ADVANCED MATERIALS 1991–1992
I. SOURCE BOOK

Jon Binner
Paul Hogg
John Sweeney

Co-ordinating editor: Nick Butler

Elsevier Advanced Technology
Mayfield House, 256 Banbury Road, Oxford OX2 7DH, UK

Copyright © 1991
Elsevier Science Publishers Ltd
Mayfield House, 256 Banbury Road, Oxford OX2 7DH, England

British Library Cataloguing in Publication Data
Binner, Jon
     Advanced materials 1991–1992.
     I: Source book.
     I. Title II. Hogg, Paul
     III. Sweeney, John
     620.11

     ISBN 1-85617-081-0

CONTENTS

FOREWORD

The *Advanced Materials 1991-1992: Source Book* comprises three main chapters which in turn provide an overview of three of the advanced materials industries (ceramics, composites and plastics) over one year.

Each month, Elsevier Advanced Technology covers all commercial and technical developments in these industries, of a commercial and a technical nature, in its international newsletters *Advanced Ceramics Report*, *Advanced Composites Bulletin* and *High Performance Plastics*. The information in this Source Book is based on twelve issues of each of these newsletters. The editors of the three chapters (Jon Binner: Ceramics; Paul Hogg: Composites; and John Sweeney: Plastics) have arranged the information to highlight significant trends that have emerged during this time. They have also added comments on the trends (which are distinguished in the text as italics).

The original *Advanced Materials Source Book* was published in 1989. In producing the 1991-1992 version we have taken into account the criticisms of both readers and reviewers of the earlier volume. One feature that was widely appreciated before was the inclusion of contact names and addresses for all the organizations mentioned including telephone, facsimile and telex numbers. Our research department has improved upon this with the production of a complete directory which lists:

* contact names for senior management, marketing and sales, technology, and production;

* company information including sales for 1989, name(s) of parent group or subsidiaries, and alternative addresses/sites; and

* details of products and/or services provided.

The *Advanced Materials 1991-1992: Directory* is supplied as a separate volume but a comprehensive index in Chapter Five of this Source Book allows the reader to refer readily to the information supplied in both. In this way, a representation of an organization's most recent ventures can be found together with the background on its main activities. Readers wishing to follow up any leads as a result have all the contact details they need.

Greater consistency between the main chapters has been sought; each chapter follows the same basic outline covering in turn market information; materials; applications; processing; equipment; testing and standards; health, safety and the environment; fundamental research; and industry news. Each chapter begins with an executive summary highlighting the major advances and trends that have emerged.

Finally, where information could have appeared in more than one chapter, the editors have tried to classify it according to the significance of the story. For instance, an item about advances in ceramic matrix composites will be found in the ceramics chapter if the major development concerns the ceramic element of the composite. This avoids repetition of information or clumsy cross-referencing. The contents pages, which list every item's title in full, and the index ensure the reader does not overlook relevant stories.

CHAPTER 1

CERAMICS

2. EXECUTIVE SUMMARY

1990 has been a year in which steady progress has been made towards the goals generated over the past decade. In particular, two main themes have begun to dominate, these being an increase in both the desire and the ability to:

• design advanced ceramics in terms of the processing-structure-property relationship; and

• process with greater control over microstructure.

These goals are intrinsically linked and are aimed at delivering more reliable materials which manufacturing industry can use with greater confidence.

The ability to design advanced ceramics involves understanding the underlying mechanisms behind the processing-structure-property relationship. The aim is to be able to base the selection of the material and exact processing route on a known ability to achieve a desired microstructure and hence a given set of final properties and component characteristics. At present the capability to achieve this is extremely crude, but already computer-based systems are available which will allow simple optimisation of given processing routes.

The search for processing routes which will yield greater control and improved economy continues. In addition to the refinement of existing process techniques, new routes are being examined. One of the major focuses of 1990 has been the use of microwave power as an alternative energy source in the processing of advanced materials. The creation of two industrial, collaborative clubs — at Nottingham University in the UK and Wollagong University in Australia — indicates the importance attached to this process. Similar clubs already exist in other countries, including France and The Netherlands.

Associated with this wish to increase the design element of ceramic processing has been the realisation that if manufacturers are to capitalise on these materials then much greater standardisation is required. The first steps toward this have been an increase in the number of national programmes aimed at generating standard characterisation techniques and the publication of a range of standards. This drive is expected to continue throughout the 1990s.

Advanced ceramics are currently regarded as one of the major growth areas in the field of materials. Millions of dollars are being spent each year on their development and market reports regularly indicate annual growth well into double figures. In 1990 many companies have looked to create joint ventures as a means of gaining new technology and markets. Electronic applications still dominate the market; however, significant growth markets have been identified in precursor materials and powders, biotechnology and, particularly, coatings.

The ceramic engine continues to be largely a goal for the future but the next decade is expected to see a significant increase in the range of components being manufactured and fitted to vehicles. The cutting tool market is growing with advanced ceramics finding increasing application as coatings on carbide and other hard material substrates. High critical temperature superconductors continue to receive considerable attention, although even here the drive for new materials has waned, attention shifting to the production of coatings, wires and cables. A particular focus of attention has been the ability to achieve flexible products and greater stability in terms of properties.

Japan and the Far East continues to dominate the development of advanced ceramics. The oriental approach leads to a greater balance between short-term development and long-term research than that achieved in the West. As a result of financial constraints, the latter is moving towards the abandonment of its long-term research base in favour of swifter profits. This is predicted to have significant consequences in the next century. A solution is to provide a more stable background for Western companies so that they can commit a higher percentage of profits to long-term research and development. Although pan-European cooperation is still some way off, and indeed may never be truly achieved, the improvements which have occurred over the past few years have succeeded in slightly increasing the European threat to the Japanese and American domination of the advanced ceramic markets.

3. MARKETS

The increasing interest being shown in advanced ceramic materials is highlighted by the positive tone of virtually all market reports published during 1990. Annual growth rates are typically predicted to exceed about 10% for most advanced ceramics markets, and this is reflected across a range of published data, from studies of individual processing routes through to predictions of overall markets.

One feature which is of note, however, is a decrease in the volume of reports published but associated with this is an overall increase in standard. Two or three years ago these reports were being produced at a rate of several per month from a range of different sources. However, they often suffered from the problem that their findings on related subjects usually had little more in common than the prediction of a substantial increase in sales and general usage of the materials. While some of these discrepancies could be accounted for by different definitions of market sectors, inadequate correlation of data did lead to many readers questioning the veracity of the information contained within some reports. In addition, there were a significant number of reports which contained obvious errors and misuse of scientific terms, indicating an underlying carelessness and/or lack of understanding of the subject matter. This situation now seems to have resolved itself to a certain degree. Some of the more questionable agencies simply seem to have not attempted to publish reports recently, while others have apparently taken to employing consultants. It should be clearly stated, however, that despite the above comments there are still a significant number of publishers of market reports who provide excellent work with detailed and plausible analyses.

The details provided here generally come from reports of a high quality, and have been divided into several categories; those dealing with international market trends for the whole advanced ceramic business, and those focussing more on one particular facet, such as materials, applications or processing.

3.1 INTERNATIONAL

3.1.1 US$2.6 BILLION EUROPEAN MARKET FOR TECHNICAL CERAMICS

The European market for technical ceramics will exceed US$2.6 billion by 1993, according to a report from Frost and Sullivan, New York, USA.

The publishers say the 400 page report, called 'The European market for technical ceramics', is based on interviews with personnel involved with the ceramics industry — including producers, suppliers, users, government officials and trade associations. The report, which costs US$5500, says the 1990 value of the market is US$1.8 billion and estimates that by 1993 it will have grown as follows:

Electronics: US$1978 million

Catalyst supports: US$325 million

Wear parts: US$267 million

Cutting tools: US$60 million

Bioceramics: US$28 million.

Frost and Sullivan claim the report, which covers the period 1987–1993, provides market analyses for three principal user countries in Europe and information on, and analyses of, eight ceramic types. These are: alumina, zirconia, beryllia, ferrites, titanates, mixed oxides, silicon nitride and silicon carbide. Prices and availability of a slightly wider range of materials are also given.

Fourteen application areas are covered, the publishers say. These include insulators, substrates, electronic packages, permanent magnets, soft ferrites, capacitors, thermistors, varistors, piezoceramics, sensors, wear parts, cutting tools, catalyst supports and bioceramics.

3.1.2 JAPANESE ADVANCED CERAMICS MARKET TO REACH ¥6 TRILLION BY 2000

The fine ceramics industry in Japan, and in particular the structural ceramics sector, can expect a high growth rate leading to a market size of ¥6.1 trillion in 2000, some five times that of the 1990 market (¥1.15 trillion), according to the US-based publication *High Tech Ceramics News*.

These are the findings of a final report from the Committee of Basic Problems for the Fine Ceramics Industries to the Ministry of International Trade and Industry (MITI), based in Tokyo.

However, the report does indicate that a number of developments are required if this figure is to be reached. These include:

• technical developments;

• a computerized database;

• more engineers; and

• standardization of materials evaluation processes.

The report also stresses the importance of active international development programmes.

3.1.3 JAPANESE ZIRCONIA MARKET DECLINES

Dai-ichi Kigenso Kagaku Kogyo has estimated that the Japanese market for zirconia powder declined during 1988 compared with consumption in 1987, according to *Roskill's Letter from Japan*.

The decrease was at just 3.1%, but is being taken seriously given the 13.9% increase in zirconia consumption which occurred during 1987 compared to 1986.

The main reason for the decrease is said to be the increasing cost of the required precursors, zircon sand, baddeleyite and the intermediate, zirconium oxychloride. This has led to a sizeable drop in demand from the electronic materials, glass and sensors industries and hence to decreased production of zirconia by the wet method, the process which largely feeds these industries.

3.1.4 NATIONAL PROGRAMMES HAVE MADE EUROPE AND JAPAN FULLY COMPETITIVE WITH USA

National programmes in technology development related to materials science and engineering have made Europe and Japan fully competitive, and in some cases able to surpass the USA, according to a US report.

Commissioned by the US National Research Council, the report is a comparative study of national materials programmes throughout the world. 'International cooperation and competition in materials science and engineering (NISTIR 89-4041)' was undertaken by Panel Three of the Committee on Materials Science and Engineering (COMMSE).

The report deals with many facets of materials science and engineering as practised in the USA and other countries. It contains information from a survey of national programmes for science and technology and materials science and engineering, and it elaborates on the administrative structures to carry out research and development.

3.1.5 INVESTMENT OPPORTUNITIES IN ADVANCED CERAMICS

Strategic Analysis Inc (SAI), based in Reading, Pennsylvania, USA, has produced a multiclient study entitled 'Global investment opportunities in high performance ceramics'. The report contains business profiles of 65 producers of advanced ceramics — made up of 30 US, 20 Western European and 15 Japanese companies. Each company has been analysed by SAI for its joint venture or technology licensing possibilities.

The report is divided into three separate volumes, one for each geographical region. Each volume is priced at US$6000, or the entire study is available for US$15 000.

3.2 MATERIALS

3.2.1 GROWTH RATE FOR CERAMICS NEARLY DOUBLE THAT OF OTHER ADVANCED MATERIALS

The average annual growth rate for advanced ceramic products in the period 1983–1990 has been nearly double that of any of the other new material categories, according to a report prepared for the UK Government.

'Manufacturing into the late 1990s' projects the average annual growth rate for ceramics to be 17.4%; the next highest being 10.4% for new glass products, while that for composites is 8.2%.

The authors, the PA Consulting Group, Cambridge, UK, argue that the international competitiveness of every sector of industry is dependent on being able to make effective use of new materials technology. "New materials give a crucial competitive advantage and leverage to manufacturers of existing product types", the report says.

Probably the greatest impact will be on production methods. With the new materials the product and material almost always come into existence at the same time. Although there are exceptions to this, the report notes, the traditional user must also become a processor to take full advantage of newer materials.

Design is also important. The design of a component in a new material must generally take into account the production process as well as the properties of the materials.

The PA Consulting Group prepared the report for the UK Government's Department of Trade and Industry. A copy of 'Manufacturing into the late 1990s' costs £20 and is available from Her Majesty's Stationery Office (HMSO), London, UK.

3.2.2 GOVERNMENT-SPONSORED CERAMICS RESEARCH

'Advanced structural ceramics' a report from TechTrends of Paris, France, surveys the major government-sponsored research and development programmes in the USA, Japan, West Germany, UK, Sweden, and France. Key areas where innovation is crucial to the successful exploitation of structural ceramics are highlighted for each region and target performance goals are identified.

Major trends and developments in the technology of monolithic structural ceramics and ceramic matrix composites are also discussed. Monolithic ceramics examined include alumina and zirconia, silicon carbide and silicon nitride, and sialons. Synthesis of ceramic powders and fabrication and processing of monolithic ceramics are also covered.

Ceramic matrix composites (CMCs) are examined, one chapter being devoted to ceramic fibre and ceramic whisker reinforcements. Toughening mechanisms, fabrication and materials applications of CMCs comprise another chapter, with attention being given to the fabrication of particulate, whisker and continuous fibre reinforced composites. All the topics are illustrated by reference to recent commercially oriented research and development activities.

The report also examines applications of structural ceramics in specific end-uses, including: automotive engine components, ceramic cutting tools, bearings and seals, wear parts, and thermal barrier coatings for aerospace applications. The market for structural ceramics is assessed in the context of other advanced materials and examined in particular for the automotive and cutting tool sectors.

Advanced Structural Ceramics is 230 pages long and costs US$500.

3.2.3 DEVELOPMENTS IN ADVANCED MATERIALS HIGHLIGHTED

'Advanced materials, technology highlights 1990', is a report from Innovation 128, based in Paris, France. It contains 215 pages of information on significant developments in metals and alloys, ceramics and glasses, plastics and polymers, composite materials, coatings and surface treatment, electronic materials and optical fibres.

The report takes the form of reports from local correspondents in Europe, the USA and Japan, and topics include: R&D projects, new products and processes, key patents, licensing opportunities and agreements, market data and trends, company activities and strategies, product applications and new research facilities/projects. Each item of information, however, only consists of an abstract of about 20 lines. The cost of the report is US$500 or £270.

3.2.4 CERAMIC PRECURSOR TECHNOLOGY

The world market for precursor-derived advanced structural and electronic ceramics will reach over US$500 million by 2000, according to a study recently completed by the Advanced Materials Group of Kline and Co of Fairfield, New Jersey, USA.

Chemical vapour deposition (CVD), chemical vapour infiltration (CVI), sol-gel and polymer pyrolysis are targeted by the company as technologies on the leading edge of new materials development, enabling ceramics to compete in new high performance applications.

The report says that the 1989 world market for precursor-derived ceramics is estimated at US$200 million, with CVD processes representing the largest and most established segment at 86%. However, But growth in use of this process is anticipated at only 9% annually, compared with growth rates of 13–17% for the other three techniques.

The report highlights the market and technological factors which Kline and Co believes are driving development and growth, these include:

- the demand for high performance fibre reinforcements for metal and ceramic matrix composites (in particular silicon carbide and alumina fibres which cannot be produced from mineral raw materials);

- the need for improved cost effective coating technology;

- desirability of moderate to low temperature processing of ceramics; and

- the increasing need for control of the purity and microstructure of raw materials for structural and electronic ceramics.

The information on which this study is based came from 90 individuals in 65 organizations, geographically spread throughout the USA, Western Europe and Japan. In addition, Kline and Co says that it assembled a panel of leading industrial and academic experts to provide technical guidance and assistance.

3.2.5 ELECTRONIC CERAMICS MARKETS ANALYSED

'Electronic ceramics in the 1990s' is the title of a report from Business Communications Co which covers markets, materials, fabrication techniques, current and emerging production techniques, new developments (including recent patents in both materials and processing technologies), current and emerging applications and a market analysis for each of several groups of electroceramics. These include: insulators, substrates and integrated circuit packages; capacitors; piezoelectric ceramics; ferrite magnets; and ceramic superconductors.

The current size and future growth of the markets are estimated for the period 1989 to 2000.

BCC says the report is based on information derived from interviews with almost all US producers, potential producers and suppliers of electronic ceramic components, research and development departments, and industry experts. The report concludes by profiling all US producers and suppliers of electronic ceramic components.

The report was published in July 1990 and costs US$2650.

3.2.6 ALUMINA MARKET STUDY

Mitchell Market Reports (MMR) has published a second edition of its market report on alumina. The first edition was published in 1987 and since then the report has been brought up to date and greatly extended, MMR says.

The report is in three volumes. The first volume covers the different forms of alumina, their sources, methods of production of powders, components and composites, world production, trade and consumption of non-metallurgical forms, and end uses. These latter include: the refractory, abrasive and chemical industries; electrical, electronic and mechanical applications; and other applications, such as catalysts and catalyst supports, filler and flame retardants, optical applications, batteries, fibres and composites, films and coatings.

Volume 2 examines alumina activities in 30 countries and by some 750 companies and organizations, while Volume 3 considers recorded exports and imports of aluminium hydroxide and non-metallurgical grades of alumina.

3.2.7 TITANIA MARKETS EXAMINED

Annual world production of titania powders is in excess of three million tonnes and has a value of US$7000 million, acording to the UK publishers of a report, 'Titania and the titanates'. However, all but US$300 million worth is used in pigmentary applications and so lies outside the scope of the report which focuses on the use of titania powders in ceramic applications.

Mitchell Market Reports of Monmouth, UK, says that 'Titania and the titanates' is a comprehensive guide to suppliers, applications and current activities in the field, covering the activities of 650 companies and research institutions in 33 countries. The report costs £600 (US$1050).

3.2.8 PITCH-BASED CARBON FIBRES

The current market for all types of carbon fibres totals US$400 million and will rise to US$600 million by 1995 and possibly US$1000 million by 2000, with pitch-based materials representing about 5% of this total, according to a report produced by two Japanese companies.

The report, by Techno Co Ltd and Shinko Research Co Ltd, is available in an English edition through Omnia of Raleigh, North Carolina, USA. It says the market share for pitch-based fibres could rise to 10% by the year 2000 if improved properties and lower production costs are realised.

The report looks at the background to pitch-based fibres, markets and applications, and compares competitive fibres. The current market is considered in detail to allow predictions up to the year 2000. Other issues, such as the international patent situation, and proposed methods for cost reduction, are explored.

The report costs US$9700.

3.2.9 DIAMOND AND DIAMOND-LIKE CARBON COATINGS MARKET TO REACH US$1 BILLION

The worldwide market for diamond and diamond-like carbon coatings should expand to US$1 billion by the year 2000, according to a study by Gorham Advanced Materials Institute (GAMI) of Gorham, Maine, USA. However, this is significantly under other industry estimates which predicts a market up to US$16 billion by the turn of the century, says the publisher.

The report, 'Diamond and diamond-like carbon coatings', says the thrust behind the rise in popularity of the materials is their combination of properties. They are hard, have high thermal conductivity, are electrically insulating, chemically inert and optically transparent over a wide range of the electromagnetic spectrum.

However, diamond has not been previously applied to any great degree as an engineering material because it has only been available in small, expensive stones with odd geometries. But, recent developments in the growth of thin film diamond have opened up a range of applications.

GAMI says that it has taken a more conservative view of the potential market because it could take decades to commercialize the coating technologies at full scale.

The study is available in four separate volumes:

* volume one is a techno-economic assessment of diamond and diamond-like carbon coating technology; while

* volumes two, three and four each profile the current market, forecast growth prospects in existing

and emerging applications, and assess business opportunities in North America, Pacific Rim and Western European countries, respectively.

The volumes can be purchased separately or in combination at a cost of US$7500 for one volume, US$11 600 for any two volumes, US$16 100 for any three, and US$17 900 for all four.

3.2.10 DIAMOND THIN FILMS PROFILED

The Business Intelligence Centre of California-based SRI International (formerly the Stanford Research Institute) has released its new TechMonitoring Service 'Profile of diamond thin films (DTF)'. The profile is SRI's evaluation for senior management of this emerging technology, focused on its potential commercialization. It is updated each month for clients, with SRI's assessments of significant developments and reviews of issues affecting the commercialization of the technology.

According to SRI, if the coating processes that make diamond and diamond-like films improve in speed and cost sufficiently, these films will appear in consumer electronics, analytical equipment, machine tools, knives and cutters, materials processing equipment, engines and motors, lenses and optical components, light and radiation sensors, communications equipment, electronic packaging and electronic circuitry. Within the next 6–7 years, says SRI, the market for diamond films could reach US$1 billion, and even US$2 billion by the turn of the century.

However, SRI warns that because of the nature of DTF technology, a protracted technical hurdle or sudden breakthrough in a major market sector could completely change all previous market estimates. According to SRI, several volatile issues currently affecting market predictions include:

adhesion problems could limit the use of DTFs to 10–20% of the potential tooling market — although if diamond films find successful use in steel tooling, the markets could expand dramatically;

few analysts apparently expect audio speaker applications to offer a large commercial market, but consumer interest could generate increased demand for a quality product in this field;

optical component coatings may see a dramatic rise in sales, especially if scientists are able to coat plastics at low temperatures. This application is thought to depend upon equipment development cycles and process improvements; and

the most important factor in the next few years is claimed to be the increased number of commercial applications in semiconductor devices and electronics. This market will probably outstrip existing markets in the early 1990s; however, fundamental research will be essential to success in this area.

3.2.11 US$1 BILLION SALES BY 1995 FOR BIOCOMPATIBLE MATERIALS

Biocompatible materials will account for sales of approximately US$661 million in 1990 and US$1 billion by 1995, according to a report from Business Communications Co (BCC) Inc of Norwalk, Connecticut, USA. End-use markets will reach nearly US$3.5 billion in 1990, rising to US$6.4 billion by 1995, the company claims.

The report, 'GB-072B: Biocompatible materials for the human body', costs US$2450 and lists applications as including implants, valves, grafts, pacemakers, bone repair and replacement devices, artificial organs, dental materials and drug delivery systems, as well as dialysis/separation/filtration systems and catheters.

Biocompatible materials — which include advanced ceramics, pyrolytic carbon, medical grade polymers, metals, composites and natural materials — are used in a wide range of medical devices marketed by an estimated 10 500 US manufacturers.

3.2.12 COMMERCIAL POTENTIAL FOR SUPERCONDUCTIVITY

Elsevier Advanced Technology, Oxford, UK, has published a report which it describes as a realistic appraisal of the markets and commercial potential for high temperature superconductors.

The publisher says the report is aimed at senior management and researchers in a range of industries: including electronics, high field magnets, and electricity generation and transmission. *'Superconductivity — a realistic appraisal of technology, markets and commercial potential'* costs £135 (US$243).

3.2.13 HARD AND SUPERHARD MATERIALS REPORT

'Hard and superhard materials — markets, technologies and opportunities: 1989-1994 analysis' considers the US market for industrial materials with a Knoop hardness greater than 1000, its US publisher says.

The report is available from World Information Technologies Inc, based in Northport, New York, and costs US$1800. Partial segments of the report are available for less than the full cover price.

3.3 APPLICATIONS

3.3.1 ADVANCED ARMOUR MARKETS SET TO INCREASE

According to a recently released report by Kline and Co Inc of Fairfield, New Jersey, USA, the combined North American and Western European market for advanced armour materials, including advanced ceramics, polymer composites, fabrics and speciality metals, is predicted to almost double by the year 2000, despite the significant improvements in East-West relations. The study reports that the market for advanced armour materials was valued at US$220 million in 1989 and this figure is forecast to exceed US$400 million by the end of the century.

James Weatherall, Vice-president of Kline's Advanced Materials Group says, "Although cuts in military programmes are a political reality on both sides of the Atlantic, the weaponry still exists to defeat current armour systems. Therefore, as new threats — including depleted uranium rounds, shaped charges, and long rod projectiles — increase the capability of defeating current military systems, development of more effective armour will need to continue."

Armour designers are currently focusing on a modular approach which enables materials to be interchanged depending on the expected threat. The report claims that many experts expect future military vehicles to contain considerable amounts of ceramic-based armour. In addition, vehicle manufacturers are looking at possibly upgrading or replacing several non-structural or semi-structural parts.

In civilian applications, armour procurement continues at a moderate pace, driven by expected increases in drug related crimes and terrorist activity. Threat levels are also expected to rise in this market.

The 350 page report is based on interviews with 290 individuals in 214 organizations in the USA and Western Europe. Titled 'Advanced armor materials 1990', it assesses material requirements for military and civilian applications, identifies critical product specifications, and analyses technical developments and future demand for materials which compete with and replace conventional metal armour unsuitable in defending against higher level threats. It details developments in over 15 advanced materials consumed in six major military and civilian end-use markets.

3.3.2 AUTOMOTIVE INDUSTRY REPORT

'Materials use in the automotive industry 1989–90' is the second report on the use of engineered materials in the international automotive industry to be published by ASM International. The first report covered the previous year.

The new report is more comprehensive, presenting more than 1100 entries that detail the research, development and commercial activities of 700 individual organizations for the period May 1989 through April 1990. The advances, trends and applications of materials in the international automotive industry are documented in the areas of ferrous and non-ferrous metals and the full range of polymers, ceramics and composites.

Each section of the report opens with tables that serve as a directory to the materials as they are used in the industry, and to the individual manufacturers, supplier companies and related organizations and institutes. The review contains concise descriptions and full source documentation of the individual projects referenced. Additional aspects of industry intelligence are included in the appendix, such as a listing of corporate sources referenced with addresses and summaries of issues, trends and developments noted in process application areas.

The report costs US$325 or £190.

3.3.3 MEMBRANE SURFACE MODIFICATION TECHNOLOGY

Membrane separation technology will continue to advance into the present decade with the advent of third and fourth generation membranes, according to a report from a US publisher.

'Membranes for the nineties: highlighting surface modification technology (C-112)', from Business Communications Co Inc of Norwalk, Connecticut, states that by 2000 it is expected that membrane sales in the USA will reach US$1.6 thousand million. Of this 44% of the membranes will be surface modified. The report costs US$2850.

3.4 PROCESSING

3.4.1 WORLDWIDE MARKET SURVEY OF HOT ISOSTATIC PRESSING

Gorham Advanced Materials Institute (GAMI), Gorham, Maine, USA, has published a business opportunities study of the hot isostatic pressing (HIP) industry.

GAMI says the study is designed for corporate executives responsible for planning for continued participation in, or entry into, the HIP industry, and coverage is worldwide.

From 1983–1988, the number of hot isostatic presses in operation worldwide increased 228%. At the same time, the worldwide annual market for materials processed by HIP tripled; it may reach US$6 billion by 1993, according to GAMI.

3.4.2 CHEMICAL VAPOUR DEPOSITION STUDY

A global technology and North American market study to assess all of the chemical vapour deposition (CVD)-based coating technologies has begun.

Gorham Advanced Materials Institute (GAMI) of Gorham, Maine, USA, says the study will also profile and forecast the leading North American markets for consumables, equipment and applied coatings.

Currently, Gorham estimates the market for CVD in North America as US$1.5 billion and it projects future expansion over the next decade to be 10% annually, on the basis of real growth.

The study's technical content will focus heavily on the CVD coatings position compared with competitive processes and materials.

3.4.3 10% GROWTH FOR PHYSICAL VAPOUR DEPOSITION BUSINESS

The value of the physical vapour deposition business worldwide is forecast to grow at 10.3% over a five year period, increasing from an estimated US$1.2 billion in 1989 to US$2.2 billion by 1995, according to Business Communications Co (BCC) Inc of Norwalk, Connecticut, USA.

'GB-133: Physical vapor deposition (PVD)' predicts that the largest growth will occur for new materials, including ceramic compounds and diamond coatings. BCC says the major factors for growth in PVD technology are:

- higher performance requirements for electronic and optical depositions and surface coatings;

- technological improvements; and

- new applications and materials.

The report costs US$2850.

3.4.4 THERMAL SPRAY MARKET TO GROW

According to a multiclient study carried out by Gorham Advanced Materials Institute (GAMI), the North American thermal spray market will grow from its 1990 value of US$610–675 million to between US$1.8 and US$2.0 billion by the year 2000 (1990 dollars).

Applications for thermal spray and other coatings have been identified in 33 industrial sectors. Of these, aircraft gas turbine applications account for 30–40% of the total North American market. This high percentage is due to the very large number of parts which need coating — some new engines have close to 5500 parts which are thermal spray coated. Growth to the year 2000 is projected at 7–8% per year on the strength of expanding applications, particularly for ceramic coatings for thermal barriers and for clearance control, and on the basis of continuing strong demand for engines into the next decade.

The report, entitled 'Thermal spray (TS) coatings', is contained in two volumes totalling more than 400 pages, with charts, tables, graphs and detailed explanations of how each market segment operates. The two volumes cost US$9000, or US$5000 and US$5500 respectively to buy them individually. Volume I covers manufacturing technology assessments of thermal spray coatings as well as assessments of competitive and complimentary coating technologies. Volume II covers market forecasts and an industry profile.

3.4.5 ION AND MOLECULAR SURFACE MODIFICATIONS

According to a study by Business Communications Co (BCC) Inc of Norwalk, Connecticut, USA, the total value of the ionic/molecular business for surface modifications is due to grow by nearly 10% by 1995.

In report 'GB-134: Ion and molecular surface modifications', BCC predicts that the total value of the ionic/molecular business will increase by 9.6% from an estimated US$145.5 million in 1989 to US$253.1 million by 1995. The worldwide value of ion implantation and molecular beam epitaxy equipment sales will account for US$108 million of the total by 1995 (up 9.4%), while conventional ion beam implantation systems, which currently account for 73% of this market, should represent less than half of the systems shipped in 1995.

The report also predicts that the major growth areas in ion implantation technology will be ion beam mixing, ion-beam assisted deposition and the development of a new plasma source system.

Meanwhile, revenue from ionic and molecular services, currently estimated at US$60.6 million, will increase at an average rate of 9.5% to US$104.3 million by 1995. Thus, says the report, it will continue to be economical for companies to contract service firms that provide ionic/molecular techniques.

The report was published in July 1990 and costs US$2850.

4. MATERIALS

This section examines some recent developments concerning materials, whether as powders for use in making ceramic components or in terms of the solid ceramics themselves. Not all materials currently of interest to the engineering community are featured, since this source book represents only developments reported during 1990. However, it is notable that there are relatively few exceptions. The materials are examined in alphabetical order.

4.1 ALUMINIUM NITRIDE

Aluminium nitride (AlN) is a ceramic which displays a relatively high thermal conductivity and is therefore being developed for use as a substrate in electronic components. Its major competitors are alumina and beryllia. The former has a lower thermal conductivity than aluminium nitride but is usually cheaper and has been in extensive use for many years. Beryllia, on the other hand, has a higher thermal conductivity than AlN but many companies are cautious about using it due to its toxicity in powder form.

Three of the reports below outline improvements to the material in terms of substrate applications, the fourth concerns the production of aluminium nitride whiskers.

4.1.1 IMPROVED ALUMINIUM NITRIDE

Ceramics Process Systems, based in Milford, Massachusetts, USA, claims to have developed an aluminium nitride which displays exceptional thermal and mechanical properties. The material is said to have a thermal conductivity of 230 $W.m^{-1}.K^{-1}$, which is significantly higher than that of alumina or even conventional aluminium nitride. While beryllia can have a higher value than this, the material has traditionally not found great favour with the advanced packaging manufacturers due to concern about its toxicity, says Ceramics Process Systems. The company has developed the improved aluminium nitride ceramic for its proprietary 'Quickset' binderless injection moulding process, which is used to fabricate advanced packages and thin film substrates for electronics applications.

4.1.2 ALUMINIUM NITRIDE POWDER INERT IN WATER

An aluminium nitride (AlN) powder is now available which does not react with water, according to its manufacturer, Advanced Refractory Technologies Inc (ART) of Buffalo, New York, USA.

The company claims to have developed a process which makes AlN powder inert to water. Previously this powder reacted with moisture, forming ammonia and creating handling and processing problems. Not only have these problems been avoided, but processing routes such as spray drying and slip casting can now be based on aqueous systems, reducing costs, ART says. The powder is available in a range of purities and particle sizes.

4.1.3 METALLIZATION PROCESS FOR ALUMINIUM NITRIDE

Japanese researchers claim to have developed a nickel metallization process for aluminium nitride which is simpler than existing procedures.

The scientists, from NEC Corp, Tokyo, Japan, say the method does not require a precursor stage, involving electroless nickel plating, or a final firing in a hydrogen atmosphere at high temperature, as was the case with previous tungsten or molybdenum metallizations.

The method is similar to the process used for thick film pastes. Firing takes place at only 1300°C in a nitrogen atmosphere. The final product is claimed to have an adhesive strength of 30 MPa; falling to 22 MPa after 500 hours ageing in a high temperature humidity test (65°C, 93% RH), *Interceram* says.

4.1.4 ALUMINIUM NITRIDE WHISKER PROCESSING

The Faculty of Engineering at Niigata University (Niigata Prefecture, Japan), has claimed to have developed an efficient processing route for aluminium nitride whiskers.

The process, developed by Professor Hotta and his group, involves mixing 5-10% of ammonia gas with nitrogen gas and aluminium vapour at a temperature of 1500°C. Under suitable conditions (not specified), aluminium nitride whiskers up to 1 mm long could be grown in about 10 minutes, while whiskers 15 mm long were formed after an hour. The efficiency of the process has been put at 40%.

4.2 BERYLLIA

Because of its high thermal conductivity, beryllia (BeO) is primarily used for engineering applications where heat needs to be removed, such as in substrates for electronic or microwave devices. To improve metallization characteristics a good surface finish is required, while a high strength provides increased resistance to thermal and mechanical shock.

4.2.1 HIGH STRENGTH BERYLLIA WITH GOOD SURFACE FINISH

The addition of small quantities of magnesia and zirconia creates a beryllia ceramic with near theoretical density, high strength, and fine grain and pore size, according to the material's UK manufacturer, CBL Ceramics Ltd from Milford Haven.

The company reports that the combination of these properties leaves a typical as-fired surface finish of 0.3 μm roughness average (μmRa). Polishing can improve this to 0.04 μmRa. The material is based on a 99.5% pure beryllia body whilst the additions improve the cross-break strength from 200 to 300 MPa. The company says that the material, 'Beramic Z' has advantages over conventional beryllias, particularly in the application of thin film metallizations on large, thin substrates (approximately 9 cm square by 1 mm thick being standard). Other potential uses are thick film substrates, helix support rods and windows in travelling wave tubes and laser capillaries.

4.3 BIOCERAMICS

Certain ceramics are finding increasing usage in biological applications, such as reconstructive surgery, because of their compatibility with bones. Within the bioceramics field attention is increasingly focusing on the bioactive materials, that is, those which generate a positive and useful response from bone. These are generally calcium phosphate-bearing materials.

4.3.1 SUPERPLASTIC HYDROXYAPATITE

Hot isostatically pressed (HIPed) bodies of ultrafine grained hydroxyapatite (HAP) powder show not only high density and mechanical strength, and good translucency, but also superplasticity, providing the grain size remains uniform and around 0.2 µm, according to the Materials Evaluation Group of the Basic Ceramic Division, Government Industrial Research Institute (GIRI), based in Nagoya, and TDK Corp, Tokyo, Japan.

The HAP ceramics are superplastic when heated to about 1000°C, while zirconia-based ceramics require temperatures around 1450°C. This allows forging moulds and apparatus designed for conventional titanium alloys to be used. The superplastic HAP can be stretched to twice its original dimensions in 20–40 minutes at 1000–1050°C, the developers say. Diffusion bonding of the superplastic HAP with other materials can also be achieved.

The developers hope this discovery will allow greater flexibility when producing complicated parts from combinations of alumina, zirconia or titanium alloys (which provide mechanical strength) with HAP (which is biocompatible).

4.3.2 CALCIUM PHOSPHATE-COATED ZIRCONIA

Japanese scientists claim to have succeeded in making calcium phosphate-coated zirconia suitable for load bearing prostheses. The researchers, from the Government Industrial Research Institute in Nagoya, use magnesium metaphosphate as a binding layer between the two ceramics. They now intend to optimize the material's biological and mechanical properties.

According to the *Japan Industrial Journal*, surgeons use both zirconia and alumina in load bearing prostheses. However, the materials are bioinert and so do not encourage the formation of new bone growth. Calcium phosphate-based ceramics, like hydroxyapatite, are bioactive and encourage bone growth, but do not have sufficient mechanical properties, such as strength and toughness, for use in load bearing applications. Previous attempts to apply a surface coating of calcium phosphate to ceramics such as zirconia have not been successful.

4.3.3 ANTI-BACTERIAL MATERIALS

A glass which is anti-bacterial has been developed by Ishizuka Glass Co Ltd, based in Aichi Prefecture, Japan.

The glass apparently contains metal ions such as silver or copper and these leach out over time, killing any bacteria with which they come into contact. In addition, the glass prevents the build up of moulds and algae. It can be used as a glass product in its own right or combined with resins or papers to form composite materials.

In a second development, Sangi Co Ltd, based in Tokyo, Japan, has formulated an anti-bacterial material based on hydroxyapatite. The material is said to be durable at temperatures up to 1200°C.

4.4 BORON NITRIDE

Boron nitride (BN) can exist in two polymorphic forms, hexagonal and cubic. The former is a very soft material and resembles graphite in many of its mechanical properties. In its cubic form, however, boron nitride is very hard like diamond, and hence it is increasingly being used for machining of metals such as steel.

A second important property for boron nitride is its chemical inertness. This has given it a number of applications where it is applied as a coating to protect a more reactive substrate material.

4.4.1 HIGH STRENGTH BORON NITRIDE

A high strength cubic boron nitride which is reported to be suitable for machining hard steels under a range of conditions has been developed by Sumitomo Electric Industries Ltd, based in Tokyo, Japan.

'Sumiboron BN250' can be used for high speed cutting and does not require the application of a cutting fluid, the company says.

Sumitomo Electric has also produced 'BN200' for continuous and 'BN300' for intermittent cutting, claiming the range offers higher efficiency with greater flexibility than can be achieved with grinding.

The company describes the hardness of cubic boron nitride as second only to diamond. It is ideal for machining hard steels, Sumitomo Electric adds, because it does not react with ferrous metals.

4.4.2 LOW TEMPERATURE COATINGS OF BORON NITRIDE ON STEEL

Radio frequency magnetron sputtering allows for steel to be coated with boron nitride at only 300°C, according to the National Research Institute for Metals, Tokyo, Japan.

The Institute, part of the Science and Technology Agency, says conventional coating processes need temperatures of about 600°C. The *Japan Industrial Journal* reports that the lower temperature has two advantages:

* gas volatilisation is reduced, allowing a higher vacuum to exist during the coating process which reduces the impurities in the film; and

* there is less degradation of the steel.

4.5 CARBON

In its various different forms, carbon offers a number of attractive properties. As a fibre it has one of the highest stiffnesses available to man, and hence carbon fibres are extensively used in composite materials. In its diamond form carbon is both the hardest material known and also an excellent electrically insulating, heat conducting material. These two properties have led to applications for both cutting tools and on electronic substrates. However, since diamond is very expensive a major research effort continues to be the formation of thin diamond coatings.

Carbon fibres are primarily made from two precursors, polyacryonitrile (PAN) and pitch. The former route generally yields the higher stiffness, while the latter is usually cheaper.

4.5.1 POLYACRYLONITRILE-BASED CARBON FIBRE

'Torayca-M60J' is a polyacrylonitrile (PAN)-based carbon fibre which offers a combination of strength and stiffness, according to its Japanese developer. Toray Industries of Tokyo says the tensile modulus is 600 GPa and the strength is 3.9 GPa. In contrast, an earlier Toray product, 'M50', has a stiffness of 490 GPa, but a strength of only 2.45 GPa. The heat treatments needed to produce high strength or high stiffness are not always compatible. However, the company claims that the superior properties of M-60J result from the disordered structure created during the heat treatment at temperatures between 2000°C and 3000°C.

4.5.2 PITCH-BASED CARBON FIBRE

Two companies from Osaka, Japan, have developed a short filament, pitch-based carbon fibre which they claim has high strength.

Kuraray Inc and Petoka Co Ltd make the meso-phase fibre from petroleum pitch and state that it has a tensile strength of 1.5 GPa and a tensile modulus of 150 GPa; they also claim to have made a calcined grade with a tensile strength of 3.2 GPa and a tensile modulus of 450 GPa. The two partners are delivering the fibre in matt, tow and chipped forms.

4.5.3 CARBOSILICON FIBRES

Researchers from Tokyo, Japan, claim to have made carbosilicon fibres with strengths of 2.5 GPa.

Members of Tadao Seguchi's team at the Japan Atomic Energy Research Institute (JAERI) calcine the fibres at 1200–1500°C in a helium atmosphere. This lowers the oxygen concentration to 0.5% and leaves a fibre with a thermal resistance to 1500°C, according to JAERI.

The Institute believes the fibres will find uses in reinforcing ceramics and metals.

4.5.4 INORGANIC CATALYST FOR DIAMOND SYNTHESIS

Transparent and colourless diamond fine powders have been successfully produced by heat treating mixtures of carbon and inorganic materials such as limestone and sodium carbonate, according to the National Institute for Research in Inorganic Materials (NIRIM), in Ibaraki, Japan.

The ratio of carbon to inorganic materials is set at 10:2 and the heat treatment conditions are 2150°C for 20 minutes under 77 000 atmospheres. Increasing the fraction of inorganic materials reduces the conditions required, NIRIM says (e.g., 80% inorganics reduces the required temperature to 1800°C and the pressure to 65 000 atmospheres); however, the size of the diamond powder particles is also substantially reduced.

The *Daily Industrial News* reports that applications for the powders include the polishing of metals or glasses.

4.5.5 CONTINUOUS PLASMA DEPOSITION OF DIAMOND FILMS

A company in Tokyo, Japan, is developing an arc discharge plasma generating system for the continuous deposition of diamond films.

Working in collaboration with scientists at the Tokyo Institute of Technology, Onoda Cement Co Ltd says it has made an apparatus which can maintain a stable plasma jet for more than 12 hours. To make diamond films, researchers spray the hydrogen and argon jet onto the substrate which is kept in a methane atmosphere.

They say they have grown a diamond film on a molybdenum substrate at a rate of 930 μm an hour. X-ray diffraction and Raman spectroscopy show that the crystallinity is good, they add.

4.5.6 INCREASED ADHERENCE OF DIAMOND FILMS

The adhesion between diamond films and tungsten carbide substrates is increased when its processing technique is used, a Japanese company reports.

Kawasaki-based Toshiba Tungaloy Co Ltd decarbonizes the surface of the substrate before recarbonizing it with 'micro-carbon'. Subsequent microwave plasma chemical vapour deposition (MPCVD) results in a diamond film with an adherence increased by as much as a factor of 3–4 when compared with conventionally produced versions, according to the *Daily Industrial News*.

4.5.7 DIAMOND-LIKE COATING

The availability of a diamond-like coating, 'Spi-Diamond 1000', has been announced by Spire Corp, based in Bedford, Massachusetts, USA.

Applied by its dual ion beam assisted deposition (IBAD) process, the coating can be deposited to substrate materials of any shape, says Spire. The key to the production of the ultra-hard, very adherant thin film is said to lie in the use of energetic ion beams which impart a graded interface structure while 'atomically' peening the coating into the substrate.

4.5.8 ISOTROPIC SPONGY CARBON

Joint research between Nippon Steel Chemical Co Ltd and Nippon Steel Corp, both based in Tokyo, Japan, has resulted in a material described as 'Isotropic spongy carbon' by its developers.

The *Nikkei Industrial Daily* reports that the advantages of the material are that it can withstand operating temperatures up to 2800°C, and has twice the bending strength, over 20 times the compressive strength and 1.2 times the thermal insulation capacity of conventional carbon felts.

4.5.9 SUPERIOR GRAPHITE

The ability to produce sheets and blocks of polycrystalline graphite which has physical properties nearly identical to those of single crystal graphite, but at a much lower cost, has resulted from research at the Matsushita Research Institute, Kawasaki, Japan.

The key to the success of the work was the use of poly(p-phenylene-1,3,4- oxadiazole), or POD, as the raw material. While normal polymers yield poor quality amorphous carbon products on carbonization (heating to high temperature), POD remains essentially unchanged until about

1000˚C. Above this temperature it turns into a black solid with a crystal structure resembling that of graphite, but containing nitrogen atoms. This has been called heterographite. Further heating to about 3000˚C, combined with pressure, results in the formation of super-graphite. In addition to the ability to form polycrystalline graphite films, by stacking sheets of the initial POD film, it is also possible to produce blocks of super-graphite since the layers bond together.

Applications for the material are thought to be very extensive. They range from the aerospace and nuclear power industries to automobile engines and gaskets for chemical plants. However, the first application has proved to be as the diaphragm in tweeters (high frequency loudspeakers). The material apparently eliminates some of the distortion found when other materials are used.

4.6 MULLITE

Mullite ($3Al_2O_3.2SiO_2$) is a single-phase aluminosilicate which has particularly good high temperature properties. It is currently used primarily in the refractories industry, however a substantial market probably exists for a successful advanced ceramic material based on high purity mullite.

4.6.1 POROUS MULLITE CERAMICS BY GLASS PHASE EXTRACTION

Japanese scientists claim to have made a porous mullite by extracting the glassy phase from a sintered body based on natural clay.

Researchers from the Ceramics Industrial Research Institute in Nagasaki Prefecture sinter a mixture of kaolinite (a clay from New Zealand) with either cobalt oxide or copper oxide. This results in a body with a substantial glassy phase which can be extracted with a caustic soda solution. The end product typically contains 50–60% porosity. The pores have a uniform size in the range of 0.5–1.0μm, controlled by the sintering temperature and the amount of oxide addition. The specific surface area of the material is 50 $m^2.g^{-1}$, the *Daily Industrial News* reports.

4.7 OXIDE FILMS

This section contains a group of items in which the central theme is the production of thin oxide films for use in electronic components. Details of applications for the films are given in the items themselves where appropriate.

4.7.1 ROOM TEMPERATURE OXIDATION OF SILICON WAFERS

The presence of copper silicide allows the oxidation of silicon at room temperature, according to scientists at IBM's T.J. Watson Research Center in Yorktown Heights, New York, USA.

IBM's scientists say they can grow a layer of silica more than 1 μm thick at room temperature when a layer of copper silicide(Cu_3Si) is present. In normal device fabrication, silica is formed by heating silicon to a high temperature (above 700˚C) in an oxygen atmosphere. It may take hours to grow a layer of silica only a few hundred nanometres thick on a silicon surface. At room temperature, no measurable growth of the silica layer occurs. Cu_3Si is a compound formed by heating a layer of copper on a silicon wafer at a temperature of 200˚C. The silicide begins to oxidize upon exposure to air, but on further examination, a layer of silica is found underneath the copper silicide.

After several days at room temperature in air, many hundreds of nanometres of silica grows between the original copper silicide layer and the underlying silicon wafer. Examination after several weeks reveals over 1 μm of silica.

High resolution measurements by transmission electron microscopy reveal the presence of very small particles (less than 10 nm across) of Cu_3Si at the interface between the silicon wafer and the growing silica layer. These particles act as a catalyst to promote the oxidation of the silicon and reform to copper silicide during the reaction, the researchers say.

The scientists do not expect this discovery to revolutionize the standard methods for oxidizing silicon. However, they say that further study should increase understanding of chemical reactions at low temperatures when a catalytic agent is present.

4.7.2 EPITAXIALLY GROWN THIN FILM TITANIUM OXIDE

Scientists in Tsukuba, Japan, claim to be able to grow films of titanium oxide 50–100 atomic layers thick using epitaxial techniques. The researchers at the National Chemical Laboratory for Industry say that they use pulsed molecular beam epitaxy to deposit the titanium oxide on a magnesium oxide substrate. The epitaxial layers have the same crystallographic orientation as the substrate and the titanium oxide forms a phase in which the titanium bonds to just one oxygen atom. This is the only phase known to be conducting and is traditionally difficult to form as a thin film.

4.7.3 LAYERED TITANIUM DIOXIDE/TIN OXIDE THIN FILMS

Taiyo Yuden Co Ltd, based in Tokyo, Japan, has announced that it has developed a method for producing alternate layers of titanium oxide and tin oxide on glass substrates.

No details have been released other than the layers are typically 1-2 μm thick, they are formed at a processing temperature of 500˚C, and the titanium dioxide crystal structure is that of rutile.

4.7.4 HIGH SPEED GROWTH OF TANTALUM OXIDE THIN FILMS

The successful development of a high speed process for producing tantalum oxide thin films has been announced by the National Chemical Laboratory of Industry, based in Ibaraki Prefecture, Japan.

The process is based on the use of excimer lasers combined with chemical vapour deposition and operates at a temperature of just 130˚C. The speed of film growth is stated to be twice that for conventional production of tantalum oxide thin films. The National Chemical Laboratory reports that if a high energy excimer laser is used the thin film produced will be crystalline, while if a low energy laser is used the film produced is in an amorphous state.

4.8 SIALON

Sialons are ceramics based on the quaternery system silicon–aluminium–oxygen–nitrogen and they display excellent mechanical properties as well as good temperature and chemical resistance. The item below reports what is claimed to be the world's first sialon fibres.

4.8.1 SIALON FIBRES

The world's first sialon fibres have been produced in Japan, according to scientists from the Colloid Research Institute, based in Fukuoka Prefecture, and the National Institute for Research in Inorganic Materials (NIRIM), Niihari.

The Japanese scientists say they used a sol-gel technique to make the fibres. They expected this research work to result in low production costs and high yields when commercial operations begin, the *Japan Industrial Journal* reports. The principal applications are expected to be reinforcement of ceramic, metal, glass and cement matrices.

4.9 SILICA

Silica (SiO2) has a wide variety of uses in addition to forming a primary constituent in a range of industrial ceramics. The item below refers to a very specific application for a silica-based material, that of an aerogel.

4.9.1 LIGHTWEIGHT AEROGELS

Silica aerogels with densities one-tenth those of previous materials of this kind have been developed by researchers at the Lawrence Livermore National Laboratory in California, USA. The scientists say the silica aerogels have densities of the order of 5 mg.cm^{-3}. The semi-transparent materials have high specific strengths; for instance, a sample of 0.06 g can support a body weighing approximately 96 g (1600 times heavier).

Possible applications include insulation for spacecraft and the collection of cosmic dust without damaging or contaminating it, according to *Nature*.

4.10 SILICON CARBIDE/CARBONITRIDE

Silicon carbide (SiC) is a refractory ceramic which can often be used at temperatures up to about 1300°C in air and even higher in an inert or reducing atmosphere. With the exception of the first item, which reports a new low temperature processing route for the material, attention has now largely focussed on the production of silicon carbide whiskers and fibres for reinforcing ceramic and metal matrices.

There has been some concern that work with whiskers may be unacceptable without precautions being taken, because of suspected health risks. However, it appears that at present concern is strongest in Europe, with the major research programmes within the USA and Japan not appearing to be seriously affected. Nevertheless, interest in fibres for reinforcement seems to be increasing, especially with the improved properties and weaving capabilities now being achieved.

4.10.1 LIQUID-BASED ROUTE YIELDS SILICON CARBIDE

Researchers at the Toyota Technological Institute in Nayoya City, Japan, claim to have developed a low temperature processing route for silicon carbide. They obtain a liquid state reaction between silicon tetrachloride, dichloromethane and a sodium naphthalene complex when the precursors are heated to about 1200°C. The *Daily Industrial News* reports that, owing to the low processing temperature, the silicon carbide formed is much cheaper than conventionally synthesized material.

4.10.2 SILICON CARBIDE WHISKERS FROM RICE STRAW AND SUGAR CANE

The production of silicon carbide whiskers from organic sources is discussed by Indian researchers in 'Silicon carbide from sugar cane leaf and rice straw', *Journal of Materials Science Letters* (Vol.9, April 1990).

It is known that silicon carbide (SiC) whiskers can be made from the carbothermal reduction of rice husks because they contain both silica and carbon. However, M. Patel and P. Kumari report that their work is aimed at producing such whiskers from alternative sources, in particular rice straw and sugar cane leaf. They describe initial results as encouraging, with evidence for whisker formation in small quantities. An external catalyst was not required.

4.10.3 SILICON CARBIDE WHISKERS WITH INCREASED THICKNESS

Tokai Carbon Co Ltd of Tokyo, Japan, says it has developed a technique which will allow mass production of SiC whiskers which have a diameter as large as 2–7 µm. The whiskers, known as 'Toka Whisker', are now available in sample quantities, according to *Nikkei Industrial Daily*.

4.10.4 REFRACTORY SILICON CARBIDE FIBRES

A research group headed by Chief Engineer Seguchi of the Takasaki Radiation Chemistry Resarch Establishment of the Japanese Atomic Energy Research Institute is claiming to have developed a silicon carbide fibre which does not lose strength on heating to temperatures as high as 1500 ˚C. The fibre is produced by firstly heating the polymeric precursor fibre under vacuum using an electron beam, followed by firing in an inert gas.

4.10.5 SILICON CARBONITRIDE FIBRE

Details about the silicon carbonitride fibre, 'Fiberamic', being developed by Rhône Poulenc Recherches in Saint-Fons, France, emerged at the 1990 European Conference on Composite Materials (ECCM).

The fibres are being developed in an attempt to produce a material resistant to higher temperatures than silicon oxycarbide versions (such as 'Nicalon' and 'Tyranno') and that would match pure silicon carbide monofilament (where the high filament diameter of some 140 µm rules out most fabric production).

The composition of Fiberamic was stated by Perez and Caix from Rhône Poulenc as being 57 wt% Si, 22 wt% N, 13 wt% C and 8 wt% O. About 5–10% of the phases present are free carbon with the remainder $SiN_xC_yO_z$. The fibre structure is described as homogeneous with the free carbon distributed as small clusters throughout the material. The fibre diameter varies within a range of 12–16 µm, with 14 µm being the average. The elastic modulus is 200 GPa and the average tensile strength is between 1500 and 2200 MPa. The fibre retains some 80% of its room temperature strength when tested at 1400˚C, and after ageing in an oxidising atmosphere retains 85% of its strength after 100 hours at 1000˚C and 10 hours at 1200˚C.

The fibre is at the pilot plant production stage of development, but agreements have already been reached with Hercules for marketing and distribution (and possibly manufacture) in the USA.

4.10.6 SILICON CARBONITRILE FIBRES CAN BE WOVEN

Rhone-Poulenc of Paris, France, claims to have developed a range of silicon carbonitrile fibres which can be woven into fabrics.

The company says it produces the fibres by polymerizing chlorosilane precursors, which it then spins, cures and pyrolises. This produces fibres with diameters of only 10–15 µm, far finer than those produced by the more conventional technology of chemical vapour deposition onto a metallic or carbon fibre core, according to *High Tech Ceramic News*.

4.11 SILICON NITRIDE

Silicon nitride (Si_3N_4) is currently the most favoured monolithic ceramic for a number of applications in heat engines due to its high strength and high temperature capabilities. Current research is focusing on improving these properties even further.

4.11.1 PRODUCTION PROCESS FOR SILICON NITRIDE POWDER

An imide decomposition method produces silicon nitride powder which has high purity and readily controlled grain boundary composition, the Japanese developer claims.

Ube Industries Ltd of Tokyo reports that the grain size is 0.2–1.0 µm, giving the powder a high level of activity. A high proportion of alpha phase makes the material easy to sinter to high density. The company has a pilot plant which can produce 100 t a year of the powder using the process. Its aim is to test-produce components, such as turbo-chargers and parts for diesel engines and gas turbines. Ube Industries is considering full commercial production for 1992/93.

4.11.2 SILICON NITRIDE HAS NEEDLE-SHAPED CRYSTAL STRUCTURE

A silicon nitride-based ceramic which has a high fracture toughness and also a high temperature strength has been developed by NKK Ltd, based in Kawasaki, Japan.

The ceramic has needle-shaped crystals in the microstructure. The *Daily Industrial News* reports that NKK claims its material has a toughness in excess of 8 $MPa.m^{3/2}$ and that values of 12 $MPa.m^{3/2}$ have been achieved. No details have been received of the measurement technique by which these values were obtained. Room temperature strengths of the new silicon nitride are said to be in excess of 1 GPa.

4.11.3 HIGH TEMPERATURE AMORPHOUS SILICON NITRIDE

A form of silicon nitride which is amorphous and remains so even on heating at 1700°C in a nitrogen atmosphere is reported to have been developed by Toa Nenryo Kogyo KK, Tokyo, Japan.

According to the *Daily Industrial News*, the tensile strength of the material is claimed to be 2.5 GPa and the modulus of elasticity to be in excess of 200 GPa.

4.12 SUPERCONDUCTORS

This section, on superconductors with high transition temperatures (T_cs) contains by far the largest number of items, a fact which is indicative of the tremendous amount of research going on in this field. There is, however, a noticeable trend away from accounts of new superconducting phases, which dominated the literature during 1988, towards more application-oriented items. In particular, emphasis has been placed on the processing of thin films and wires of superconductors, reflecting the current and widespread viewpoint that it is in these forms that ceramic superconductors are most likely to be commercialised in the near future.

There is simply no possibility of every news item on these materials from 1990 being featured here — there have simply been too many. Rather, therefore, an attempt has been made to select key news items to give an overview of the developments being made.

Items start with an interesting report from the US Department of Energy suggesting that these materials may not be ceramic after all - with a hint of the inevitable repercusions concerning patents if true. We then consider processing, both in terms of powder production for subsequent use in making high-T_c superconductors and in the processing of the materials themselves. The section concludes with a look at some of the latest achievements in terms of properties which have arisen as a result of composition or processing. Further information can be found in the applications section.

4.12.1 HIGH TEMPERATURE SUPERCONDUCTORS MAY BE METALS

Tests performed by researchers at the US Department of Energy (DoE) indicate that high temperature superconductors may be metallic in nature.

Early tests of these materials found non-metallic electron levels. However, the recent experiments at the DoE were performed at liquid helium temperatures which give results that are easier to interpret. If the findings are confirmed it will not only affect current theories explaining the effects but may also lead to battles over patent rights.

4.12.2 MOISTURE RESISTANT POWDER

A moisture resistant powder for fabricating superconductors has been developed and commercialized by Dowa Mining Co of Tokyo, Japan, the *Nikkei Industrial Daily* reports.

The powder is produced by adding 5–10% of silver oxide to the yttrium barium copper oxide system. Purity is quoted as 99.9% and the powder is produced with four different mean grain sizes; 1, 3, 10 and 50 µm.

4.12.3 LEAD SUBSTITUTION STABILIZES SUPERCONDUCTOR

A more stable and homogeneous material results from the addition of 1% lead to a thallium-based high temperature superconductor, Japanese researchers say.

The usual metallic composition of such superconductors is 10% thallium, 30% calcium, 20% barium and 40% copper. However, the company substituted the lead for some of the thallium, resulting in a material which has a critical current density of 10 800 $A.cm^{-2}$ at −196°C in the absence of a magnetic field, the *Japan Economic Journal* reports.

Sumitomo Electric says it has produced superconducting tape 3 mm wide and 0.15 mm thick from the material, adding that such tapes are usually less than 0.1 mm thick.

4.12.4 US PATENT FOR THALLIUM–LEAD SUPERCONDUCTORS

A US company claims to have obtained the first US patent for a family of high temperature superconductors based on thallium–lead compositions.

From its headquarters in Wilmington, Delaware, Du Pont Co says that the materials are thallium lead strontium calcium copper oxides. They differ from standard thallium-based superconductors in that they contain lead and strontium in the place of barium. The compounds have superconducting transition temperatures in the range 120–125 K, among the highest known, according to Du Pont. The company adds that the thallium–lead materials may have advantages over other high temperature superconductors. For instance, they may be easier to process and less prone to flux creep, a problem which lowers a superconductor's resistance to magnetic fields.

Dr M.A. Subramanian conducted the research on thallium–lead superconductors at the company's Central Research and Development Department which has led to Du Pont's fourth US patent in the field.

Du Pont has signed separate co-operative agreements with Hewlett-Packard Co and the Los Alamos National Laboratory, the Argonne National Laboratory, and the Oak Ridge National Laboratory to develop high temperature superconductors.

4.12.5 FINE COPPER POWDER RAISES TRANSITION TEMPERATURE OF THALLIUM SUPERCONDUCTOR

The fineness of the copper powder used to make thallium-based superconductors affects the critical transition temperature (T_c) of the final material, according to the Japanese *Daily Industrial News*.

Scientists at Dowa Mining Co Ltd in Tokyo say the T_c of a superconductor made with copper particles in the range 10–20 nm is 120 K, compared with a material made using a more usual 1 µm powder which has a T_c about 15 K lower.

4.12.6 ULTRAFINE COPPER OXIDE POWDER

Dowa Mining Co Ltd of Tokyo, Japan, says it has devised a method for producing ultrafine copper oxide powder which it has used to make a high temperature superconductor. The company claims that the copper oxide powder is 10–20 nm in diameter and that the resulting thallium-based superconductor has a transition temperature 15 K higher than materials made using conventional 1 µm powder.

4.12.7 TEN MINUTE PREPARATION OF YTTRIUM SUPERCONDUCTOR

High quality bulk 1-2-3 phase yttrium barium copper oxide ($YBa_2Cu_3O_{7-x}$) is synthesised by fusing stoichiometric amounts of yttrium and copper nitrates with barium hydroxide, in air, in a process which takes only ten minutes, the US inventors claim.

The use of barium hydroxide for this purpose is not new. However, the scientists from the Center for Materials Research at Temple University in Philadelphia,' Pennsylvania, say that in the past these syntheses necessarily involved the use of water to remove the hydroxide, which ultimately interfered with the direct synthesis of the superconductor. Further, control of stoichiometry was also difficult.

In the new process, the researchers use barium hydroxide both as a flux and reactant. This exploits the fact that barium hydroxide has two hydration states which differ greatly in their melting points. The low melting octahydrate (78°C) acts as a solvent for the nitrates of the copper and yttrium. The higher melting anhydrous form (408°C) serves as a medium and reactant, allowing the mixing and subsequent reaction of the three metal ions. The barium hydroxide is thus itself consumed as a reactant.

The above reactions take about 10 minutes. However, complete decomposition and solid state reaction occur during subsequent thermal processing. This stage, in common with other bulk processing routes, takes an additional 18–24 hours.

The researchers say that the existence of an intermediate liquid phase allows homogeneous doping with minute amounts of controlled impurities, which may prove to be the biggest benefit of this technique.

The final material produced is better than 99% pure, with copper oxide (CuO) as the only other phase present. It exhibits a transition temperature of 92 K and a 15.5% perfect diamagnetic response (field cooled), 76% (zero field cooled).

4.12.8 FASTER PROCESSING FOR 1-2-4 PHASE YTTRIUM SUPERCONDUCTOR

A new process takes only three hours, one-tenth the time taken by conventional methods, to make a 1-2-4 phase yttrium barium copper oxide superconductor, according to its inventors, Kobe Steel Ltd, Kobe, Japan.

The technique uses hot isostatic pressing (HIPing) at 1000°C under 250 atmospheres of oxygen. This, claims Kobe Steel, reduces energy consumption considerably. The material produced is stable to 800°C.

The *Nikkei Industrial Daily* quotes a critical temperature of 90 K for a material partially substituted with calcium.

4.12.9 SOL-GEL ROUTE TO YTTRIUM SUPERCONDUCTORS

A sol-gel route to the production of films of yttrium-based high critical temperature (T_c) superconductor has been developed by the Japanese company Mitsubishi Cable Industries, based in Tokyo. The process requires the use of the acetic salts of yttrium, barium and copper in the appropriate ratio (1:2:3). These are mixed together and then dissolved in propionic acid solution. The product is coated onto silver tape 0.1 mm thick and fired at 900°C. The properties of the resultant film are quoted as a critical temperature (T_c) greater than 88 K and critical current density (J_c) of 6700 $A.cm^{-2}$.

The main advantage of the process is claimed to be the cheapness of the raw materials compared to the more conventional alcoxide precursors.

4.12.10 UNIAXIAL HOT PRESSING FABRICATES SUPERCONDUCTORS

High temperature ceramic superconductors are being fabricated by a uniaxial hot press which has been developed as a result of a joint development study between Tokyo-based researchers.

Mitsubishi Heavy Industries Ltd and Assistant Professor Yoshizaki at Tsukuba University claim they have used the press to produce a bismuth-based material with a critical current density, J_c, of 10 000 A.cm^{-2} at liquid nitrogen temperatures (77 K). They predict that they will be able to improve this so as to exceed 100 000 A.cm^{-2} within a year. Further, the body produced requires no subsequent processing to achieve the appropriate oxygen levels in the material, according to the *Daily Industrial News*.

4.12.11 ONE-STEP PREPARATION OF SUPERCONDUCTING THIN FILMS

A one-step method for synthesising and depositing superconducting materials on a substrate makes use of a high frequency electrical discharge to produce a plasma with temperatures up to 10 000 K, according to the US developers. High deposition rates are achieved, only atmospheric pressures are required and no post oxidation is needed, the team claims.

Scientists from the University of Minnesota in Minneapolis say that the process involves atomising stoichiometric quantities of mixed aqueous nitrate solutions of yttrium, barium and copper. They introduce the aerosol of liquid precursors into the radio frequency thermal plasma using oxygen both as a carrier and as the plasma gas. This eliminates the need for post-oxidation.

The 1-2-3 phase yttrium barium copper oxide ($YBa_2Cu_3O_{7-x}$) superconductor is deposited as thin films on gas cooled substrates of polycrystalline calcium oxide stabilized zirconium dioxide (ZrO_2). Subsequent characterisation shows the films to have transition temperatures (T_cs) above 80 K. X-ray diffraction analysis also reveals that the films grow with their c-axis perpendicular to the substrate.

4.12.12 3 nm THICK SUPERCONDUCTIVE FILMS

Superconducting thin films based on the bismuth series of superconductors measuring just 3 nm (30 Å) thick have been sucessfully produced for the first time, according to the Research Institute for Superconductors, based in Tokyo, Japan.

This has been achieved using metal organic chemical vapour deposition (MOCVD) on magnesia substrates. The crystallinity of the films is said to be good; however, a plot of critical temperature (T_c) versus film thickness indicates that these thin films have a value of only about 50 K. A T_c of around 80 K is not achieved until the film thickness is in excess of 10 nm (100 Å). Nevertheless, the *Japanese Industrial Journal* reports that the Institute believes these films could ultimately lead to bipolar superconductive transistors having far higher operating speeds than currently available.

4.12.13 THICK FILM SUPERCONDUCTORS FORMED AT ATMOSPHERIC PRESSURE

Pyrolysis of organic acid salts can produce long tapes of thick film, high temperature superconductors at atmospheric pressures, according to Showa Electric Wire and Cable Co Ltd of Tokyo which claims to have produced a tape of a 1-2-4 phase yttrium-based superconductor.

Researchers at the company spread the raw materials on the surface of a backing tape and pyrolysed them before processing the resulting product in oxygen for several tens of hours at atmospheric pressure. The *Japan Industrial Journal* reports that the final tape has a transition temperature of 80 K and a critical current density of 20 000 A.cm^{-2}, at liquid helium temperatures.

4.12.14 INCREASED FLEXIBILITY FOR YTTRIUM-BASED SUPERCONDUCTORS

US researchers have developed a high temperature superconductor which they claim is flexible enough to withstand bending and is capable of carrying a current ten times greater than household wiring.

Argonne National Laboratory, Illinois, says that the scientists bond the brittle 1-2-3 phase yttrium barium copper oxide to a silver foil backing. Prior to firing, the resulting ribbon is formed into tight coils quite easily, the Laboratory says.

The scientists mix the superconducting and silver powders with a wetting agent. They then squeeze the paste onto the silver ribbon before firing it at 950°C, just below the melting point of the metal. The Argonne team says that if microcracks appear in the ceramic when it is stressed the silver helps to carry the current past the break.

4.12.15 SUPERCONDUCTOR FILMS ON SEMICONDUCTOR SUBSTRATES

US researchers claim to have made a thin film of a high temperature superconductor on a silicon substrate.

The scientists from Xerox Palo Alto Research Center, Stanford University, and the University of Santa Clara say the breakthrough may have a considerable impact on supercomputer and other high technology applications.

Many research groups are trying to form high temperature superconductor films on a silicon substrate. If this can be done it would allow semiconductor devices to exploit the properties of superconductors and lead to superior performance and function.

However, one obstacle has been the adverse chemical reaction which occurs when a superconducting film comes into direct contact with a silicon layer during growth. This leads to poor quality films which do not effectively carry electric current.

The group focused on the yttrium barium copper oxide superconductor and to separate it from the silicon created a buffer zone, less than 100 nm thick, of yttria stabilised zirconia. A patent application has been filed.

4.12.16 SUPERCONDUCTIVE TAPE IN LENGTHS UP TO 100 m

Joint research and development between Hitachi Cable Ltd and Hitachi Ltd, both based in Tokyo, Japan, has resulted in the ability to produce superconducting tape in lengths of up to 100 m.

The product consists of the thallium-based superconductor encased in silver. The manufacturing route involves simultaneous rolling and pressing of the constituent materials.

The tape has a critical current density (J_c) of 10 000 A.cm^{-2} at liquid nitrogen temperatures; however, a maximum value of 22 000 A.cm^{-2} can be produced at the expense of other properties.

4.12.17 SUPERCONDUCTING FIBRES

According to a report in the Japanese *Daily Industrial News*, Tomoko Goto and her research group at the Nagoya Institute of Technology, Fujisawa, have developed a method for making high temperature superconducting fibres. Their process involves extruding a viscous polymer suspension of the oxide powder. The fibres produced are then pyrolysed.

4.12.18 CONTINUOUS RODS OF HIGH TEMPERATURE SUPERCONDUCTOR

Research performed at the University of Houston, Texas, USA, under the guidance of Professor Paul Chu, has resulted in a continuous process capable of producing rods of high temperature superconductor. Work to date has focused on the yttrium barium copper oxide material. The process involves the use of a zoned oven through which the material passes.

By regulating the temperature of each zone precisely, a green rod inserted at one end is sintered and correctly annealed by the time it reaches the other end. The oven is sufficiently well designed and built to permit reproducible material to be produced with a relatively high critical current density. This varies between 7500 and 20 000 $A.cm^{-2}$ depending on the strength of the magnetic field imposed during measurement.

4.12.19 SHOCK WAVES INCREASE CURRENT CARRYING CAPACITY

Exposure to shock waves substantially increases a high temperature superconductor's capacity for carrying current, US researchers claim.

The scientists, from the Lawrence Livermore National Laboratory in California, use a gas powered gun to fire lightweight plastic projectiles at powder samples of bismuth strontium copper oxide. The shock wave produced deforms the material, creating tiny defects in its crystal structure, they say.

Measurements show that the treated materials carry higher superconducting currents when exposed to magnetic fields. The researchers believe this is because the defects pin the lines of magnetic flux passing through the superconductor.

One of the team, William J. Nellis, says scientists have tried other techniques to create similar defects, such as neutron irradiation.

However, the shock wave approach is suitable for mass production, unlike these alternatives. The team now plans to try the technique on the yttrium barium copper oxides.

4.12.20 CURRENT DENSITIES REACH PRACTICAL VALUES

Australian scientists claim they can routinely make bulk high temperature superconductors with current densities large enough for practical applications.

Using their patented process, Dou Shixue and his team, from the University of New South Wales, Sydney, also form the material into wires and tapes.

According to the materials scientists, the bulk bismuth lead strontium calcium copper oxide superconductor has a critical current density of more than 12 000 $A.cm^{-2}$. The team also claims to have made four coils of tape. each about 1 m long, with current densities of about 2000 $A.cm^{-2}$.

Shixue admits that US scientists have reported bulk materials with higher current densities of around 17 000 $A.cm^{-2}$, but adds that the sample concerned was relatively short, about 1 cm long.

The team states that its success results from the development of a sintering process which produces a pure single phase of the material and from a method for incorporating silver into the ceramic, the malleability of the metal compensating for the brittleness of the superconductor.

"These coils have now reached the stage where they may possibly be used in an industrial application", according to Shixue. To test his claim, the scientists have sent some of their coils to the University's Department of Electric Power Engineering where Colin Grantham and co-workers will examine their suitability as fault current limiters.

4.12.21 YTTRIUM-BASED SUPERCONDUCTOR PASSES CURRENT OF 225 A

The largest steady state DC current ever to pass through a high temperature superconductor at liquid nitrogen temperatures (77 K) was recorded in a test at the Naval Research Laboratory (NRL), Washington DC, USA, according to scientists there.

A cylindrical pipe of $YBa_2Cu_3O_{7-x}$, measuring 25.4 cm (10 inches) long with a cross-sectional area of 1 cm^2, is said to have passed a current of 225 A. ICI Advanced Materials, Wilmington, Delaware, USA, made the pipe, and the current measurement test was performed at NRL by a research group headed by Dr Donald Gruber, Superintendent of the Materials Science and Technology Division, as part of a collaborative research effort.

According to ICI, the results have important implications for the use of high temperature superconductors as current leads in power engineering applications — such as motors, power convertors, underground transmission lines and energy storage systems.

4.12.22 SILVER-COATED TAPE HAS HIGH CRITICAL CURRENT DENSITY

Sumitomo Electric Industries Ltd, Osaka, Japan, has produced a bismuth-based superconducting wire with a critical current density of 25 000 $A.cm^{-2}$ at liquid nitrogen temperatures (77 K), a world record, the company claims. It appears that the bismuth strontium calcium copper oxide superconductor is coated with a layer of silver to form a tape of 0.13 mm thickness. The critical current density achieved is a result of crystal orientation and the "improvement on tight particle-to-particle boundaries, by means of a machining process", the company says in the *Japan Industrial Journal*.

4.12.23 CRITICAL CURRENT INCREASED IN TAPE CAST SUPERCONDUCTOR

Cooperative research between Asahi Glass Co Ltd and the National Research Institute for Metals, both of Tokyo, Japan, has resulted in improvements to the critical current density (J_c) of high temperature superconductors.

Using the 2-2-1-2 phase bismuth strontium calcium copper oxide system, they tape cast 15 μm thick layers of the ceramic, 3 mm wide by 50 mm long, onto silver foil. They then heated the material to 890°C, partially melting it, and then cooled it slowly to 870°C over a period ranging from 1 to 10 hours. This product is claimed to have displayed a critical current density of 140 000 $A.cm^{-2}$ at liquid helium temperature and under a 'high' magnetic field.

Meanwhile, researchers at the Radiation Chemistry Research Establishment of the Japan Atomic Energy Research Institute, based in Takasaki, claim to have increased the J_c of both yttrium and bismuth series high temperature superconductors by up to 50%. This has been achieved by irradiating the ceramics with a 3 MeV electron beam.

4.12.24 HIGH CRITICAL CURRENT DENSITY DISPLAYED BY GADOLINIUM-CONTAINING SYSTEM

A gadolinium barium copper oxide superconductor which displays a critical current density (J_c) of 40 000 A.cm^{-2} under a magnetic field as high as 4 T, has been developed by Nippon Steel Corp, based in Tokyo, Japan. The material is made by a melting process, followed by oxygen annealing.

Meanwhile, Sumitomo Electric Industries has developed superconducting wire based on the bismuth series materials which has a J_c of 11 000 A.cm^{-2} at liquid nitrogen temperature under a magnetic field of 1 T. The tape measures 0.15 mm thick by 4 mm wide and is made by sheathing the superconductor in a silver pipe before firing.

4.12.25 FILM WITH 8 000 000 A.cm^{-2} CRITICAL CURRENT DENSITY

Japanese researchers have made a film of a high temperature superconductor which they claim has a critical current density of eight million A.cm^{-2} at liquid nitrogen temperatures (77 K).

The Tokyo-based companies, Sumitomo Electric Industries Ltd and Tokyo Electric Power Co Inc, say that in their studies they formed a film of an yttrium-based superconductor on a magnesia substrate using laser-stimulated vapour deposition. The film has a critical transition temperature of 91.5 K, according to the *The Japan Industrial Journal*.

4.12.26 CRITICAL TEMPERATURE OF 130 K REACHED BY TWO SUPERCONDUCTORS

The developement of high critical transition temperature (Tc) superconductors with Tc values of about 130 K has been announced by two Japanese companies.

Hitachi Ltd, based in Tokyo, announced recently that it has produced a high temperature superconductor with a Tc of 130 K (−143°C), which it claims is the highest value to date. The composition is known to contain vanadium rather than copper.

Since the Hitachi announcement, Nippon Telegraph and Telephone Corp, Tokyo, has announced the synthesis of a compound that also exhibits superconductivity at a temperature of 130 K. Once again the composition contains vanadium rather than copper, the other elements present being bismuth and strontium.

4.12.27 PROCESS INCREASES RESISTANCE TO MAGNETIC FIELDS

Japanese researchers claim to have devised a process for making high temperature superconductors which have a high resistance to magnetic fields. Scientists at Sumitomo Electric Industries Ltd and Kansai Electric Power Co Ltd, both of Osaka, collaborated to produce a thin film of a bismuth-based superconductor. They say that the film's critical current only dropped from 240 000 A.cm^{-2} to 220 000 A.cm^{-2} when they applied a magnetic field of 1 T at liquid nitrogen temperatures. Usually the critical current drops to between one quarter and one hundredth of its original value. The two companies are now working to create wires from the material, the *Asian Wall Street Journal* reports.

4.12.28 WIRES RESISTANT TO HIGH MAGNETIC FIELDS

Cooperative research in Japan has produced a high temperature superconducting wire which has a critical magnetic field as high as 40 T, the partners say. As a result of independent work, a third group is making similar claims.

The two Tokyo-based organisations, Asahi Glass Co Ltd and the National Research Institute for Metals, make their wire from a bismuth-based superconductor. At present the length of the wire is limited to 10 cm and current research aims to increase this, according to the *Nikkei Industrial Daily*.

Kobe Steel Ltd of Tokyo has not disclosed the maximum field under which its wire remains superconducting, but says the material is a 1-2-4 phase yttrium barium copper oxide with a critical current density of 5000 $A.cm^{-2}$ at liquid helium temperatures (4 K). Kobe Steel is now working to increase the wire's critical current density by improving processing.

4.12.29 SUPERCONDUCTOR DISPLAYS ELECTRO-OPTIC PROPERTIES

Researchers at the Applied Electronics Laboratories of Nippon Telegraph and Telephone (NTT) Corp, based in Tokyo, Japan, have found that thin films of a yttrium-based superconductor will change their voltage output when exposed to certain wavelengths of electromagnetic radiation.

The workers deposited a 100 nm thick layer of the superconductor onto a strontium titanate substrate and applied a voltage across the component at liquid helium temperatures. When exposed to infra-red light with a wavelength of 1.3 Çmm, the voltage changed by 5 V in 5 ns. Applications for the discovery are being considered.

4.12.30 SUPERCONDUCTING PASTE

A high temperature superconductor paste which is suitable for producing circuit patterns on wafers is now available from Hayashi Chemical Industry Co Ltd, Kyoto, Japan.

Until now, the company says it could not commercially produce the paste because of the large grain size of the ceramic powder. However, the use of a co-precipitation mixing process, involving oxalic acid and ethanol, reduces the particle diameters by an order of magnitude to 0.3–0.8 μm.

Hayashi Chemical makes the paste by mixing nitric acid with yttrium, barium and copper, in the ratio 1:2:3. It then uses the co-precipitation method to produce the powder which it kneads together with a solvent binder and a viscosity promoter, and hardens. The paste has a viscosity of 6000 cP.

4.13 TITANIUM DIBORIDE

Titanium diboride (TiB_2) is gradually finding increasing applications, particularly as new methods have been found to produce the material. A major application at present is for the ballistic industry where it is used as a very effective armour. However other applications, including as an electrode in the metal smelting industry, are possible.

4.13.1 THERMITE REACTION YIELDS TITANIUM DIBORIDE

Pure, heat resistant titanium diboride (TiB_2) powders have been produced by self-propagating thermite reactions, according to Georgia Tech Research Institute of Atlanta, Georgia, USA.

To produce the powder, a heat resistant crucible is filled with powders of titania (TiO_2), boron oxide (B_2O_3) and a reducing agent, either aluminium or magnesium. The crucible is located inside a protective chamber and heated to between 600 and 1000°C with a hot wire. This hot wire causes the precursor materials to react exothermically, the heat of reaction ensuring that it goes to completion, and temperatures of 2000–3000°C have been reached, Georgia Tech says. The product is titanium diboride uniformly dispersed throughout an alumina or magnesia matrix. This latter can simply be dissolved to yield the TiB_2.

Alternatively, in one step an homogeneous, lightweight foam composite can be shaped into a crucible or other desired product.

By infiltrating the foam with metal, the material could be useful for various applications such as mechanical components, say the researchers.

The TiB_2 produced will withstand temperatures up to 3000°C, has a particle size of less than 1 μm, and conducts electricity while resisting acids and molten metals. In addition, the self-propagating nature of the process yields a particularly pure material, with no traces of the carbides found with the more conventional production processes.

Applications are said to include molten metal pipelines and processes such as aluminium smelting. The research team has already tested titanium diboride electrodes as potential replacements for carbon, currently used to smelt aluminium.

4.14 ZEOLITES

Zeolites are materials which have particularly fine and controllable pore structures which can be used to advantage in the catalyst industry. This item considers a report by a US company which focuses on the potential for zeolites in the production of speciality chemicals.

4.14.1 ZEOLITES FOR FINE AND SPECIALITY CHEMICALS

According to a recently released study by Catalytica, based in Mountain View, California, USA, zeolites could be the basis of a new generation of clean processes for the manufacture of fine and speciality chemicals. *Catalytica Highlights* reports that the study, 'Zeolites for the production of fine and speciality chemicals', shows how the unique properties of zeolites can be advantageous to organic synthesis. It points out the common pitfalls in zeolitic catalysis of organic syntheses and precautions that must be taken to avoid them. The study also provides a critical review of recent progress in specific types of reactions and concludes with a look at future directions that are likely to receive the most research and commercial attention.

4.15 ZIRCONIA

During the early 1980s zirconia (ZrO_2) was possibly the most widely researched material within the field of advanced ceramics. At the time it seemed as if every issue of every journal published carried at least one paper on some aspect of the processing or application of this ceramic. Unfortunately, some of the claims researchers made for its properties proved to be exaggerated, leading to disillusionment by many industrialists. Following a quieter period in the late 1980s interest is once again beginning to rise, particularly as a result of the excellent work which has been ongoing throughout this whole period by the Australians.

4.15.1 AUSTRALIAN RESEARCH COULD LEAD TO LOW COST ZIRCONIA

Research at the University of Melbourne, Australia, aimed at producing pigment grade titania from the mineral ilmenite, could be developed through a chemically related approach to offer a comparatively low cost route to pure zirconia.

A research team from the University's Chemical Engineering Department recently filed a patent on the titania process estimated to be worth about Aus$20 million a year to the University in technology licensing fees and royalties. The team have developed the process under contract to a potential commercial licensee. The group's latest contract is to scale up the technology to mini pilot plant status; however, Professor David Wood, Head of the Chemical Engineering department, has said that by using a somewhat different, but chemically related approach, the team had performed some basic experiments which had shown that this general area of inorganic chemistry could also offer a comparatively low cost route to pure zirconia.

Given that about 60% of the world's zircon is produced in Australia, the process could multiply Australia's earnings from zircon by providing the basis of a multi-million dollar value-added export industry.

4.15.2 BINDER-FREE ZIRCONIA

A new 3 mol% yttria stabilized zirconia powder which can be isostatically pressed and then green machined to a high surface finish without the need for binders has been developed by Z-Tech Pty Ltd, part of the ICI group, based in Melbourne, Australia. This offers major advantages to the partially stabilized zirconia ceramics industry.

The elimination of the need for a binder means that:

- firing times can be reduced since there is no need to heat very slowly during the early stages when binder burnout normally occurs, this will in turn mean that;

- significant increases in furnace productivity can be achieved with less complex firing schedules;

- there should be fewer problems with distortion and cracking of components as a result of binder burnout;

- higher sintered densities should be achievable since there will be no pores arising from the presence of the binder;

- recycling of scrap from the green maching stage should become easier to handle; and

- pollution in the form of furnace emissions arising from binder burnout will be eliminated.

This is a very important development and could significantly change, and improve, the processing of zirconia ceramics. Hopefully the technology can also be applied to other advanced ceramics.

4.16 OTHER

4.16.1 ALUMINIUM BORATE WHISKERS

Japanese researchers claim to have made aluminium borate whiskers with superior mechanical properties to silicon carbide and alumina competitors, at only a fraction of the cost.

Takao Kitamura, of the Government Industrial Research Institute in Takamatsu, led the research group which also included scientists from Shikoku Chemicals Corp, Marugame. The group's process uses aluminium sulphate and boric acid precursors, adding a further sulphate as a fusing agent; for instance, the researchers say the use of potassium sulphate yields whiskers with a uniform length and thickness. The team adds that whisker lengths of 10–300 μm with thicknesses of 0.5–10 μm are possible.

The researchers say that because the whiskers are 86% aluminium, they should have mechanical properties comparable with silicon carbide and alumina whiskers. Furthermore, the monocrystalline aluminium borate whiskers are likely to be stronger and more resilient than polycrystalline alumina whiskers and a fraction of the cost of silicon carbide whiskers.

The researchers claim their achievement contrasts with programmes mounted in the USA during the 1960s which unsuccessfully attempted to synthesise aluminium borate as a possible replacement for asbestos.

4.16.2 TRANSPARENT POLYCRYSTALLINE YTTRIUM ALUMINIUM GARNET

A Japanese company claims to have succeeded in making a transparent polycrystal of yttrium aluminium garnet (YAG).

Kurosaki Refractories Co Ltd from Kitakyushu says that the process involves sintering powder compacts under atmospheric pressure. The method could reduce manufacturing times to one tenth that of conventional techniques and also reduce costs significantly.

5. APPLICATIONS

This section contains a series of items that are concerned with product development — that is, the use of advanced ceramics in practical applications. Two points should be noted, however, while reading this section. Firstly, the broad diversity of fields in which advanced ceramics are finding applications; and secondly, the geographical distribution of the items. The vast majority originate in Japan and the Far East, with the USA holding second place and the rest of the world trailing significantly behind.

5.1 AEROSPACE

Apart from fibres for reinforcing metal-based composites and sensors and electronic components (of which the former lies outside the scope of this chapter and the latter two are dealt with in later sections) advanced ceramics currently see relatively few applications in the aerospace industry. There is considerable speculation in some market reports that ceramics may be used in aircraft engines, however they are not yet considered reliable enough in many instances for use in automobile engines, so extensive use in aircraft must surely be some considerable way off.

Set out below, however, are two items from the Battelle organisation in the USA in which advanced ceramic fabrics are discussed in terms of potential applications in space.

5.1.1 LIGHTWEIGHT FABRICS FOR SPACECRAFT PROTECTION

Scientists at the US Department of Energy's Pacific Northwest Laboratories (PNL) in Richland, Washington, have developed new materials that could be used to protect spacecraft from damage by micrometeoroids and orbital debris. The materials could also reduce the costs to deploy and operate the spacecraft by millions of dollars because of their light weight.

The advanced ceramic fabrics are made by weaving fibres of ceramics and metals, such as carbon and silicon, together. They are being developed and tested at PNL in cooperation with the US National Aeronautics and Space Administration (NASA).

The fabrics are claimed to have a strength comparable to steel and aluminium, but are only one-tenth the weight. In addition, the pliable fabrics can be shaped into a variety of forms and thickness for a range of applications. For protective shielding the fabrics would make up the outer layers of a multilayed shield designed to encompass a spacecraft or space station.

Laboratory testing of the materials conducted by the NASA Johnson Center in Houston, Texas, has shown the fabrics effectively destroy simulated micrometeoroids at high speeds. During this Summer's space testing, the materials will be exposed for three days. They will then be analysed as to their ability to resist ultraviolet radiation, high speed impacts and different types of radiation found in space, according to Brent Webb, a senior research engineer at PNL.

In addition to shielding, Battelle researchers, who operate PNL for the Department of Energy, are developing the materials for other uses in space. These are expected to include piping, radiators and fuel storage tanks for spacecraft.

5.1.2 FABRIC-BASED RADIATOR FOR SPACE APPLICATIONS

Using fabrics made from advanced ceramic fibres, US researchers claim to have developed a prototype, lightweight radiator for cooling the power supplies on spacecraft.

The researchers from Batelle's Pacific Northwest Laboratories, Richland, Washington, call their device the 'Rotating Bubble Membrane Radiator'. They say they make it from a range of advanced ceramic fabrics — based on fibres of metal oxides, carbon and silicon — as well as plastics and lightweight metals.

The radiator takes the form of a bubble. The power supply's liquid coolant (e.g. water, ammonia, nitrogen or oxygen) is pumped to the centre of the bubble and is sprayed onto the inner wall. The liquid cools as it comes into contact with the bubble's wall and the heat radiates from the outer surface into space. The cooled liquid is collected and returned to the power source.

Only a fraction of the energy generated by a power supply can be converted into usable electricity, according to Battelle. The remaining energy is in the form of heat which must be removed for the power system to function properly. As power levels increase, the amount of this excess heat also increases and so the size and weight of the required radiator becomes larger relative to that of the power system and spacecraft.

5.2 AUTOMOBILE

The automobile industry is predicted as being a major user of advanced ceramics within the next decade by many market reports. Indeed some components, such as turbochargers, are already being made out of ceramics and are appearing in production model automobiles. However, extensive use of ceramics is probably still some way off, largely because of an inability to make components with a high enough production yield and a guaranteed lifetime. Nevertheless, considerable effort is being expended in this direction as the following items show.

5.2.1 UK PROGRAMME ON ENGINE CERAMICS — RESULTS ANNOUNCED

The results of a £5.5 million series of projects aimed at more efficient vehicle applications have been presented to the Consortium on Ceramic Applications in Reciprocating Engines (CARE). The projects are the result of a programme set up by the UK Department of Trade and Industry (DTI) in October 1986 to manage collaborative research into ceramic engine components. Participating organisations received up to 50% support from the DTI. The research focused on four main areas:

- materials substitution;

- thermal insulation;

- turbocharging; and

- materials development.

Summaries of the results for each of these four areas are as follows.

Materials substitution

Work on materials substitution has included a study of the basic tribology of ceramics operating in valve train and cylinder component applications. Benefits in terms of reduced levels of friction and wear were found to be affected by the degree of the surface finish. Ceramic valve train components running in diesel engines were found to be capable of operating with minimal lubrication without any deterioration in performance. In a high performance petrol engine, ceramic bucket tappets exhibited the dual benefits of a low wear rate and a low mass; the latter feature offers the possibility of reducing friction or uprating the valve train performance. The fourth materials substitution project has investigated possible benefits of ceramics in the pre-combustion chamber of an indirect injection diesel engine.

Thermal insulation

The installation of a thermally insulated engine into a vehicle gave the anticipated advantages of reduced cooling system size, improved vehicle packaging and hence increased installed efficiency. In more fundamental studies, test rigs were developed to apply thermal shocks, more severe than those experienced in use, and hence to evaluate in an engine the properties of glass ceramics applied as insulating piston caps. In other fundamental studies using a 130 mm bore single cylinder diesel engine, small improvements in thermal efficiency were demonstrated after re-optimization, although at the expense of excessive NO_x emissions. The fourth thermal insulation project was concerned with the development of aluminium titanate port liners suitable for incorporation into a cast iron cylinder head.

Turbocharging

Two projects were undertaken by the turbocharger section, one concerning an automotive unit and the other a turbocharger suitable for a medium speed engine. Silicon nitride and silicon carbide rotors for the automotive turbocharger were manufactured by green machining and also by injection moulding. The latter is believed to provide a potentially viable manufacturing route. The ceramic bearings for the large turbocharger operated entirely satisfactorily without perceptible wear.

Materials development

Within the materials section, zirconia powders were manufactured by the potentially more economic electro-fusion route, as opposed to the usual chemical precipitation method. The resulting material was shown to be suitable for plasma sprayed coatings, which were applied to the exhaust valve of a medium speed engine. The remaining materials project concerned the design from first principles of novel thermal barrier coatings; at least one of these coatings is being followed up with a view to commercial applications. A list of participating organizations can be obtained from the Department of Trade and Industry. Details of individual projects will be published individually by the organizations concerned at their commercial convenience.

5.2.2 DESIGN FINALIZED FOR CERAMIC GAS TURBINE ENGINE

The New Energy and Industrial Technology Development Organization (NEDO), Tokyo, Japan, is developing a ceramic gas turbine engine intended to produce about 300 kW of power. The design has been finalized and work is underway towards a completion date in 1996.

NEDO are currently at the stage of studying a metallic prototype engine to determine its characteristics while, simultaneously, starting to build the ceramic version.

According to the *Nikkei Industrial Daily*, the ceramic engine will be finished by the end of 1993, leaving three years to study and develop it.

5.2.3 CHINA'S FIRST CERAMIC ENGINE

In common with most of the rest of the developed world, China has been working towards the design and manufacture of a ceramic-based engine which requires little cooling.

According to the *Japan Economic Journal*, the Chinese have produced a prototype engine which has successfully completed the 1500 km trip from Shanghai to Beijing. With a capacity of 6200 cm^3, the engine generates 130 horsepower and needs no water coolant.

It was jointly developed by the Chinese Academy of Sciences and the China State Ship-building Corp.

5.2.4 JAPANESE AIM FOR LOW POLLUTION ENGINE

Ceramics will play an important part in the Japanese Government's initiative to reduce pollution from automobile engines, according to the *Nikkei Industrial Daily*.

The Tokyo-based Ministry of International Trade and Industry (MITI) will lead the project, due to begin in 1990, but both government and private enterprise will be involved. One of the project's aims is to develop a gas turbine engine with reduced pollution. This will involve the use of ceramic components to allow higher temperatures to be used, the paper reports.

5.2.5 ISUZU DEVELOPS CERAMIC DIESEL ENGINE

Japanese researchers claim that their ceramic diesel engine has 30% higher output and fuel efficiency than similar engines made with conventional materials.

The Ceramics Research and Development Center of Tokyo-based Isuzu Motors Ltd says its V6 diesel turbo-compound engine has a maximum output of 180 horsepower.

Isuzu uses silicon nitride for the turbocharger rotor, valves, piston rings and cylinder liners, and a combination of silicon nitride and aluminium titanate for the turbine scroll which reduces the amount of heat emitted from the outer walls of the combustion chamber, according to *High Tech Ceramics News*.

Isuzu Motors formed the ceramics center in 1988, three years after it developed its first ceramic adiabatic engine. Since then the group has worked hard to introduce an increasing number of ceramic components into engines.

5.2.6 LOW POLLUTION AIM FOR ISUZU MOTORS

Isuzu Motors of Tokyo, Japan has started the development a new ceramic engine with the specific aim of lowering pollution levels.

Isuzu says it has already made ceramic engines which can run for 5000 hours, and have reduced NO_x and soot emissions, coupled with fuel consumptions which are lower by 30%. The company has given the project a budget of ¥6000 million, according to the *Nikkei Industrial Daily*.

Isuzu is among the world leaders in the development of ceramic engines. As early as the 1989 Tokyo motor show the company demonstrated a V6 engine with ceramics used for the piston rings, the cylinder liner, head liner and ceramic surfaces on the camshaft.

5.2.7 TOYOTA INSTALLS CERAMIC TURBOCHARGER

Toyota Motor Corp, Tokyo, Japan, is in the process of making changes to some of its models by installing ceramic turbochargers as standard.

According to the *Nikkei Industrial Daily*, the 'Celica GT-4' underwent the change in September 1989, and the 'MR-2' in October 1989.

Toyota is now considering making the same move with the 'Soarer' and 'Supra' models.

5.2.8 KYOCERA STARTS ROTOR PRODUCTION

Kyocera Corp, based in Kyoto, Japan, has started commercial production of silicon nitride ceramic turbocharger rotors and is producing about 10 000 pieces per month for the Toyota Motor Co (Aichi Prefecture). This production rate is expected to increase to 30 000 per month by the summer of 1991. Kyocera are currently manufacturing 5.2 cm diameter rotors, but the *Nikkei Industrial Daily* reports that they have spent ¥6000 million on this project and are hoping to be able to make rotors 3.5–13 cm in diameter in the future.

5.2.9 CHRYSLER LEADS RESEARCH INTO ROLLER BEARINGS

The US car producer, Chrysler Corp, has been investigating the use of ceramic roller bearings in cooperation with the Sullivan Mining Corp of San Diego, California, USA.

The bearings will be for Chrysler's 2.2 and 2.5 litre, four cylinder engines and, according to the companies, nitride bearings have already undergone 50 000 mile durability tests with virtually no measurable wear. This is attributed to the fact that the ceramic made by Sullivan Mining has reduced brittleness compared to other silicon nitrides.

A further benefit to come from the use of the ceramic is that the bearings are said to be cheaper than their metal counterparts.

5.2.10 CERAMIC–COATED EXHAUST MANIFOLD

Spray coating with a patented ceramic produces exhaust manifolds made from pressed sheet steel which are lighter and have superior thermal resistance to their rivals, the US manufacturer says. Cyb Tool and Die Manufacturing Inc, based in Canton, Michigan, says each manifold weighs only 4.1 kg (9 pounds) rather than 17.3 kg (38 pounds) for a cast iron counterpart. The higher thermal resistance results in lower temperatures under the bonnet, which in turn means reduced deterioration of neighbouring parts. The principal disadvantage of the 0.5–0.8 mm (0.020–0.032 inch) ceramic coating is that the extra heat retained within the manifold degrades the performance of catalytic converters. According to *Automotive News*, both Chrysler Motors and Navistar International Corp have tested or will be testing prototypes of the manifold on truck engines. Other car and truck companies have also expressed an interest.

5.3 BIOCERAMICS

The use of certain ceramics for biomedical-based applications, including reconstructive surgery and dentistry, is increasing as developments in the materials themselves occur. The major advantage that ceramics possess over metallic rivals is that many of them display a signficant degree of biocompatibility. Indeed some (particularly those based on calcium phosphate) are described as bioactive, that is they encourage natural bone and tissue regrowth, rather than simply existing passively within the body.

The number of items in this section is significant in itself in that in the first volume of the source book (1988) there were only a few items on bioceramics in the 'Materials' section and none at all in the 'Products' section. The range of ceramic materials being considered is also growing; however it is particularly noteworthy that hydroxyapatite (a synthetic version of natural bone) now features in nearly half the items.

5.3.1 CARBON FOR IMPLANTS

Cardio Carbon Ltd, based in Swansea, UK, has announced the availability of a highly biocompatible material, Turboform Carbon, for biomedical implants. The ceramic-type material is a form of vitreous carbon which can be moulded to give complex three-dimensional shapes with excellent surface finishes. Applications for the material are seen in many fields including renal, dental, maxillo-facial, orthopaedic, neurosurgical, percutaneous access, micromachines, long term and targeted drug delivery, and cardiovascular.

The company also offers a complete manufacturing design service.

5.3.2 DIAMOND-LIKE COATINGS FOR MEDICAL IMPLANTS

Spire Corp, based in Bedford, Massachusetts, USA, has announced a diamond-like coating (DLC) for medical orthopaedic implants. The low temperature dual ion beam assisted deposition (IBAD) process produces a graded film interface which results in a highly adherent coating.

Characteristics include high fretting and wear resistance, a low coefficient of friction, high electrical resistivity, extreme hardness and chemical inertness. The DLC is claimed to be well suited for orthopaedic implants, bearings in ventricular assist devices and prosthetic systems with potential in-vivo problems.

5.3.3 CERAMIC COATINGS EVALUATED FOR METALLIC IMPLANTS

Scientists at Kyushu University are studying the *in vivo* biocompatibility of metal implants coated with ceramics.

The research team, led by T. Kanemanru from the Department of Orthopaedic Surgery, is investigating alumina, titanium oxide, titanium nitride and hydroxyapatite coatings. The scientists believe the composites have a potential use in the stems of total hip prostheses.

5.3.4 HYDROXYAPATITE COATING FOR TITANIUM TEETH ROOTS

Pure titanium teeth roots which are then coated with a layer of hydroxyapatite to give them biocompatibility have been developed by researchers in the Medical and Dental departments of Tokyo University, Japan.

Existing artificial teeth roots are usually either uncoated metal titanium or, increasingly, alumina. However, both are regarded as having disadvantages, with the the primary problem being the attachment of the roots to the jaw. Usually this is done by use of a bone cement, but the heat given off during setting of the cement can cause necrosis (death of the adjacent cells), resulting in subsequent loosening at the interface of the bone and the implant.

The use of a hydroxyapaptite (HA) coating should reduce this problem. HA is the ceramic version of the mineral constituent of bone and as such has far greater biocompatibility than the bioinert titantium or alumina. Since HA is bioactive, bone will grow onto the ceramic directly, eliminating the need for a cementitious adhesive.

The *Daily Industrial News* reports that the coating has been achieved by covering the titanium teeth roots with a mixture of a calcium-based organometallic salt and an organic phosphate and sintering the system at 600–800°C. The result is a 1 μm thick coating of HA. This process is then repeated five times to provide a final coating some 5–6 μm thick.

5.3.5 ARTIFICIAL TOOTH ROOT

A company from Kyoto, Japan, has developed an artificial tooth root called 'POI' which, it claims, is thinner and more flexible than conventional implants and so will simplify dental operations.

According to the *Japan Economic Journal*, Kyocera Corp makes the root from ceramics, including titania. The company is to market the material, together with equipment for the operations, at a cost of ¥23 000–28 000 a unit.

5.3.6 ARTIFICIAL BONE

An artificial ceramic bone has been introduced by NGK Spark Plug Co Ltd, based in Aichi Prefecture, Japan. The ceramic, which has the trade name 'Ceratite', is a hybrid of hydroxyapatite and tricalcium phosphate and is sold in 117 different shapes ranging from balls to pieces similar in shape to a thigh bone. Chugai Medicine Co is acting as the sole agent for the products in Japan, and the material costs about ¥10 000 per gram.

5.3.7 HYDROXYAPATITE PASTE FOR BONES

US researchers are developing an hydroxyapatite paste for injection into bones that that have been weakened by disease. The scientists, from Norian Corp, based in Mountain View, California, USA, call the material SuperBone.

The paste is expected to be used to strengthen damaged bones, for example, by osteoporosis. When injected into the centre of a bone from which the marrow has been removed it solidifies in two to three hours. Then cells normally present in the bone and marrow can tunnel into the SuperBone implant, regenerating the bone, the *Washington Post* reports.

5.4 CUTTING TOOLS

Cutting tools are predicted by many market reports to be one of the fastest growing fields of application for advanced ceramic-based materials. Within this field the current trend is very much towards the use of coatings on substrates, as illustrated by the items below.

5.4.1 CUTTING TOOLS FOR STEEL HAVE TITANIUM CARBIDE/TITANIUM NITRIDE COATING

Sumitomo Electric Industries Ltd, based in Osaka, Japan, has developed a ceramic-coated cutting tool for steels, according to the *Nikkei Industrial Daily*. The tool is made of tungsten carbide and Sumitomo Electric has added a coating of titanium carbide/titanium nitride. The coating is achieved by vacuum evaporation and is said to double the life expectancy of the cutting tool tips.

5.4.2 SAW CHIP LIVES EXTENDED BY TITANATE COATING

A ceramic titanate film endows saw chips with wear resistance and gives them a Vickers hardness of 2000, according to a Japanese company. Nippon Metal Industry Co Ltd from Tokyo is now marketing a chip saw called 'Stainless Line' for cutting stainless steels. The chips in the saw have an original design which, together with the coating, helps to prevent fusion of the blade with the alloy. This gives the saw a longer life than its competitors, the company says.

5.4.3 SURFACE COATINGS AVAILABLE FOR CUTTING TOOLS

A series of ceramic coatings are now available from the Dutch company, Hauzer Coating Centrum. Hauzer has extended its coatings from titanium nitride (TiN) to include titanium carbonitride (TiCN), titanium aluminonitride (TiAlN) and chromium nitride (CrN). Each of these offers advantages over TiN; for example, TiCN coatings display superior hardness and coefficient of friction, whilst TiAlN has better heat and corrosion resistance. CrN coatings are intended primarily for the polymer processing industries, where the coatings display excellent wear properties. All the coatings can also offer improved product lifetimes over TiN-coated parts, according to *Materials Edge*. The coatings are applied to metal cutting tools using physical vapour deposition (PVD) techniques and large scale production is available. At the present time, Hauzer is focusing on industrial drills and cutting tools.

5.4.4 DISPOSABLE CERAMIC TOOLING

Mitsubishi Metals Corp, based in Tokyo, Japan, claims to have developed cubic boron nitride (CBN) tooling which is sufficiently cheap that it can be considered disposable. The CBN is in the form of a coating on a substrate and Mitsubishi Metals have succeeded in reducing the thickness of coating needed to about one third of that previously used. This has been achieved by the development of a special solder, according to the *Nikkei Industrial Daily*. The reduction in thickness of the coating has resulted in a similar drop in the price of the CBN tool. These are now available at a cost of ¥4300 per piece.

5.5 ELECTRONICS

Many ceramic components have made notable in-roads into the electronics industry over the last decade. Often this has been due to the inherent properties of the materials of which they are made. It has been said that this is currently the only section of the advanced ceramics industry to be showing a profit in terms of money expended on research and development and income from sales. While this is an oversimplification, it is certainly true that the markets for ceramics in the electronics industries are the largest by a considerable margin.

This section contains a diverse range of materials and applications. It starts with a consideration of dielectric and related types of materials, then examines the use of diamond coatings before considering substrate materials and then silicon technology.

5.5.1 PIEZOMOTOR

Researchers at the German car manufacturer, Daimler Benz, have developed an electric motor which they say has no moving parts and which produces high torque even at low speeds. The motor is based on piezoelectric ceramics, which change shape when a potential difference is applied to them.

The motor consists of a stator which comprises a flat, stationary disc divided into segments and edged on its lower surface with piezoelectric elements based on lead zirconate titanate. By applying a potential difference to each of the segments in turn, the stator changes shape, developing a moving waveform on its surface. Because the rotor is clamped to the stator with the appropriate surface-to-surface contact forces, it is forced to rotate by the friction between it and the travelling waveform on the surface of the stator.

The travelling waves are actually produced by the superposition of two separate, stationary waveforms and it is this, combined with the nature of the contact between stator and rotor, which results in a motor which produces very high torque, even down to extremely low speeds. Daimler Benz has not revealed details of the contact region, except to say that it involves a polymer coating on the rotor.

The prototype versions of the motor are said to be quiet and lightweight compared to conventional motors (although it does produce significant amounts of noise at ultrasonic frequencies — a feature which needs improvement). It can be operated at slow speeds and will reverse swiftly. In addition, Daimler has claimed an operational life of up to 15 000 hours.

So far, applications are said to include winding car windows; however, the company believes that there are many more potential applications.

5.5.2 LEAD ZIRCONATE TITANATE BUBBLE DETECTOR

Japanese researchers say that they have developed a device which uses ultrasound to detect gas bubbles as small as 0.5 mm in a variety of fluids.

Murata Manufacturing Co Ltd of Kyoto developed the detector which uses lead zirconate titanate (PZT), a piezoelectric ceramic. The company intends it for applications where the presence of bubbles can be dangerous, such as in intravenous drips or automotive brake systems.

The detector is a cylinder, 10 mm in diameter and 10 mm in length, which contains the PZT element. The latter is sensitive to ultrasonic waves and detects any attenuation in an ultrasonic signal passed through the liquid.

The *Japan Industrial Journal* says that Murata Manufacturing's device can be used in opaque tubes, unlike conventional optical bubble detectors.

5.5.3 CERAMIC OSCILLATOR CAN BE SURFACE MOUNTED

Murata Manufacturing Co Ltd of Kyoto, Japan, has made a ceramic oscillator which it says can be surface mounted and measures 5 x 4 x 0.5 mm. The oscillator's design allows it to operate without a spring terminal, permitting the thickness to be reduced to 0.5 mm. The casing is thin alumina.

The resonant frequency is 200 kHz and the operation range is 50–600 kHz. The device has a faster rise time than a quartz oscillator, Murata Manufacturing reports.

5.5.4 LAMINATED CAPACITORS WITH HIGH DIELECTRIC CONSTANT

A ceramic which can be sintered at low temperature and has a dielectric constant twice that of conventional materials has been developed, NEC Corp of Tokyo, Japan, claims.

NEC says it is using the ceramic to make laminated capacitors with volumes just one seventh those of previous models. The capacitors are as small as tantalum chip versions with the same capacitance and require just one quarter the area for installation. In addition, the insulation of the ceramic capacitor does not breakdown at high temperatures.

5.5.5 CERAMIC ANTENNA FOR WEAK COMMUNICATION SIGNALS

The development of a ceramic antenna which enables very weak signals to be received from communication satellites using a unit one-sixth the size currently needed has been announced by Murata Manufacturing Co Ltd, based in Kyoto, Japan.

This has been achieved by making the antenna out of a ceramic material with a high dielectric constant — the composition of the material has not been released. The dimensions of the antenna are: width 70 mm, height 70 mm and thickness 20 mm. The shape is described as that of a homebase plate, as used in the game of baseball. According to the *Japanese Industrial Journal,* Murata believes that the antenna can be successfully used in global positioning sytems and that this will be the first time a ceramic antenna has been used for this application.

5.5.6 10 MHz, LOW LOSS DIELECTRIC RESONATORS

A Japanese company has commercialized a range of 10 MHz dielectric resonators which are made from low-loss ceramics. Murata Manufacturing Co Ltd, Nagaokakyo, says the resonators form part of their 'Resomix R' series.

According to the *Japan Industrial Journal,* the principal application for the resonators is as oscillator elements and filters for satellite broadcast receivers. Production was 50 000 pieces a month in January 1990, each resonator costing ¥150–700.

The dielectric is composed of barium zinc tantalates, carefully processed to avoid the formation of lattice dislocations to ensure that losses stay low, the company says.

5.5.7 DIAMOND FILM-BASED TRANSISTOR

A Japanese company claims to have developed a transistor based on a thin film of diamond.

Sumitomo Electric Industries Ltd, Osaka, says the transistor is composed of a boron-doped diamond film, coated on either side by titanium and aluminium electrodes respectively. Coating was achieved by chemical vapour deposition (CVD). The *Nikkei Industrial Daily* says the principal limitation of the device is its inadequate current carrying ability.

5.5.8 DIAMOND DIODES

Research performed at Tohkai University in Tokyo, Japan, under the guidance of Professor Iida, has resulted in the development of a diamond-based diode.

The diode was produced by depositing n-type synthetic diamond in a layer on a silicon substrate, followed by a layer of p-type diamond, the electrical resistivity of which was about 100 ohm.cm. The major advantage of the diode is claimed to be its ability to withstand high temperature or high radiation environments. For example, an upper limit in excess of 500°C has been reported.

5.5.9 HIGH POWER HYBRID PACKAGES

High power hybrid packages are now available from Ceradyne Inc of Costa Mesa, California, USA.

The company is a supplier of custom brazed ceramic–metal assemblies for aerospace markets and says it is now applying this technology to the packages, which have:

- hermetic alumina ceramic feedthroughs and copper lead pins with improved reliability and current ratings over glass seals;

- high temperature (780°C), all brazed construction which permits users to safely employ high melting solders to attach devices; and

- injection moulded tungsten–copper or molybdenum–copper housings that provide low thermal expansion, high thermal conductivity, and reduced cost.

5.5.10 LOW DIELECTRIC CLOTH FOR PRINTED CIRCUIT BOARDS

Printed circuit board manufacturers can now use a glass cloth made from a silica-rich fibre which has a low dielectric constant and is suitable for high speed electronics and high density wiring, according to the Japanese developer.

Unitika Ltd of Osaka says it uses Akzo NV's 'Silica Fiber' to make the cloth, samples of which are now available. The company is examining the potential uses of the material, particularly at high frequencies, such as in portable telephones or receivers for direct broadcasting.

5.5.11 POWER HYBRID PACKAGES BONDED TO ALUMINIUM NITRIDE

A process which allows power hybrid packages to be bonded to aluminium nitride substrates, in addition to the more usual alumina substrates, has been developed by ATI of San José, California, USA.

The advantage of using aluminium nitride is its superior thermal conductivity over alumina. Working jointly with Powerex Inc, ATI have produced double sided, metallized aluminium nitride substrates which show greater wettability and adhesion to both the underside of a high power microwave hybrid package on one side and to a heat dissipating copper bus-bar on the other. ATI believe that this development opens the way to the production of a new family of ready-to-solder aluminium nitride substrates).

5.5.12 POLISHED ALUMINA SUBSTRATES

Polished alumina substrates with a surface finish of less than one microinch (less than 25.4 nm) are now being offered by Materials Research Corp's (MRC) Hybrid Products Division. Called 'MRC Ultragrade Superstrates', they are produced by using proprietary lapping and polishing techniques on tape cast 99.6% alumina. The resulting polished substrate is claimed to provide the smooth surfaces needed by manufacturers of thin film microwave and resistor networks with lines and spacings of less than one microinch.

5.5.13 STEEL–CERAMIC SUBSTRATES

Sophisticated electronics can take advantage of the strength of steel as a result of the development of a steel–ceramic substrate, according to Heraeus Silica and Metals of Weybridge, UK. The company says it has replaced ceramic substrates with multilayer materials in which steel–ceramic bases are coated with 'Heramic', a brand name dielectric. The materials can also be used for strain gauges and power resistors.

5.5.14 LOW MELTING POINT GLASS TO REPLACE GOLD–TIN SOLDER

A glass which has a melting point of only 360°C has been developed by the Japanese company, Nippon Electric Glass (NEG) Co Ltd, based in Shiga Prefecture.

The *Daily Industrial News* reports that the glass, 'LS-1401', is intended to replace gold–tin solder for use in ceramic packages such as pin grid arrays (PGAs) or leadless chip carriers (LCCs). The price of the glass, which comes as a frit, is ¥12 000 a kilogram.

5.5.15 CHEMICAL VAPOUR DEPOSITION OF SILICON AND GERMANIUM CRYSTALS

A procedure for depositing crystals of silicon and germanium with a speed some 50 times that currently available has been developed by NEC Corp, based in Tokyo, and its subsidiary Anelva Corp of Fuchu, Japan.

The technique uses an ultra-high vacuum chemical vapour deposition (CVD) method, in which the pressure needs to be reduced to just 10^{-10} torr and the temperature is 650°C. According to the *Japan Industrial Journal*, the companies believe that the process will open the way to the production of thin films (typically about 50 nm) of these materials for the next generation of bipolar large scale integrated circuits.

5.5.16 LARGE SILICON SINGLE CRYSTALS

Large single crystals of high purity silicon have been sucessfully produced by the continuous Czochralski process using a double crucible system by NKK Ltd, based in Kawasaki City, Japan. The crystals measure six inches (15.24 cm) in diameter and are 2 m long. NKK started shipping samples to device manufacturers during 1990.

5.5.17 LARGE AREA SILICON SOLAR CELL

A solar cell with an area of 225 cm^2 made from polycrystalline silicon has a photoelectric conversion efficiency of 14.5%, according to its Japanese developer.

Kyocera Corp of Kyoto says it achieved this efficiency by:

- reducing defects in the material;

- reducing reflection from the surface by coating the device with a silicon nitride film using chemical vapour deposition (CVD); and

- using screen printing to form very fine electrodes.

The company carried out the work as part of the Ministry of International Trade and Industry's 'National Sunshine' project, under an agreement reached with the New Energy and Industrial Technology Development Organization (NEDO). NEDO now plans to produce a 100 cm^2 cell with an efficiency of 18%, and a 225 cm^2 device with 17% efficiency by 1992.

5.5.18 AMORPHOUS SILICON SOLAR CELL

Sanyo Electric Co Ltd of Tokyo, Japan, has developed a single layer, 100 cm^2 solar cell made from amorphous silicon which it claims has an optoelectronic energy conversion efficiency of 10.2%.

The company says the amorphous film is formed in a high vacuum so that impurities — such as oxygen, nitrogen and carbon — are reduced to one-tenth their usual levels. Sanyo Electric has also designed the cell to improve conversion efficiency.

The surface area of a house roof in Japan is about 30 m^2 on average and, if it is facing south, the conversion efficiency of solar cells must exceed 10% to generate the 2 kWhours of electricity typically used, according to the company. Conventional single crystal, polycrystal and amorphous laminated cells have been produced which meet this requirement, but this is the first single layer device to do so.

5.5.19 HIGH SATURATED FLUX DENSITY FOR MAGNETIC FLUID

Japanese researchers claim their magnetic fluid has the highest magnetism of any such material. The fluid has a saturated magnetic flux density of 0.17 T, four times that of its conventional rivals, say the scientists from the National Research Institute for Metals (NRIM) in Tokyo.

Magnetic fluids are usually based on iron oxide, however, the NRIM group uses powders of iron nitride. The higher saturated flux density could lead to practical applications of the material in magnetic fluid dampers and actuators. The team are now looking at potential uses with the assistance of outside enterprises, according to the *Japan Industrial Journal*.

5.6 FILTERS

Filters for industrial, food, biotechnology and medical applications are increasing in usage. The advantage of ceramics in such applications is their chemical inertness and temperature resistance allied with the ability to clean and reuse them. The demand for increasingly fine pore sizes is reflected in the development work being carried out.

5.6.1 ULTRAFINE FILTER

NGK Insulators Ltd, based in Nayoya City, claims to have developed a ceramic ultrafine filter which has pore sizes in the range of 5–50 nm. The filter is made of alumina, the *Nikkei Industrial Daily* reports.

5.6.2 ION SEPARATION MEMBRANE

Scientists at the Leningrad Institute of Cytology, USSR, say that they have made an artificial membrane which is able to select out certain ions from all others. The membrane has been incorporated into a computerised system for purifying a variety of materials, according to *Soviet News and Views*

5.6.3 HOLLOW CERAMIC FIBRES TO BE USED AS FILTERS

NOK Corp, based in Tokyo, Japan, has developed a range of filters made from hollow ceramic fibres.

The filters consist of hollow α-alumina fibres having a diameter of 1–2 mm and a wall thickness less than 300 μm. The filters will remove fine particulates from high temperature gases, acids and alkalies.

NOK intends to use the filters in applications such as membrane reactors, processing of industrial wastes, cleaning of high temperature waste gases and in the biotechnology, medical and food industries. The *Japan Industrial Journal* reports that the fibres were originally developed as a result of cooperative research between NOK and Professor Morooka of Kyushu University.

5.6.4 FILTRATION MODULES

A series of ceramic microfiltration and ultrafiltration membrane modules which have nominal pore sizes ranging from 5 to 200 nm are the first commercial products of the CeraMem Corp, based in

Waltham, Massachusetts, USA. The membranes are applied to high surface area porous honeycomb ceramic monoliths. According to the *Membrane Quarterly*, the modules are relatively small in size, measuring 2.54 cm in diameter by 30.48 cm long and contain just 1300 cm^2 of membrane area. However, industrial scale modules are under development which will contain 12.1 m^2 of membrane area and should be commercialized during 1991.

The current 'Series LM' ceramic microfiltration and ultrafiltration membrane modules are to be introduced to the beverage, chemical, dairy, food and pharmaceutical markets.

5.7 OPTICAL

5.7.1 SINGLE MODE INFRA-RED FIBRE

'Irguide SM1' is the world's first single mode infra-red optical fibre, according to the French company which has produced it. Le Verre Fluoré from Vern sur Seiche makes the fibre from high purity fluoride glasses. The company says the fibre offers a wide window of transparency and is designed for sensor applications and the development of telecommunication links in the middle of the infra-red range. The diameter of the core is 11 μm and that of the cladding 125 μm. The fibre is single mode at 2.2 μm and has a window of transparency which extends to 4 μm, Le Verre Fluoré says. The minimum attenuation is less than 50 dBkm^{-1} and occurs at 2.6 μm.

Le Verre Fluoré has made fibres up to 1 km in length, claiming they are the longest yet produced in fluoride glass. The company believes that the low attenuation in the middle of the infra-red range (2.55 μm) makes the fibre suitable for intercontinental telecommunications. At this wavelength, the attenuation is ten times less than that of silica fibres currently in use, according to Le Verre Fluoré. Experimental links at 2.55 μm have been built and the company has started work on the study of propagation phenomena within single mode fluoride fibres, particularly with respect to non-linear effects.

5.7.2 FAST CARBON COATING PROTECTS OPTICAL FIBRES

A 40 nm thick film of carbon protects an optical fibre from chemical attack, according to the US scientists who claim they have developed a method for applying a coating less than 1 s after the fibre is formed.

Researchers at the AT&T Bell Laboratories in Princeton, New Jersey, say the carbon film offers protection against moisture and hydrogen, the two greatest corrosion threats, without degrading the optical properties. The carbon coatings perform better than polymer films in many hostile environments, such as those encountered in some underwater installations or at the tip of an oil drill, *Ceramic Industry* reports.

5.7.3 THIN LIQUID CRYSTAL DISPLAY

Kyocera Corp, based in Kyoto, Japan, claims to have succeeded in producing a liquid crystal display (LCD) which is only 1.9 mm thick.

The display consists of two thin (0.7 mm) glass substrates with the liquid crystals sandwiched in between. One of the substrates incorporates an aluminium wiring pattern and thin film integrated circuit mounted on its surface.

The display measures 140 mm (width) by 123 mm (height) and contains 640 x 480 dots. Black, blue and yellow colours are available using super twisted nematics (STN). According to the *Japan Industrial Journal*, Kyocera is producing 2000 pieces per month initially, rising to 10 000 pieces per month by the end of 1991. The primary application for this product will be palm-top computers.

5.7.4 GREEN DIAMOND LASER

A diamond laser which generates green light could lead to the development of an optical memory device capable of storing high densities of information, according to its Japanese manufacturer.

Osaka-based Sumitomo Electric Industries uses electron irradiation and heat treatment to create nitrogen impurity–vacancy centres in the diamond crystal. It is these complexes which cause the laser action.

In tests, the crystal lased at a wavelength of 530 nm when excited by a pulsed laser operating at 490 nm. However, Sumitomo Electric says it can modify the complexes and, therefore, the emission wavelength by altering the treatment conditions.

The achievement could lead to the development of optical memory devices based on synthetic diamond, a material with a high resistance to heat and radiation.

5.8 REFRACTORIES

The use of advanced ceramics in refractories has been steadily increasing for many years now, with ceramic fibres playing a dominant role. Initially these fibres were made of silica; however, first alumino-silicate and then alumina fibres have been developed and used in commercial products. The driving force is to be able to use these highly efficient thermal insulators at ever higher temperatures and in more flexible applications.

5.8.1 1800°C INSULATION BOARDS

The Japanese company Taiyo Chemical Co Ltd, based in Tokyo, has developed refractory insulation boards for use in laboratory scale furnaces which, it claims, can be used up to 1800°C.

Taiyo Chemical makes the boards from alumina fibres produced by ICI Advanced Materials, Runcorn, UK. The fibres are mixed with binding agents and pressed into boards 0.3–9 cm thick with an area of 90 x 60 cm.

According to the *Japan Industrial Journal*, the density of the boards is only 0.525 g.cm^{-3}, leading to high thermal efficiency.

Production at the start of 1990 was about 50 boards a month. The cost of a board 4 cm thick, weighing 12 kg, is ¥200 000.

5.8.2 CERAMIC FIBRE INSULATION MODULES

'Anchor-Loc' is a range of ceramic fibre insulation units which are quick and easy to install, and can be used up to 1350°C, according to Carborundum Resistant Materials of Rainford, UK.

The company makes Anchor-Loc products from folded pleats of 'Fibrefrax' spun 'Durablanket 1400' for use at temperatures up to 1350°C. To allow for continuous use up to 1200°C, Carborundum Resistant Materials uses Fibrefrax spun 'Durablanket S' to make modules with thicknesses from 100–300 mm.

The Anchor-Loc range comes in three basic types: 'Power-Loc', 'Thread-Loc' and 'Screw-Loc'. Power-Loc modules are principally for maintenance and repair work on the insulation of ladle pre-heaters, forge and heat treatment furnaces, ceramic kilns, covers, and doors. Installation times are under a minute per module, Carborundum Resistant Materials says.

The main application for Thread-Loc units lies in furnace linings in the chemical, petrochemical and ceramic industries, and all types of heat treatment equipment. Screw-Loc modules offer an alternative method of repair and maintenance in smaller installations — such as laboratory furnaces, craft kilns, ladle covers and pre-heaters, doors, and covers.

The range became available throughout Europe in April 1990 with the company producing the Fibrefrax spun Durablanket at a newly commissioned ceramic fibre plant in Rainford (see 'US$9 Million Ceramic Fibre Plant Commissioned', in the Industry News section).

5.9 SENSORS

Advanced ceramics are finding rapidly increasing uses in sensors, including those intended for temperature measurement or the detection of gases such as oxygen or carbon dioxide, humidity and infra-red emissions. Many sensor types only operate because of the particular properties of the ceramic material involved. The principal aims now seem to be to improve performance and reduce costs, rather than develop entirely new sensors.

5.9.1 HUMIDITY SENSOR WORKS OVER WIDE TEMPERATURE RANGE

A hygrometer which is capable of monitoring humidity over the temperature range $-40°C$ to $+350°C$ has been developed by Murata Manufacturing Co Ltd, based in Kyoto, Japan. The sensor is constructed of laminated multiple layers of ceramic discs with a composition based on barium, calcium and neodymium. The electrical resistance of the module, which measures 9 mm in diameter by 8 mm high, varies proportionally with the absolute humidity of its surroundings. The *Japanese Industrial Journal* reports that applications for the device are now being developed; in one of these, the sensor is expected to be used in conjunction with heating elements.

5.9.2 THERMOMETER FOR MOLTEN METAL

A thermometer with a protective tube made from zirconium boride can measure temperatures up to $1600°C$ and survive in molten metal baths for more than 40 hours, according to its joint Japanese developers. Nippon Steel Corp and Asahi Glass Co Ltd, both of Tokyo, say the zirconium boride has a melting temperature of $3060°C$.

The measurement of temperature in molten metal baths has always been expensive in the past, according to *Interceram*, since thermocouples do not last for long under such conditions.

5.9.3 LOW RESISTIVITY TITANIA FILM YIELDS OXYGEN SENSOR

A semiconducting titania film doped with niobium has a low enough electrical resistivity to give it potential applications as an oxygen sensor and humidity detector, according to the Japanese developer.

The Government Industrial Research Institute at Nagoya says the processing route is based on sol-gel technology. Scientists dissolve niobium and titanium alcoxides in ethyl alcohol solution and add diethanolamine as a stabiliser. The final gel forms as a thin film covered with silicon oxide. Particles are 0.6-1.5 μm in diameter, according to the *Japanese Industrial Journal*.

5.9.4 CHEAPER CARBON DIOXIDE GAS SENSORS

Researchers at the Musashi Institute of Technology, Tokyo, Japan, claim to have developed two types of compact sensors for carbon dioxide which are one-hundredth the cost of existing devices.

In one sensor, the scientists deposit layers of zeolite interleaved with gold or silver paste on an aluminium substrate. They say that as the carbon dioxide gas concentration in the surrounding environment increases the electrical resistance of the zeolite decreases.

The other sensor has a sensitivity ten-times higher and is made to a similar design, but has an apatite-based material as the ceramic phase. In this case, the electrical resistance increases with increasing carbon dioxide concentration.

According to the *Nikkei Industrial Daily*, sensitivity is improved in both cases by purifying the air to remove dust and/or smoke prior to measurement.

5.10 SUPERCONDUCTORS

Practical applications for high temperature superconductors are limited by the need to improve the properties of the materials, particularly the current density. However, this statement should not be regarded as surprising given that only about four years have passed since high temperature superconductors were discovered. That any applications at all are being considered at this stage is testimony to the massive amounts of effort which have gone into developing these materials. At the moment attention seems to be firmly focused on the production of wires and thin films for electronic components.

5.10.1 SUPERCONDUCTING MAGNETIC SHIELD TO BE COMMERCIALIZED

Scientists from Tokyo, Japan, claim to have made the world's largest shield of magnetic fields based on a high temperature superconductor and are now aiming to produce a commercial device.

Professor Tachikawa, of the Engineering Educational System at Tokai University, and researchers from Nippon Kokan KK say the shield is a cylinder 30.5 cm (12 inches) long and 15.2 cm (6 inches) in diameter.

The cylinder is made of copper and coated with a 100 µm layer of an yttrium-based high temperature superconductor. Nippon Kokan says that it will begin commercial production of a liquid nitrogen-cooled, superconducting magnetic shield in the near future.

At liquid nitrogen temperatures (77 K), the superconductor has a critical current density, J_c, of 3000 $A.cm^{-2}$ and is capable of shielding out magnetic fields as low as 0.6 gauss. According to the *Japan Industrial Journal* and the *Asian Wall Street Journal*, this allows the researchers to shut out magnetic fields from the environment and study extremely weak fields, such as those emanating from human brain cells.

5.10.2 SUPERCONDUCTOR RESONATORS IMPROVE CIRCUITS

A UK company and a US firm have jointly developed a family of radio frequency and microwave resonators using high temperature superconductors which, they claim, have superior characteristics to conventional devices.

ICI Advanced Materials, Runcorn, UK, and AT&T Bell Laboratories, Murray Hill, New Jersey, USA, say that in circuits operated at liquid nitrogen temperatures (77 K) the microwave resonators generate about 100 times less noise, yield better frequency stability and have at least five times lower insertion loss than conventional, all copper devices. Resonators constructed for radio frequencies, which are lower than microwave frequencies, exhibit even greater performance improvement. The resonators have been successfully used in several electronic circuit applications, including oscillator stabilisers and supergenerative receivers.

Research on the design of the resonators was conducted by AT&T using a yttrium barium copper oxide ($YBa_2Cu_3O_{7-x}$) developed by ICI. The cavity resonators are transverse electromagnetic (TEM) mode coaxial and helical devices, constructed from copper tubes with inserts of superconducting rods and helical wires. In conventional resonators, normal helical wires and rods are major contributors to heat loss and cavity inefficiency. Using high temperature superconductors for these components accounts for the improvement in resonator performance. Further, additional performance improvements are expected if the outer tube is also made from superconductors.

Several existing radio and microwave frequency circuits can benefit from the use of high temperature superconductor resonators. Applications such as low frequency communications equipment may now incorporate high temperature superconducting helical resonators to reduce noise and channel spacing, and improve selectivity. Search acquisition and Doppler radars can operate with better target discrimination and greater sensitivity. However, such applications must be compatible with device operation at low temperature (77 K).

5.10.3 WOVEN SUPERCONDUCTING CABLES SUPERIOR TO WIRES

Flexible high temperature superconducting filaments woven into thicker cables could transmit electrical power or form the windings in motors, according to the two US signatories of an agreement to develop such materials. The Superconductivity Pilot Center in Argonne, Illinois, is now working with HiT_c Superconco of Tullytown, Pennsylvania, to make flexible filaments of high temperature superconductors about 10 µm thick. The idea is that the cable will be more flexible and carry more current than simple wires of these materials. The project also aims to develop contacts to connect the cables to normal electrical conductors.

5.10.4 SUPERCONDUCTING LOGIC CIRCUIT DEMONSTRATED

The basic operating movements of a logic circuit have been demonstrated in a device made from a high temperature superconductor, a Japanese company claims. Tokyo-based Hitachi Ltd says it hopes to build a working circuit within three years.

5.10.5 ENERGY STORAGE SYSTEMS EVALUATED

A joint US research project is assessing the technical and commercial feasibility of small superconducting magnetic energy storage (SMES) systems using high temperature superconductors.

Combustion Engineering Inc of Windsor, Connecticut, and the Superconductivity Pilot Center at the Department of Energy's Argonne National Laboratory, near Chicago, say that if shown to be feasible, small SMES systems could be used by industries, hospitals, office complexes and shopping centres to reduce electricity costs during peak demand periods. SMES systems could store electricity during periods of low demand and release it during periods of high demand, a concept known as load levelling.

Small SMES systems might also be used as power backup to protect sensitive equipment, such as large computer or telecommunication systems, against sudden power failures.

The joint research project will assess the use of high temperature superconductors in SMES devices up to approximately room-size. Such devices would use superconducting magnets to store 20–300 kW.h of electricity; 300 kW.h would power an average home for 1–2 weeks.

Combustion Engineering says it is examining market needs to identify performance requirements for small SMES systems, and will study systems data for the SMES magnetic storage coil, coil restraint structures and conditioning equipment to deliver power to and from the coil. The firm will also estimate manufacturing costs for a number of coil and system designs, and compare costs to those of other energy storage technologies.

Argonne is providing information on the properties and performance of existing high temperature superconductors and those likely to be developed in the near future. It will also assess the required size of the superconducting magnet and its containment structure, examine various combinations of high and low temperature superconductors in system components, and analyse the system's heat transfer requirements.

Both partners in the venture will contribute US$60 000 to the one year project.

5.11 WEAR

The extreme hardness of many structural ceramic materials contributes to their generally very good wear and abrasion resistance. However, the ability to fabricate components with all the required properties — including strength and toughness — has led to limitations in the applicability of these materials, which are inherently brittle. Nevertheless, the items below indicate that work on introducing ceramics to wear components is active.

5.11.1 VACUUM PUMP FOR FUSION REACTORS

A prototype ceramic vacuum pump is ideal for use in fusion reactors, reports its developer, the Japan Atomic Energy Research Institute (JAERI), Tokyo.

The pump is suitable because it is capable of operating in high magnetic fields, at high temperatures, and in radiation and corrosive gas environments, JAERI claims. This is because the pump is made entirely of ceramic materials. For example, the 16 rotors, which measure 640 mm long and weigh 19 kg each, are made of silicon nitride. In addition, they are levitated on gas bearings rather than the conventional oil-lubricated or magnetic bearings.

According to *Asia Technology*, while the pump is not yet fully developed — JAERI believes it will be some years before such pumps are commercialized — other applications besides fusion are already foreseen. With minor modifications, JAERI thinks the pump could be used in the manufacture of semiconductors.

5.11.2 DIES FOR EXTRUSION OF COMPLICATED SECTIONS

A research programme which aims to develop extrusion techniques for making ceramic dies for the production of metal components with complicated sections has won funding of 2 million European Currency Units (MECU).

The funding comes from the first round of the Basic Research in Industrial Technologies for Europe/European Research in Advanced Materials (BRITE/EURAM) scheme. The programme involves 14 companies in five different European Community (EC) countries. It will last three or more years. Approximately one third of the money will go to the two prime contractors, both located in the UK; AEA Technology in Harwell and BNF Metal Technology Centre of Wantage.

Ceramic dies are already used for the extrusion of metals since they often provide much greater wear resistance compared to more traditional metal dies. However, current dies are limited in the complexity of shape which can be extruded — dies are generally circular in cross-section. The project aims to develop dies which will allow far more complicated sections to be extruded. It is hoped that this will be achieved by the enhancement of ceramic properties and the improvement of die housing designs, to allow the dies to survive greater thermal and mechanical stresses during use; and the development of electro-conductive, wear resistant ceramics so that electro-discharge machining can be used to produce the dies.

AEA Technology will work on the development of the electro-conductive ceramics, concentrating on zirconia and silicon nitride. Both these materials are used in current dies. BNF Materials Technology Centre will examine the stress on dies under loading using finite element analysis and a 750 t extrusion press.

5.12 OTHER

5.12.1 BURNER DEVELOPED FOR 100 000 kW GENERATOR

The successful conclusion of firing tests on a ceramic burner marks the latest stage in the development of a 100 000 kW gas turbine generator, according to Tokyo Electric Power Co Inc and Toshiba Corp, both based in Tokyo, Japan. Tokyo Electric Power is also involved in the production of a ceramic starter with Mitsubishi Heavy Industries Ltd, and a ceramic rotor with Hitachi Ltd. According to the *Daily Industrial News*, the companies expect to build and test the generator by 1995. It will be 43–47% more efficient in energy terms than current conventional generators.

5.12.2 CERAMIC-COATED PRINT ROLLER

A hydrophilic ceramic coating on the roller of an offset printing press ensures a uniform transfer of water to the printing plate and results in better quality without the need to use isopropyl alcohol, according to its Tokyo-based developers, Nippon Steel Corp and Akiyama Printing Machinery Manufacturing Corp.

Offset presses rely on the immiscibility of oil (the ink is oily) and water. The parts not to be inked on the printing plate must be uniformly wetted. In the past, printers achieved this by sucking up a mixture of water and isopropyl alcohol for transfer to the plate. However, isopropyl alcohol is expensive and has associated health risks, the two companies claim.

6. PROCESSING

The items appearing in this section relate to developments in processing technology, rather than to the processing of individual materials. Items concerning the latter generally appear in the Materials section.

The reports themselves can be conviently broken down into a series of subgroups dealing principally with such topics as the design of processing parameters to optimise product characteristics; powder processing to yield superior starting powders and improved techniques to yield coatings and thin films. In addition, 1990 saw research focused on investigating new and potentially exciting processing routes such as the use of microwaves as an alternative energy source for obtaining heating effects. Each of these themes is examined separately below.

6.1 DESIGN

It is becoming increasingly clear that if we wish to exploit advanced ceramic materials to the full we will not be able to rely on empirical determination of the processing conditions required to obtain the optimum properties in the final product. Precise design is needed at every stage, from choosing the most appropriate microstructure through to selecting the best balance of product attributes. It is only in the last few years, however, that sufficient expertise has existed to allow such design work and even now it is often only possible at a very crude level compared with our ability to fabricate components from other materials such as metals. Nevertheless, with the current emphasis on this stage of the production process, giant steps forward are expected over the next decade.

6.1.1 SOFTWARE DESIGNED TO HELP SELELCT OPTIMUM BALANCE OF RAW MATERIALS

A personal computer software package is available to help decide the best balance between as many as 20 product characteristics, its US developer says.

'Improvelt Version 2.0' allows users to select the optimal combination of ingredients and/or process variables to obtain the best balance of product attributes, such as cost, strength, colour, viscosity and processability, according to Battelle of Columbus, Ohio.

Version 2.0 is an upgrade of the original Improvelt program which was available during 1988/9 only through a Battelle multiclient programme. 18 subscribing companies used Version 1.0 to optimize a variety of products' including coatings and ceramics.

The software, which operates on an IBM or IBM-compatible personal computer is designed to assist with statistical experimental design, data analysis and the complex mathematics involved in property balancing, Battelle reports.

6.1.2 SOFTWARE PACKAGE DESIGNS STRUCTURAL COMPONENTS

'Ceram' is a software package for analysing the reliability of structural ceramic components, according to the Swiss-based Battelle-Geneva, part of the Battelle Institute.

Battelle developed the package as part of its research programme, 'Computerized methodology for the design of ceramic components'. It calculates the failure probability of two and three dimensional structures using what the institute describes as advanced techniques in statistical fracture mechanics.

Ceram determines the reliability of the component. If this is considered insufficient the geometry is altered until an acceptable value is reached. Should this prove impossible, the programme decides that a new material must be sought. Ceram determines the required material properties and gives the design and operating conditions. Battelle says material specifications are expressed as conventional test parameters for comparison with existing databases. If no suitable material is found the operating conditions must be modified.

The programme uses both the 'Weibull' and 'multi-axial elemental strength' models to compute the reliability. The latter combines fracture mechanics with a fundamental statistical analysis (the so-called elemental strength approach). It incorporates a non-planar fracture criterion at the level of microstructural flaws.

Battelle is also offering a ceramic design service which uses Ceram. Starting from a component geometry and loading conditions specified by the client, the institute claims to design a reliable component. There are four stages to this process:

- determination of materials data;

- determination of the stress distribution in the component;

- determination of failure probabilities using Ceram; and

- interpretation of failure probability computations and possible improvement of the design.

Depending on the needs of the client, Battelle says the service includes one or more of these four tasks. Cost and duration depend on the complexity of the programme and the number of these tasks needed.

6.1.3 TECHNOLOGY LICENSING/PRODUCT DEVELOPMENT PROGRAMME

A US company is offering a technology licensing and product development programme to those manufacturing, or considering the manufacture of, advanced ceramic parts, and the suppliers of raw materials.

By means of a 12 month laboratory research programme, Gorham Advanced Materials Institute (GAMI), Gorham, Maine, USA, offers to develop the optimum process technology for any one of five materials groups selected by each sponsor; alumina, silicon nitride, aluminium nitride, partially stabilized zirconia, and silicon carbide. A non-exclusive, royalty-free license will be given for the process technology used. Expertise developed will be transferred to the licensee and on-site training will be provided.

As the first step in the design of an appropriate programme, GAMI will assist an individual company in selecting the specific material for a particular product or product area. The major research effort to follow would be a 12 month, pilot plant/laboratory programme aimed at developing specific commercial products based upon GAMI's technology. If the sponsor wishes, the programme may continue beyond 12 months. The programme fee for any one of the five ceramic material areas is US$120 000.

GAMI considers this programme to be a natural follow-on to its three year study on developing and commercializing sintering/hot isostatic pressing, high pressure reactive sintering, and sintering processes, which was completed in February 1989. This US$2 million project produced fully densified, near net shape, advanced ceramic parts, GAMI claims, with excellent physical and mechanical properties. The processes were optimized for the five material groups.

6.1.4 EVALUATING DESKTOP MANUFACTURING

'Rapid Prototyping Technology' is a multiclient programme to help companies understand the impact of desktop manufacturing, and to choose systems best suited to their needs, according to Battelle of Columbus, Ohio, USA. The idea behind desktop manufacturing is to use computerized rapid prototype systems which can design and produce a prototype in a matter of hours rather than days or weeks, Battelle says.

6.1.5 RATIONAL DESIGN FOR CATALYSTS

A simple process for making catalytic thin films containing immobilized, highly dispersed metal atoms that do not reduce catalytic efficiency by forming clusters is currently being refined by scientists at the Georgia Institute of Technology, Atlanta, Georgia, USA.

Based on an original computer programme and rational design principles, it is intended that the research will result in more efficient and durable catalysts to produce gasoline, alcohol and other useful substances — without generating undesirable byproducts. A further advantage of the work is said to include the ability to use the raw materials more effectively, eliminating waste.

By experimenting with unconventional solvents, the Georgia scientists have been able to make catalytic thin films consisting of non-porous silica covered by a single layer of metal atoms using standard chemical procedures. Silica and copper were stirred into a non-aqueous solvent, the metal dissolved and stuck to the silica substrate in a thin film.

Analysis showed that the copper atoms were spaced at least 0.5 nm apart and positioned to bond tightly with atoms on the substrate. At the same time, energy effects which normally cause the metal atoms to cluster did not occur.

6.1.6 NATURE SHOWS THE WAY TO TOUGHER CERAMICS

US researchers are examining shellfish in an attempt to produce ceramics tougher than any made so far.

According to scientists at the Pacific Northwest Laboratories (PNL), in Richland, Washington, the Materials Science and Engineering Department at the University of Washington, the Washington Technology Center, and Case Western University in Cleveland, Ohio, shellfish produce their shells by exuding a mixture of calcium carbonate together with chitin, a sort of polymeric glue. The result is a natural ceramic which is much tougher than any synthetic ceramic material yet made — and produced entirely at ambient temperatures.

The researchers wish to mimic this approach. However, at PNL they are using advanced polymers instead of chitin, and titania (TiO_2) or iron oxide instead of calcium carbonate. They aim to construct a 'brick wall' consisting of microscopic building blocks of the ceramic phase, with the polymer acting as a mortar. This is achieved by producing a honeycomb template of the polymer, and precipitating the ceramic phase into it which then grows into crystals, filling up the voids.

So far the researchers have created thin films suitable for lithographic patterns for use in the electronics industry. However, the range of potential applications is extensive. For example, the work at Case Western is directed towards trying to grow artificial bones which will not be rejected by the human body.

6.2 POWDER PROCESSING

The development of advanced ceramics over the past half-century has been characterized by a steady shift of the focus of research backwards along the processing route. That is, sintering processes were among the first to be studied followed by the green forming routes. In each case it was realised that the ability to make substantial improvements in terms of removing defects from the final product was limited by the defects already present from the previous processing stage. Finally this culminated in the realisation that while individual processing techniques could be improved, the final properties of the component were inextricably linked to the properties of the precursor materials. This has led to gradually increasing emphasis being placed on powder processing.

6.2.1 CERAMIC POWDERS STUDY

'Ceramic powders for structural components and electronic applications : (GB-102)' is a report which reviews the technology for reproducibly making the uniform quality powders needed to fabricate reliable advanced ceramic components, its US publisher says.

Business Communications Co Ltd from Norwalk, Conneticut, says the report discusses the emergence of various synthesis techniques for ceramic powder production and makes an assessment of the major current areas of research to achieve uniform size and evenly distributed powder particles. The report costs US$2650.

6.2.2 FINE CERAMIC POWDERS BY PLASMA SYNTHESIS

A device described as a triple DC torch plasma reactor (TTPR) can synthesise fine ceramic powders, according to researchers at the University of Minnesota in the USA.

The production of homogeneous, fine powders by vapour phase reactions requires the complete evaporation of the solid precursor particles. However, there is a problem with this route, in that the relatively short time the raw material spends in the plasma region necessitates the use of very fine initial particle sizes, the scientists say. This creates problems with feed into the plasma stream because of the high viscosity of thermal plasmas.

The researchers say the TTPR consists of three identical DC plasma torches, each of which can operate at a power of 3–30 kW. The jets from these torches form a converging plasma volume into which the precursor powder is fed. So far they have studied the synthesis of aluminium nitride (AlN) and zirconium carbide (ZrC) powders. The former used Alcoa aluminium powder, median size 1–2 μm as a precursor, while the latter used a liquid organometallic precursor, $(C_4H_9O)_4Zr$. Single, hexagonal phase AlN and porous, spherical ZrC were produced. The scientists consider the latter material to be of great potential use for producing thermal spray coating powders.

6.2.3 COMPUTER AIDED METROLOGY

The French company Saviphar, based in Orléans, has developed a high speed powder dispenser which takes a completely new approach to the problem of mixing powders and liquids on an industrial scale.

Unlike conventional techniques, Saviphar's approach involves measuring only small quantities at a time, but at high speed and under computer control. Using a local area network, up to 99 balances are connected to a central computer. Each balance can measure a different product in varying quantities. Once the doses are measured they are all fed directly into the same micro-mixer.

The company says the key is a special dispenser which combines the accuracy of traditional dosing balances with the speed of volumetric dispensers. It can measure up to 100 mg with an accuracy of ±0.3 mg, or up to 10 kg with an accuracy of ±10 g, in 2–5 s.

Traditional metrological techniques involve weighing large quantities at a time, collecting them in a hopper, and then pouring them into a mixer. This often leads to handling problems and to a lack of homogeneity in the final mix because of gravitational settling. In some cases there may also be a risk of ignition or of an explosion occurring when bulk quantities are handled.

6.2.4 FAST MIXING ROUTE FOR POWDERS

A fast mixing and homogenizing route for ceramics has been developed and patented by Dena Enterprises Ltd, based in Leeds, UK.

The invention has been developed primarily for mixing engineering ceramics, semiconductors, screen printing materials for electronic components and surface coating technologies; however, the technique is claimed to be applicable anywhere where conventional mixing techniques are inadequate.

The principle of the process is to pump initial feedstocks through a series of specially designed tubes, baffles and constrictions at moderate to high pressure, thus inducing turbulent flow, shear mixing, etc. Dena claim that the process can handle roughly mixed feedstocks, and seperate feeds, agglomerates, partially or fully separated emulsions and high viscosity components. Generally, the material is circulated through the mixing units several times using a batch processing technique until the required degree of homogeneity is achieved. However, continuous processing is said to be possible.

According to Dena, a typical application such as mixing a ceramic slurry for use in tape casting multilayer capacitors can be achieved in as little as 60 minutes, compared to the 24 hours that conventional processing requires. Dena also claim that the final product can be much more homogeneous and have reduced agglomerate problems. Low viscosity slurries can be mixed in very short times.

Dr Sulaiman, Managing Director of Dena, has won a UK Department of Trade and Industry (DTI) SMART award to develop the process further, and at present it is being scaled up to deal with large quantities of material and continuous rapid mixing. Special surface coatings and treatments are also being investigated to allow corrosive and abrasive materials to be safely mixed without the risk of contamination from the mixing units. Dr Sulaiman is currently offering free tests with agreed criteria for assessing the results to any company or individual with a serious interest in mixing materials, using the facility he has created in the School of Materials, Division of Ceramics, Glasses and Polymers at Sheffield University.

6.2.5 DIRECT EXTRUSION LEADS TO CERAMIC HONEYCOMB STRUCTURES

Direct extrusion of honeycomb ceramics is possible, according to a French engineer. Dr Jean-Pierre Trotignon says the technique results in:

• cheaper production of alveolar plates and tubing;

• new cellular geometric shapes with larger specific surface areas; and

• a wide variety of shapes, profiles, dimensions, and products.

The cells run perpendicularly to the component's surface, as in a traditional honeycomb; however, some inclined faces have been specially adapted to better withstand the shear forces. This also increases the structure's specific surface area. The honeycomb-shaped structure can be extruded directly in plates or in tubes with their final dimensions; without any cutting or joining of their component parts. Trotignon claims this technique makes it possible to directly extrude plates or 'sandwich' tubes (the alveolar structure between two outer coverings).

Applications for the technology include:

• components of pollution treatment systems;

• economic catalytic exhaust pipes;

• heat exchangers;

• structural tubing or plates; and

• filtration systems.

6.2.6 LOW TEMPERATURE FIRING OF ELECTRONIC SUBSTRATES

Nihon Cement Co Ltd of Tokyo, Japan, claims to have developed a low temperature firing method for mass producing multilayer substrates.

The company says it makes the substrates by firstly mixing alumina (particle size 1–2 µm) and glass powders. The glass has a relatively low melting point, allowing the subsequent firing to take place at 850°C, about half the temperature needed for conventional alumina substrates, according to Nihon Cement. The low firing temperature means that circuit patterns on the substrates can be fabricated from silver. The laminated sheets can also contain blind holes, facilitating the surface mounting of electronic devices at a later stage, the company says.

6.3 MICROWAVE, LASER AND ELECTRON BEAM PROCESSING

New ceramics and ceramic-based composites are continually being developed and improved to meet the ever increasing demands of modern technology. While these materials have potentially exciting properties they can be difficult to processes by conventional means, so there is a rapidly growing interest in developing novel processing routes. This is reflected by the nature of the items appearing in this section. In particular, the use of microwaves to obtain heating effects in ceramics is rapidly becoming a major research focus, although relatively little is currently being published due to proprietary interests.

6.3.1 MICROWAVES USED TO SINTER TITANIUM DIBORIDE

US researchers claim to have used microwave energy to sinter titanium diboride to within 82% of theoretical density.

The scientists, from the Los Alamos National Laboratory in New Mexico, say that to completely prevent oxidation during the sintering process, they had to keep the oxygen partial pressure below 10^{-18} atmospheres. In addition, they kept the temperature below 1860°C to stop voids forming near the surface of the sintering material.

According to *High Tech Ceramic News*, these voids arise because boron oxide (B_2O_3) evolves rapidly after the loss of a titanium dioxide (TiO_2) protective film which melts at 1860°C.

6.3.2 MICROWAVE JOINING OF CERAMICS

Researchers at the QuesTech Inc Research Laboratory in Falls Church, Virginia, USA, claim to have succeeded in joining ceramic rods using microwave energy.

According to the team, the samples were silicon carbide cylinders measuring 9.5 mm in diameter by 6.4 mm in length with an unspecified metallic braze interlayer. The mating surfaces were rough cut with no polishing. Joining was performed using a microwave frequency of 2.45 GHz in a single mode rectangular cavity producing 200–250 W. The joints formed after heating at 1400°C for 5–10 minutes with a butting pressure of approximately 5 MPa.

Several samples have been produced and these are currently undergoing scanning electron microscopy examination at the Los Alamos National Laboratory. So far, the results have indicated that good wetting has been achieved with an homogeneous interfacial layer of about 50 μm in width.

However, independent and preliminary fracture studies have indicated that the metallic interlayer is brittle. Work is currently underway to optimize the mechanical properties of the joints.

6.3.3 LASER PROCESSING

Laser processing techniques are now used in many industrial sectors including aerospace, automotive, appliance, chemical, electronics and packaging, says a report from TechTrends of Paris, France.

'Industrial laser materials processing' provides a treatment of the technical and economic aspects of laser technology and reviews both current and emerging applications, the publiser says. The report includes an international directory of lasers and laser systems suppliers and contract laser materials processing services. It costs US$500.

6.3.4 ELECTRON BEAM PROCESSING

A "lightning machine" developed in Canada is among the 100 most significant inventions of 1989 and could result in improved ceramics, amongst other applications, according to the US-based *Research and Development* magazine.

The equipment, known as the 'Impela', was developed by a research team at Atomic Energy of Canada Ltd, and is the world's most powerful industrial accelerator. The machine accelerates 'bundles' of electrons along a 3 m long tube reaching the speed of light partway along it. After this the electrons gain in mass as they continue to absorb energy on travelling up the rest of the accelerator.

A high energy electron beam is produced and when it comes into contact with a material it rearranges the internal structure in a way which conventional heat treatments simply cannot achieve and in only a fraction of time.

According to the research team, a primary use for the device is in strengthening and toughening materials, such as ceramics and plastics and even creating entirely new materials. Other uses are expected to include sterilization of waste food products for use as soil conditioners or mulch.

6.4 JOINING

Joining ceramics with adequate properties in the joint has always been very difficult to achieve and considerable research is currently underway throughout the world towards developing a range of alternative techniques. Two of the items below focus on work at The Welding Institute in the UK, the third being concerned with the development of a new adhesive for ceramics. Attention is also drawn to an item on microwave joining of ceramics appearing in the previous section.

6.4.1 CERAMIC TO METAL JOINING

To meet industry's need for greater exploitation of advanced ceramics, The Welding Institute (TWI), located near Cambridge, UK, has made a significant investment in resources for developing reliable joining technology for ceramic to metal joints. So far TWI has produced joints using solid phase and liquid phase processes.

According to TWI, friction welding, a solid phase process, has shown promise for joining aluminium to a range of ceramics, including Si_3N_4, SiC and ZrO_2. The process is simple and fast and could be used as a manufacturing route in the automotive industry, claims TWI. Diffusion bonding and hot isostatic pressing are also solid phase processes under investigation.

Liquid phase bonding of ceramics to metals creates two major problems, says TWI. Firstly, most ceramics are not readily wetted by conventional brazes, although this difficulty can be largely overcome by sputter coating the ceramic's surface prior to brazing to promote wetting and chemical reaction. Alternatively, an active metal braze can be used which contains elements which react with the surface of the ceramic during brazing. TWI has produced Si_3N_4 to steel and ZrO_2 to cast iron joints using this one-step liquid phase method.

The second problem is caused by large differences in the coefficients of thermal expansion of ceramics and metals. In its work to overcome this, TWI has been developing novel interlayers which accomodate excessive strain built up during bonding.

In addition to the work outlined above TWI is involved with:

- joining ceramics to plastics for applications in the electronics industry;

- innovative techniques such as using microwaves to join ceramics (TWI has joined the Materials and Microwave Processing Group based at Nottingham University — see 10.1.7);

- spray coatings — because of their potential for the automotive and aerospace industries; and

- surface and sub-surface modification of coatings and superconducting ceramics by laser.

6.4.2 CERAMIC–METAL JOINING YIELDS CERAMIC-FACED TAPPETS

A process for manufacturing ceramic-faced car engine tappets has been developed by The Welding

Institute (TWI), based in Abington, Cambridge, UK. The development is the result of a two year project funded by the French and British Governments and industry, and was part of the European Advanced Materials Programme (EURAM).

Working jointly with a major French car company, TWI sought to find a practical means of joining the silicon nitride to steel. Possible techniques investigated included brazing a copper-silver alloy with added active ingredients, such as titanium, between the two materials. This was not successful, because brazing at a temperature of 850°C caused large thermal expansion in the steel. The result was thermal mismatch and the ceramic cracked.

The solution was found to be to braze a thin layer (1 mm thick) of novel design with a compliant metallic material as an interlayer between the silicon nitride and the steel components of the tappet. The joint is formed in one operation and the metallic material accomodates the strain so that the ceramic is not unduly stressed.

The new tappet will be tested initially in an electrically driven engine and could significantly improve performance.

6.4.3 HIGH TEMPERATURE ADHESIVE

'Ultra Temp 516' is a zirconia-based single component liquid adhesive which can be sprayed, injected, or applied with a spatula or brush, according to Meclec Co from Shoeburyness, UK.

Drying and curing takes only two to four hours at 94°C, the company says, and afterwards the adhesive has a thermal resistance up to 2425°C. Meclec also reports that the material is an electrical insulator able to withstand 250 V across a thickness of 25 μm.

The adhesive can bond diverse materials such as ceramics, graphite, quartz, silicon and metals, according to *Ceramic Industry International*.

6.5 COATINGS AND THIN FILMS

For the purposes of this section the term coating is considered to be applicable when the properties of the bulk of the material are largely desirable or acceptable except at the surface, neccessitating the use of a different surface layer. In contrast, such a layer will be considered to be a 'thin film' when it is the properties of the layer which are of prime importance and the primary function of the material below is to act as a substrate.

6.5.1 ADVANCED SURFACE COATINGS

The optical and electrical properties, as well as the wear, corrosion, erosion, fatigue or heat resistance, and fire and flame resistance of materials can all be improved by surface coatings, according to a report published by TechTrends, Paris, France.

'Advanced surface coatings' also covers surface modification technologies as used in the aerospace, automotive, nuclear, textile, paper and mining industries. The report, which costs US$500, reviews a number of techniques: including thermal spraying, chemical vapour deposition, physical vapour deposition (ion plating, etc), the 'Toyota' diffusion process and ion implantation, TechTrends says.

Major applications are illustrated and technical and commercial limitations explained, the publisher adds. An international directory of suppliers of equipment is appended.

6.5.2 EXPLOSIVE FORMATION OF CERAMIC COATINGS

The explosive bombardment of hard metal substrates leads to ceramic coatings with peeling strengths over 200 MPa, claims the Japanese subsidiary of a US company.

Union Carbide Services KK, which is a subsidiary of Union Carbide Coating Services and is located in Yokohama, says the bombardment involves a new fuel gas, in addition to oxygen and ethyne (acetylene). This increases the explosive energy, resulting in better bonding and film density.

Union Carbide expects sales of ¥1000 million within two years,according to the *Japan Industrial Journal*.

6.5.3 CERAMIC COATINGS WHICH DO NOT PEEL AT HIGH TEMPERATURES

Ceramic coatings on metals are less likely to peel off at high temperatures if they are applied with a method developed by the Japanese National Research Institute for Metals(NRIM), Tokyo, the Institute, claims.

According to NRIM, the principal barrier to successful coatings is the presence of sulphur which at high temperatures concentrates at the interface. The solution proposed by the Institute is the introduction of small quantities of rare earth elements to the substrate. These additions combine with the sulphur, preventing it from diffusing.

Cerium and lanthanum are usually chosen, the *Japan Industrial Journal* says. For example, a 0.15% lanthanum addition to a stainless steel resulted in the sulphur concentration at the surface increasing to only 2% on heating, compared with 20% without the lanthanum. The Institute reports that the steel retained its ceramic coating to 827°C.

6.5.4 POWDER METALLURGY METHOD COATS NITRIDE CERAMICS

Israeli researchers claim to have developed a powder metallurgy method for coating nitride ceramics.

Their technique involves heat treating the ceramic while in a powder bed of the desired metal or metals, according to I. Gotman and E.Y. Gutmanas of the Department of Materials Engineering Technion, Haifa. The coatings produced consist of two layers, the outer one a metal rich nitride and the inner a metal rich silicide. Thickness, morphology and sometimes position of the coatings depended on the duration and temperature of the heat treatment as well as on the metal powder's particle size.

Silicon nitride was used in the bulk of the experiments reported which involved a variety of metal powders. However, in their paper in *Powder Metallurgy International* the scientists say they also produced a titanium nitride layer on the surface of boron nitride, indicating that the technique may be more generally applicable.

If the layers were kept thin, by reducing the heat treatment time and/or temperature, then they generally bonded well to the silicon nitride substrate. As thicker layers were produced transverse cracks appeared, caused by the differences in the material's thermal expansion coefficients. Despite this, the layers remained well bonded to the ceramics, the Israeli's report. Only very thick coatings had a tendency to spall.

Treating the Si_3N_4 in powder beds containing two metallic elements resulted in preferential diffusion of the metal with a higher affinity to nitrogen (e.g. vanadium or chromium) towards the ceramic's surface. As a result, coatings produced in this way consisted almost entirely of nitrides and silicides of the metals with high affinities. Multilayer coatings were also produced by successive treatments in different metal powders.

6.5.5 SPUTTER COATING CENTER OPENS

The opening of a Sputter Coating Center for exotic materials has been announced by Spire Corp, based in Bedford, Massachusetts, USA. Advanced ceramics, polymers and metals (including beryllium) can be deposited by magnetron, radio-frequency and ion beam sputtering in the facility. Coating thicknesses range from about 1 nm to hundreds of micrometres. Substrates with diameters up to 200 mm (7.9 inches) can be accomodated in the facility.

Coating applications include optical components, medical devices, bearing surfaces, tools, and nuclear components.

According to Spire, sputter etching and ion beam texturing can be routinely performed in the facility, and the sputtering process follows OSHA approved guidelines, permitting the processing of a broad range of substrate and coating materials.

6.5.6 ION BEAM ASSISTED DEPOSITION SERVICES

Spire Corp, based in Bedford, Massachusetts, USA, is now offering the use of its 'IBAD 9000' (ion beam assisted deposition) system for the production of thin film coatings, in addition to its range of other ion beam surface treatment equipment.

The IBAD 9000 is manufactured by Nissin Electric in Japan, and is claimed to have demonstrated breakthrough performance in applications where substrates and coatings have been incompatible previously, or where specific films have been difficult to attain at low temperatures. The new equipment employs a 1 m by 1 m vacuum chamber, 2 KeV and 40 KeV dual ion sources and a four crucible electron beam evaporator to deposit very adherant films of materials such as alumina, silicon nitride, titanium carbide, diamond-like carbon and pure metals, as well as others. The principle feature of the IBAD coatings is the graded nature of the interface achieved between coating and substrate.

6.5.7 SPUTTERING TARGETS AND EVAPORATION MATERIALS

A range of sputtering targets and evaporation materials is on offer from Megatech Ltd, Havant, UK.

The company says it can supply targets fabricated from elemental metals, composite alloys, dielectrics, semiconductors, refractory metals, magnetic materials and stoichiometric oxide high temperature superconductors. Standard and custom geometries and sizes are available.

Evaporation materials for thermal resistance or electron beam evaporation are available in the form of bulk powders, granules or pellets, and fabricated shapes such as rods, cubes, slugs and cones. Purities are said to range from 99.5% to 99.999%.

6.5.8 SPUTTER DEPOSITION MAY LEAD TO PRACTICAL SUPERCONDUCTING WIRES AND TAPES

A sputter deposition method could produce wires or tapes of high temperature superconductors suitable for practical applications, say the two US partners in a project to develop the technology, the Superconductivity Center in Argonne, Illinois, and Microelectronics and Computer Technology Corp (MCC) of Austin, Texas.

MCC has developed a proprietary sputtering apparatus specifically for these materials. The device uses targets efficiently and increases deposition rates 1000 times compared with conventional machines, the company claims. With further work the process could lead to mass production of practical superconducting wires and tapes.

To this end, the Superconductivity Center and MCC have each contributed US$228 000 to a one year project which began in 1990. No funds are changing hands. MCC are trying to reduce materials and production costs by examining various substrates and processing methods, while Argonne are building a vacuum chamber for the device. The latter will also study novel processing methods, and the microstructures and properties of the sample wires and tapes.

6.5.9 DEPOSITION PROCESSES FOR FLEXIBLE SUPERCONDUCTIVE TAPE

A US Government laboratory and a US company are cooperating in an attempt to make thin films of high temperature superconductors on flexible ceramic substrates.

The Department of Energy's Oak Ridge Laboratory in Tennessee and Corning Inc, based in Corning, New York, will spend US$245 000 on the programme.

Oak Ridge is concentrating on the deposition processes — investigating such techniques as magnetron sputtering, co-evaporation and laser ablation — and the subsequent characterization of the films produced, including critical temperature (T_c) and critical current density (J_c) measurements.

Corning's role is to provide the flexible ceramic substrates and to characterise these in terms of composition, strength, electrical properties, surface roughness, etc.

According to *Lightwave*, two superconductors are being examined; a 'standard' yttrium barium copper oxide and a proprietary silver-doped variation which Corning has developed. The aim of the project is to produce these films with a critical current density in excess of 1000 $A.cm^{-2}$.

6.5.10 LASER DEPOSITION TECHNIQUES YIELD HIGH RATES

Researchers at Los Alamos National Laboratory in New Mexico, USA, claim to have achieved rates as high as 15 $nm.s^{-1}$ for depositing superconducting thin films on substrates with laser deposition techniques. The films have high crystallinity and yield very high quality thin films, according to the *American Ceramic Society Bulletin*. The speed of deposition enhances the commercial viability of the technology, claim the researchers.

7. EQUIPMENT

As the advanced ceramics industry grows in size there is a steadily increasing need for equipment to process and subsequently characterise the products. The items included in this section cover both these equipment types and are roughly subdivided according to the function of the equipment.

The first group of reports cover the processing of ceramics before firing, that is the production of powders and their formation into green bodies. The next group of items relates to equipment for firing, both with and without the use of pressure. The machining of ceramics to obtain accurate dimensional control is the focus of the next section and there follows a small number of items with the production of coatings and thin films as their theme. Finally, this section concludes by examining a range of new characterisation techniques.

7.1 POWDER PROCESSING

The first stage in the manufacture of most ceramic products involves the processing of powders. In this section three reports are presented, the first deals with a new furnace intended for the processing of powders, the second with the ability to measure out large quantities of powder accurately and the third is concerned with the fabrication of green bodies by isostatic pressing.

7.1.1 POWDER PROCESSING FURNACE

The development of a furnace which has been specifically designed to provide a variety of operational features useful in the processing of powders has been announced by Lenton Thermal Designs Ltd, based in Market Harborough, UK.

The Rotary Incline Tube Furnace (RITF) is a heat treatment system which has a maximum operational temperature of 1600°C and incorporates a work tube of 80 mm inside diameter. This may be rotated smoothly and without vibration at up to 10 rpm and may be inclined with the furnace assembly to any angle within 0-10 degrees to the horizontal. In this way, the passage of the powder is assisted through the tube during processing, Lenton claims. A number of other options are also available. These include:

• a variable amplitude electromechanical vibrator, to prevent the accumulation of powder on the inside surface of the work tube;

• a hopper screw feed, to assist sample loading;

• and facilities for processing in an inert atmosphere.

7.1.2 ANALYTICAL BALANCES WITH LARGE CAPACITY

Mettler-Toledo Ltd of High Wycombe, UK, claims to have set new standards in terms of analytical balances with the introduction to two models, the 'AT460 DeltaRange' and the 'AT400'. Both balances offer a weighing range of 450 g combined with analytical level accuracy, an automatic calibration system, an intelligent 'print' key and an automatic draught shield. In addition, Mettler claims that the AT460 DeltaRange provides greater versatility, by effectively being two balances in one. The DeltaRange provides the user with a 62 g fine range, moveable anywhere over the balance's weighing range. This enables the user to tare out a heavy container or bulk constituent and still weigh to an accuracy of 0.01 mg. This is a particularly useful feature when considering making small additions, such as sintering aids, dopants, etc., to precursor ceramic powders.

7.1.3 50 t COLD ISOSTATIC PRESS

A 50 t capacity isostatic press designed for ease of use and maintenance has been introduced by Simac Ltd, based in Rugby, UK. According to the company, the 'Densomatic 50' range has been built with flexibility and accessibility in mind. For example, says Simac, there is substantial space behind the tooling, the intensifier is easy to reach and the pumps are located near the motors, separate from the tank.

The press features automatic load and unload and is capable of producing several different types of component, including balls, rods, cones, nozzles and tubes, typically in free-flowing ceramic powders such as alumina, tungsten carbide and other materials. Offering a maximum work rate of 4 strokes per minute, i.e. 16 parts per minute using four cavities, the Densomatic 50 range features seven machine options. These include three different maximum pressures and a choice of cavities and cavity diameters.

7.2 FURNACES

With the exception of a few chemically processed ceramics, such as the cements, ceramics require a heat treatment which allows diffusion to occur and the conversion of a weak green body into a hard, strong polycrystalline ceramic component. This section of the source book considers some of the latest developments in furnace technology.

7.2.1 HIGH CAPACITY VERTICAL CHAMBER FURNACE

Lenton Thermal Designs Ltd, based in Market Harborough, UK, has announced the introduction of a high capacity vertical chamber furnace with, it claims, a maximum programmable temperature of 1800°C and the convenience of an electrically-operated hearth to simplify sample loading.

According to the manufacturer, the 'EHF' furnace has been designed for heat treatment processes requiring stable, high temperatures under programmable control. It has a vertical work area within a chamber of high purity graded ceramic fibre, heated by vertically suspended molybdenum tungsten disilicide elements.

Based upon a standard modular design, the chamber may be constructed with a capacity specified by the user, and has the option of being fitted with an alumina liner. This, claims Lenton, allows the EHF to be tailored to a wide variety of applications. For example, in the production of fibre quality glass an ancillary furnace may be incorporated in the hearth to assist glass flow and a stirrer fitted to ensure maximum homogeneity of the melt.

7.2.2 TILTING TUBE FURNACES

The 'UTF Series' tube furnace, developed by Lenton Thermal Designs Ltd, Market Harborough, UK, operates in a vertical, horizontal or angled position.

Lenton Thermal Designs says the furnaces embody state-of-the-art insulation technology and are available with maximum operating temperatures of 1200°C or 1600°C. The furnaces have an optional atmospheric control system.

In the standard form, a UTF furnace is suited for use with a reaction tube of up to 100 mm bore. However, alternative versions for larger reaction tubes can be custom built. Provision is made for remote programmable control of both the work area temperature and profile.

7.2.3 HEATING ELEMENTS FOR 2000°C FURNACES

A US-based company claims to have developed zirconium-based heating elements which enable furnaces to operate at over 2000°C. The Swedish furnace company Kanthal Furnace Products from Hallstahammar has acquired the developer, Artcor Inc of Costa Mesa, California, and says it intends to launch the technology on the international market.

Kanthal Furnace says the zirconium-based elements work up to 2280°C in air. To date, the company has been manufacturing heating elements for temperatures up to a maximum of 1900°C.

Kanthal Artcor Inc is offering complete furnaces in five sizes, with temperatures over 2000°C. The furnaces are insulated by zirconia sheets and the largest model as a chamber volume of 150 x 150 x 175 mm. The heating elements are electrically preheated to achieve an ignition temperature of 800°C, after which they can be used to heat the furnace. Previously, Artcor had mainly marketed the furnaces in the USA.

The furnaces are primarily intended for materials testing in research and development laboratories and, for example, the production of special glasses. Applications are forecast in the electronic and military sectors, as well as in waste destruction.

7.2.4 OPERATION RANGE OF GRAPHITE FURNACES RAISED TO 2500°C

A UK company says it has uprated its graphite resistance furnaces to allow their operation at 2500°C.

Consarc Engineering, Strathclyde, says the furnaces have forced gas cooling systems and binder removal facilities. The sizes of the furnaces range from laboratory models with working diameters starting at 350 mm and heights of 450 mm to commercial versions 2 m in diameter and 2.2 m high.

Previously Consarc's resistance furnaces worked up to 2200°C and found applications in the sintering of silicon carbide and silicon nitride, the company reports.

7.2.5 RAPID CYCLING VACUUM FURNACE

Yamato Scientific Co Ltd of Tokyo, Japan, a manufacturer of chemical analysers and testing machines, has developed a high temperature vacuum furnace which it claims is capable of rapid temperature cycling. The furnace has a capacity of 1.5 litres and is heated by carbon fibre reinforced carbon heating elements. These elements allow the furnace to reach 2000°C in only six minutes, and the construction of the equipment allows it to cool back to room temperature in just 40 minutes, Yamato Scientific says. The furnace is being sold under the trade name of 'Newtonian' and the price is about ¥4.5 million. This includes the furnace, a controller and a vacuum pump. Yamato Scientific expects to sell about 50 units per year.

7.2.6 PATENT FOR HIGH PRESSURE FURNACE

A convection-free hot zone for high pressure furnaces has been developed by Vacuum Industries Inc, Somerville, Massachusetts, USA. According to the company, the design eliminates convection currents between the inside and outside of the hot zone, allowing for better temperature uniformity, a longer hot zone life and higher efficiency.

In the design, convection currents are minimized by bringing electrical feedthroughs for the heating element through a horizontal spacer ring which separates a pressure body and cover. The design allows a relatively small volume of unheated atmosphere to support convection currents stemming from necessary clearances around the feedthroughs where they enter the hot zone.

7.2.7 SMALL SCALE HOT ISOSTATIC PRESSES WITH VARIABLE ATMOSPHERES

Two small-scale hot isostatic presses (HIP) are available which the Japanese developer says can work with various atmospheres or under vacuum, and even as cold isostatic presses (CIP). Futic Furnace Co Ltd is producing the HIPs which are being marketed by Dowa Mining Co Ltd, based in Tokyo. They cost ¥4.8 million, the *Daily Industrial News* reports.

7.3 MACHINING

The ability to machine hard ceramics accurately is becoming increasingly important as these materials find their way into ever more demanding applications.

7.3.1 PRECISION MACHINING FOR OPTICAL COMPONENTS

A three-axis diamond turning, grinding, polishing and measuring centre has been developed by Cranfield Precision Engineering Ltd, of Cranfield, UK.

It is claimed to incorporate all the latest advances in nanotechnology and to be capable of achieving high rates of production of optical quality surfaces in a wide range of materials for non-conventional optical components.

The machine is said to efficiently machine optical glasses, ceramics and other brittle materials in the ductile mode. This capability is apparently due to the machine design which incorporates hydrostatic bearings, a T-bed base and bridge configuration which all go to provide a loop stiffness which is greatly in excess of existing aspheric generating type machine tools, says Cranfield Precision Engineering. High speed servo controls to resolution options down to 1.25 nm and, for the rotary axis, 0.1 arc seconds, gives the machine the ability to take sub-micron cuts. These are essential to achieve good form accuracy and minimum sub-surface damage.

7.3.2 COMPUTER NUMERIC CONTROLLED PROFILING MACHINE

Fluid Developments Ltd, based in Cranfield, UK, says it has developed the first computer numeric controlled (CNC) profiling machine based on the 'Diajet' cutting technology of its parent company, BHR Group Ltd.

The CNC profiling machine combines computer controlled manipulation with hydro-abrasive jetting technology. Leaving no heat affected zones, stress, shock and minimum second operation work, the Diajet profiler can cut any material, including ceramics, glass, composites, metals, rubber, plastics, concrete and glass reinforced plastic.

The profiler is operated from a remote control console and is programmed from its own computer keyboard or on-line with full screen graphics. Cutting is performed within a safety enclosure with the swarf and spent abrasive collected below the table in a jet catcher tank.

7.3.3 ELECTRO-DISCHARGE MACHINING UNIT

Toray Industries Inc, based in Shiga Prefecture, Japan, has developed an electro-discharge machine which it says has dimensions of only 5 x 5 x 7 cm. The basis of the machine is the use of a piezoelectric ceramic element as an actuator for accurate positioning of the discharge electrode. The research leading to the machine was sponsored by the Japanese Reseach Development Corporation (JRDC) on the basis of original work performed by Professor Higuchi of the Institute of Industrial Science at the University of Tokyo. The next development is said to be an even smaller, cylindrically shaped machine, measuring only 1.3 cm in diameter by 7 cm in length.

7.3.4 PRECISION GRINDER

Japanese researchers claim to have developed a technique for grinding ceramics to within a surface roughness of 0.1 µm. The scientists from the Osaka Prefecture Industrial Technology Research Institute use a specialised diamond grinder which they can move with a precision of the order of nanometres, the *Nikkei Industrial Daily* reports.

7.3.5 ION PLATING MACHINE FOR COATING CUTTING TOOLS

The Japanese company Nachi Fujikoshi Corp, based in Toyama Prefecture, has commercialised an ion plating machine for coating cutting tools with titanium nitride. According to the *Daily Industrial News*, the adhesion strength of the titanium nitride film is twice that obtained by conventional methods. The machine costs ¥100 million and Nachi Fujikoshi expect to sell 15-20 machines a year.

7.4 DEPOSITION, METALLIZATION AND COATINGS

The use of coatings and thin films on substrate materials is becoming of increasing importance as technology places ever greater demands upon materials. Very often a single material, or even composite, is simply unable to provide all the required properties. This section examines four pieces of equipment which exist simply to place a thin layer of one material on the surface of another. This may be to provide a coating to alter the surface characteristics in some way, such as in the first item, or a thin film which requires the substrate largely for mechanical support, as in the last three items.

7.4.1 METALLIZING FURNACE

A fully automatic metallizing furnace has been designed and manufactured by Torvac Furnaces, a division of Cambridge Vacuum Engineering Ltd based in Cambridge, UK, with cooperation from Siemens AG of Munich, Germany. Purpose designed for ceramics applications, the Torvac unit is claimed to provide improved production efficiency, reliability and quality control.

The furnace is designed to produce the molybdenum/manganese coating on high purity ceramic substrates which is required to provide a better surface for subsequent deposition of additional metallic layers, or for vacuum brazing hermetically sealed components such as vacuum switches, capacitors, klystrons and electronic valves. The batch loaded vacuum controlled atmosphere furnace was adapted from Torvac's standard compact 80 V design and comprises a cold wall, top loading unit with sequential control of process gases (usually dry and wet nitrogen/hydrogen mixtures) in parallel with a variable time/temperature programme. The hot zone and all radiation shielding is constructed of molybdenum.

7.4.2 METALLO-ORGANIC CHEMICAL VAPOUR DEPOSITION SYSTEM FOR SUPERCONDUCTING FILMS

A metallo-organic chemical vapour deposition (MOCVD) system has been designed specifically for research into thin films of high temperature superconductors. The 'MOCVD 105' has independently heated evaporation sources, each with independently heated supply lines, according to its manufacturer, Archer Technicoat Ltd, High Wycomb, UK. The sample platform can maintain a temperature between 100 and 950˚C, and the reactor has a separately controlled background heater to prevent premature condensation of the reactants. The evaporation sources are interchangeable. For instance, a mixture of yttrium, barium and copper β-diketonates will produce an yttrium barium copper oxide film in the presence of oxygen. All of the precursors have low vapour pressures, but with controlled heating of the evaporation sources it is possible to achieve reasonable transport rates without causing decomposition of the reactants, the company says.

7.4.3 CHEMICAL VAPOUR DEPOSITION FURNACE

Denkoh Co Ltd, Tachikawa, Japan, has introduced a vertical, low pressure, chemical vapour deposition furnace designed to produce films of gallium arsenide (GaAs).

The company says the main features are:

* a load mechanism which isolates the atmosphere inside the process tube, thereby increasing yield through improved film uniformity and thickness;

* high reproducibility and safety, as a result of automation;

* a throughput of 20 pieces an hour; and

* a film thickness distribution of ± 5%.

7.4.4 SCREEN PRINTER FOR THICK FILM SUPERCONDUCTORS

Two Japanese companies, a materials producer and a printing machine manufacturer, have cooperated to produce a desktop screen printing machine and related equipment for making thick films of high temperature oxide superconductors.

Dowa Mining Co and Murakami Screen Co, both based in Tokyo, say they have adapted existing machines produced by Murakami for the production of hybrid integrated circuits and printing soldering pastes. The machines are now commercially available, according to *The Japan Industrial Journal*.

7.5 CHARACTERIZATION

There is now a wide range of characterization techniques available ranging from (relatively) simple mechanical and physical property measurements through to electron and many other different types of spectroscopy. As the demand being placed on materials and their performance increases, so too does the need to understand and control materials' properties.

Two features of this section which should be noted are firstly the shear number of items which are presented below which indicates the growing importance attached to characterisation and secondly the geographical distribution. Throughout this chapter of the source book, items emanating from Japan have formed the largest group. However it is noticeable that this trend is not continued here, possibly indicating that to date Japanese characterization equipment has not penetrated the Western market extensively.

7.5.1 INFRA-RED MICROSCOPE DETECTS INTERNAL DEFECTS

Direct observation of internal defects in silicon integrated circuits is possible with its infra-red microscope, a US company claims.

Based in Piscataway, New Jersey, Research Devices says the 'Model F' microscope can detect internal defects and stresses, check wire bond quality, and observe emission from infra-red generators. It can also inspect solid state lasers, light emitting diodes, minerals, pigments and organic materials.

The Model F operates in transmission, reflection, polarization, and infra-red fluorescent modes. It is sensitive to radiation from visible wavelengths up to 1.8 μm, the company reports.

7.5.2 ANALYSIS OF SUPERCONDUCTOR IMPURITIES

US researchers have developed a device for detecting surface impurities in high temperature superconductors which they claim is cheap and simple to operate.

One of the team, D. Wayne Cooke, from the Los Alamos National Laboratory, New Mexico, believes the portable instrument offers several improvements over existing models. "Sample preparation is easier, analysis is quicker, and the operation does not require a skilled technician", says Cooke.

Analysis takes 85 s and can detect impurities at concentrations of less than 10 parts per million (ppm). The device can measure impurities within 1μm of the surfaces of bulk samples as small as one square millimetre. The scientists can also use the method to screen raw materials for impurities before they try to make the superconductor.

The instrument is similar to those used to read the dosimetry badges worn by people who may be exposed to radiation. The scientists irradiate the sample with x-rays, gamma rays or ultraviolet light for about three hours. They then place it in the detector where it is heated.

According to the team, when treated in this way insulators give off considerable light (luminescence). However, pure superconductors do not luminesce. "The intensity of a sample's luminescence as revealed by our device is directly proportional to the amount of insulating material at or near the surface of the sample being examined", Cooke says. "The absence of luminescence from a sample correlates with low levels of surface impurities and indicates that the material may be a good high temperature superconductor."

Los Alamos applied for patents in October 1989 and is now seeking private companies interested in licensing the technology.

7.5.3 AUTOMATED FLUORESCENT X-RAY ANALYSIS

A fluorescent X-ray system which can automatically weigh, crush and shape specimens, and perform multi-element and ignition loss analyses, has been developed by NGK Insulators Ltd, Nagoya, Japan.

The system can simultaneously analyse constituents from a wide range of materials including ceramics, superconductors, cement, ferrite, slag, and sintered materials. 29 elements can be measured simultaneously, ranging from boron (atomic number 5) to uranium (atomic number 92), according to the company.

7.5.4 X-RAY MICROANALYSIS

An X-ray system for microanalysis applications, the 'Advanced Performance Detector (APD)', has been introduced by Kevex Instruments, based in San Carlos, California, USA.

According to Kevex, this detector combines the following features:

- maintenance free — the APD, says Kevex, requires no liquid nitrogen for operation, or any other consumable coolant;

- it is available with either the 'SuperQuantum' window for light element detection or the 'SuperBeryllium' window for enhanced sensitivity down to fluorine; and

- it includes a standard, remote controlled motor drive for automatic positioning of the detector probe. In addition, the APD is protected from damage by power outages, either intermittent or prolonged. Its weight is said to be only one third that of conventional liquid nitrogen cooled detectors. The spectra produced by the unit are claimed to be identical to those obtained with more conventional detectors.

7.5.5 INDUCTIVELY COUPLED PLASMA MASS SPECTROSCOPY SYSTEM

Perkin-Elmer Ltd's plasma mass spectroscopy (ICP/MS) system, the 'Elan 5000', has a radio frequency power supply, ICP/MS interface (which the company describes as an improvement over that of the previous model, the 'Elan 500'), mass spectrometer, vacuum system and controlling electronics. The Beaconsfield, UK-based company says the 40 MHz radio frequency generator is the first to be specifically designed for use with an ICP/MS system.

In addition, the turbomolecular pumping system can establish a vacuum in less than three minutes, significantly reducing operating times, states Perkin-Elmer. The Elan 5000 is controlled by an industry standard computer with a multi-tasking operating system. Specialized software includes five application modules for a range of analysis requirements.

7.5.6 SEQUENTIAL INDUCTIVELY COUPLED PLASMA ANALYSIS

Two pieces of equipment which perform sequential inductively coupled plasma (ICP) analysis, the 'Plasma 400' and '400 WMA', are available from Perkin-Elmer Ltd, based in Beaconsfield, UK.

Based on the 'Plasma 40' and 'Plasma 40 WMA', they provide economical, reliable and high performance analysis, according to the company.

Both include an integral 40 MHz free-running radio frequency generator, a plasma monitoring system and a thermostatted monochromator.

The system is controlled by an industry standard computer, which can use third party software. Software features include wavelength tables, ID/Wt files and Wcal for easy method development.

7.5.7 ULTRAMICROTOME FOR ANALYTICAL ELECTRON MICROSCOPY

Research and Manufacturing Co Inc (RMC), based in Tucson, Arizona, USA, has introduced the 'MT-7 Ultramicrotome' for the thin sectioning of specimens for analytical electron microscopy.

Normally used in conjunction with a diamond knife, the MT-7 will produce single or serial sections from 10 nm up to 10 μm in thickness in 1 nm increments. With a powerful driven cutting stroke and microprocessor controlled mechanical specimen advance, the MT-7 will section ceramics and composites as well as bone and teeth, according to RMC.

7.5.8 *IN-SITU* PORE STRUCTURE ANALYSIS OF GELS

Nuclear magnetic resonance (NMR) relaxation measurements at low fields on interstitial fluids can probe the *in-situ* pore structure during sol-gel processing, according to the technique's US pioneer.

Douglas Smith, a researcher at the University of New Mexico in Albuquerque, says that by varying the aging and drying conditions during processing, the physical/chemical structure of the dried gel, and hence its properties, may be significantly altered. Using his method, Smith is able to follow the evolution of pore structure during processing, instead of having to infer changes from the final dried material, allowing dynamic control of final properties.

Smith has studied changes in pore structure with processing conditions (i.e. pore fluid, pH, temperature and time), using both base and acid catalysed silica gels. The initial pore structure of the wet gel, the structure after various treatments during drying and of the resulting xerogel were all obtained and compared.

In addition, the evolution of surface area in growth and aging (at different pH) of colloidal silica suspensions were studied by NMR. By combining the results with magnetic resonance imaging (MRI) techniques, Smith was able to study the spatial distribution of pore fluid and pore structure during drying.

7.5.9 GAMMA RAY DENSIMETER

The non-destructive evaluation of pressed and sintered ceramic bodies is possible with their gamma ray densimeter, West German researchers claim. Sintermetallwerk Krebsöge GmbH from Radevormwald and Hermann-Löns-Weg of Telgte say the instrument is suitable for use in the production environment. The technique is based on the absorbtion of gamma rays in solid materials, the absorbtion coefficient depending only on the material's composition and the energy of the radiation. This allows the porosity to be determined. Further details are given in *Ceramic Forum International* (1990, No.1/2, p.11–15). According to the paper, the potential range of applications has yet to be fully determined.

7.5.10 LASER FLASH APPARATUS MEASURES THERMAL DIFFUSIVITY

A fully automated laser flash apparatus which measures thermal diffusivity has been developed by Holometrix Inc, based in Cambridge, Massachusetts, USA. The instrument is designed to determine the thermal properties of advanced materials at temperatures up to 2000°C. Only a small test sample, 10 mm in diameter by 1–3 mm thick, is required.

The 'Thermaflash 2200' comes with a built in microprocessor which controls the vacuum, valve sequencing and laser systems, in addition to data acquisition and analysis. The system therefore requires no operator attendance once a specimen is loaded and the desired test temperatures have been keyed in.

Test specimens are irradiated uniformly on one surface by a laser beam pulse, the momentary temperature rise at the opposite surface being measured using a non-contact infrared (InSb) detector. The resulting data are then analysed by the computer and used to calculate thermal diffusivity, and hence to derive the other thermal properties.

This instrument is one of three products brought out by Holometrix recently which has applications for advanced ceramics amongst other materials. The other two are: a fully automated thermal conductivity analyser which can test both solids and non-solid materials, and an automatic quantitative adiabatic calorimeter for measuring bulk specific heat.

7.5.11 MANUAL FOR HIGH TEMPERATURE THERMAL ANALYSIS

Thermogravimetric (TG) specialist Setaram, Saint Cloud, France, has produced a manual for all types of high temperature thermal analysis.

The 70 page, A4 manual has chapters on TG, differential thermal analysis (DTA), differential scanning and other forms of calorimetry and dilatometry. These are followed by application notes under the broad headings of ceramics, metals and alloys, and inorganic materials.

Called 'Thermal analysis at high and very high temperature', the manual is free. It is also available from Setaram's agent Roth Scientific Co Ltd, Farnborough, UK.

7.5.12 ULTRA-HIGH TEMPERATURE DILATOMETER

A dilatometer which operates up to 3000°C is now available from Centorr Furnaces of Suncook, New Hampshire, USA. The system features Centorr's 'Series 10' graphite tube furnace, which has a vertical double walled chamber made of stainless steel. The graphite heating element is of a one-piece cylindrical design. Thermal insulation is provided by a graphite felt blanket wrapped around a graphite sleeve. According to Centorr, the furnace design assures that an isothermal zone surrounds the specimen. For operation at 3000°C, the dilatometer can accommodate up to a 2.54 cm^3 (one cubic inch) specimen. A linear variable displacement transducer (LVDT) with high resolution controls the coarse adjustment, while the fine adjustment is accomplished with a micrometer head. This arrangement allows for thermal expansion and ensures accuracy and repeatability to 0.0001, Centorr claims. The Series 10 furnace with the dilatometer accessory ramps from ambient temperature to 3000°C at 25°C a minute, and cools back to ambient in about two hours. This allows users to make two runs a day, according to the company. The instrumentation includes an Ircon 'Maxline' pyrometer system that allows control from 93°C to 3100°C.

7.5.13 AUTOMATIC INSPECTION AND EVALUATION FACILITY

A sample evaluation facility will allow potential users of its automatic inspection system to assess its suitability free of charge, according to the UK company which has set up the operation.

London-based Image Automation Ltd says that its inspection system employs a retro-reflective laser scanner to detect defects in transparent or highly reflective materials with a resolution better than 100 μm. It detects all faults that cause a surface disturbance or affect the transmissivity of the material, with defects being classified on the basis of their optical 'signature', together with a parametric description of their shape — including area, length and width.

7.5.14 CONTROL SYSTEM FOR MATERIALS TESTING

A control system for materials testing has been introduced by MTS Systems Corp, Minneapolis, Minnesota, USA.

Called 'TestStar', the system is based on a workstation and includes a personal computer, a digital test controller, a load unit control panel and the software. The computer interfaces with the load frame through the digital controller and a load unit control panel. Data collection, function generation and closed loop control are controlled by 32 bit microprocessors. The controller also conditions the transducer and drives the valve.

The load unit control panel is mounted near the load frame and provides actuator and hydraulic controls as well as a graphics display of specific test information. The TestStar system is designed for use with MTS load frames. Software for the system provides flexibility and ease of use, the company says.

A brochure is available from MTS which provides details on the system and its components.

7.5.15 NON-CONTACTING SURFACE PROFILER

Manufacturers applying thick films are able to profile surfaces and measure their thicknesses using a non-contact scanning capacitance microscope developed by Wentworth Laboratories, UK. The device has a vertical resolution of 0.1 µm and a total vertical range of 5 mm. Each measurement covers a spot in the horizontal plane which is about 10 µm square. The microscope works by holding a probe around 10 µm from a surface and measuring the capacitance between the two. As the sample is scanned underneath, a servo loop controls the vertical position of the probe so that the capacitance is kept constant. The output from a linear variable differential transformer connected to the probe is then proportional to the sample's surface topography, Wentworth Laboratories says. The company adds that the device can measure the profiles of wet, sticky or soft surfaces of any material. The sample holder can accommodate specimens up to 150 x 150 mm, though larger samples can be tested with some reduction in accessibility.

7.5.16 RAPID RESPONSE FURNACE FOR TENSILE TESTING

Rolls Royce plc of Derby, UK, says that its newly developed materials testing furnace can cycle from room temperature to 800°C and back to hand hot in under eight minutes. Conventional furnaces take about three hours. The furnace is intended for tensile tests on ceramics, composites and alloys. Each test typically lasts only 30 s. The faster cycling better represents the conditions faced by the materials when in use in an aero engine, especially on start up. Designed by graduate trainees, the furnace concentrates the energy from a pair of quartz halogen lamps onto the test specimens using chromed elliptical reflectors. This results in a very low thermal mass, since only the sample itself is directly heated. The bulbs are cooled by passing air across them with specially shaped ducts. The design won the first gold medal to be awarded in four years at the annual Engineering Design Competition, sponsored and run by the Worshipful Company of Turners in the UK. Rolls-Royce have applied for a patent and are now evaluating the business potential of the furnace.

7.5.17 NON-CLAMPED DYNAMIC MECHANICAL ANALYSIS

A UK company has introduced a dynamic mechanical analyser which offers non-clamped, as well as traditional clamped, modes of operation. Perkin-Elmer Ltd, based in Beaconsfield, UK, reports that the 'DMA 7' allows analysis of different sample types from low to high modulus. The device can perform fequency scanning and stress, strain and creep measurements, as well as displaying tan delta plus loss and storage modulus.

7.5.18 PARTICLE SIZE ANALYSIS OF MIXED POWDERS

A particle size analyser which is capable of accurately measuring the sizes of different ceramic powder particles, even when they are present as a mixture, has been developed by the Government Industrial Research Institute at Nagoya, Japan. The analyser operates on a system known as diffuse reflection infra-red fourier transformation (DRIFT), a kind of infrared spectroscopy. The intensity of the spectrum obtained varies in accordance with both diameter and the nature of the material, thus allowing the diameters of individual components in a mixture to be determined.

As reported in the *Japan Industrial Journal*, the Institute says that to date it has obtained satisfactory results for the individual diameters of alumina and silicon carbide powders which had been mixed together. There is no information, however, on how discriminating the technique is, i.e. whether the equipment could handle mixtures of ceramic powders which are relatively similar, such as alumina and zirconia.

8. STANDARDS

One of the major problems currently facing potential users of advanced ceramics is the wide diversity of similar materials available from different manufacturers and the range of properties claimed for them. In addition, often the techniques used to obtain the property values are either not provided or differ between manufacturers, leading to extreme difficulty in assessing which material should be selected.

The solution to the problem is the creation of first national and then international standards which outline how properties should be measured and reported. Although it has to be said that ceramic manufacturers have not always been entirely dedicated to this task, recent developments have seen improvements in this field. Further progress is expected over the next few years.

8.1 GENERAL

8.1.1 JAPANESE CREATE STANDARDS COMMITTEE

The creation of a standards committee in Japan has been announced by a division of the Science and Technology Agency. The committee will look into the creation of a system for evaluating advanced materials and will be in existence for two years. The focus will be on materials such as high temperature superconductors, synthesised diamond, and functionally gradient materials.

8.1.2 HARMONIZING INTERNATIONAL STANDARDS ON HIGH MODULUS FIBRES

A subcommittee of the American Society for Testing and Materials' (ASTM) D-30 High Modulus Fibers and Their Composites' committee, has been formed to pursue the international harmonization of standards in the area. The group, D30.02, will begin by reviewing current standards to determine where non-compatibilities exist. These may occur in relatively minor aspects such as terminology, specimen dimensions and conditioning, or in areas of major importance, such as the types of test used to measure a property, the data reported and the materials used.

Attempts will be made to accommodate minor differences by adjustments to ASTM standards where possible, but major changes will require dialogue between key personnel on an international basis. The committee is therefore looking for volunteers, particularly from ASTM members overseas with contacts and affiliations with standards bodies in their own country. The committee will seek to encourage ASTM round-robin activities and the organizations that participate to become more international in their scope.

8.1.3 RECOMMENDED TEST METHODS FOR CARBON FIBRES

The US-based Suppliers of Advanced Composite Materials Association (SACMA) has issued a set of five recommended methods for testing carbon fibres. These are:

* SRM 13-90, mass per unit length of carbon fibres;

- SRM 14-90, sizing content of carbon fibres;

- SRM 15-90, density of carbon fibres;

- SRM 16-90, tow tensile strength of carbon fibres; and SRM 17-90, twist in carbon fibres.

SACMA is not a standards issuing body, but, as it represents the composites raw materials industry, its recommended methods are tantamount to *de-facto* standards. SACMA is now collaborating with ASTM and Society of Automotive Engineers (SAE) in the USA to assist in further evaluation of the methods.

8.1.4 UK STANDARDS FOR ENGINEERING CERAMICS

The standardization of engineering ceramics is an area which has, to date, received relatively little attention. Consequently, although design engineers are now becoming more aware of the attractive properties of engineering ceramics, they often turn away from using these materials due to the low levels of accepted, consistent, accurate and genuine property data, which only standardization can provide. In the UK, this shortfall is now being addressed by the British Standards Institution (BSi) which is in the process of publishing a series of standards in five parts, with 22 sections, detailing:

- 'Sampling' (Part 0);

- 'General and textural properties' (Part 1);

- 'Mechanical properties at room temperature' (Part 2);

- 'Thermo-mechanical properties' (Part 3);

- 'Thermo-physical properties' (Part 4); and

- 'Resistance to chemical corrosion'(Part 5).

9. HEALTH, SAFETY AND THE ENVIRONMENT

Concern continues to be expressed about the inherent safety of ceramic whiskers because of their dimensional similarity to asbestos. The latter is known to result in significant damage to the lungs in certain individuals and early laboratory tests on whiskers pointed to similar results. To this end, the American Society of Testing and Materials established a sub-committee to examine the issue. Set out below is the assessment of the committee's chairman.

9.1 HEALTH RISKS

9.1.1 HEALTH RISK FROM SILICON CARBIDE WHISKERS ASSESSED

Concerns regarding the health risks associated with whisker reinforcement of ceramic materials prompted the American Society of Testing and Materials (ASTM) to establish a sub-committee of its E34 committee on Occupational Health and Safety to examine the issue during 1989. The Chairman of that committee, Sam Weaver, has now given his assessment of the risks associated with silicon carbide whiskers in an article in ASTM's *Standardisation News*.

Whiskers in general, and silicon carbide whiskers in particular, are regarded as potentially dangerous due to similarities with asbestos. Information received by ASTM E-34.70 suggested that silicon carbide whiskers of similar dimensions to asbestos will probably provide similar health problems and *in-vitro* studies have confirmed an analogy in the effects of the two materials.

However, Weaver indicated that the situation is not as bad as it initially seemed and that there are a number of positive factors:

1) In three separate studies involving hundreds of animals only one cancer was reported after silicon carbide exposure, which is not statistically significant;

2) The greatest health risk from whiskers is the development of mesothelioma which is caused by very fine whiskers with diameters less than 0.2 μm. The diameter of silicon carbide whiskers can be controlled during manufacturing to ensure that this size range is avoided;

3) Silicon carbide manufacturing operations lend themselves to controls to eliminate airborne whiskers (unlike the mining of asbestos); and

4) The composites manufactured from SiC whiskers bind the whiskers into the structure and even extreme treatment such as machining with cutting tools does not lead to individual whiskers being released into the environment.

Weaver's conclusion, therefore, is that providing sensible precautions are taken during manufacturing, the risks from using SiC whiskers are minimal. However, no details are provided on how much of the research was performed by independent laboratories and how much by research and development laboratories associated with companies who manufacture such whiskers. For example, regarding point (4), given the wealth of literature on studies of machining and abrasion waste, it is difficult to believe that no whiskers are released into the environment if a whisker containing composite is machined.

10. RESEARCH INITIATIVES

This section examines some of the research initiatives instigated during 1990. The order of the reports is largely geographical in nature and a point of note is the number and magnitude of initiatives created in Australia which is rapidly becoming a world leader in some aspects of advanced ceramic technology, notably zirconia-based materials. This is a reflection of Australia's dominant position as a supplier of zircon, the principal raw material used in the manufacture of zirconia ceramics.

10.1 GENERAL

10.1.1 JAPANESE ESTABLISH INTERNATIONAL EXCHANGE FUND

In the Autumn of 1991, the Tokyo-based Ceramic Society of Japan intends to establish an 'International Exchange Fund' as part of its centenary celebrations. The Fund will have a budget of ¥160 million to send young Japanese researchers abroad or to invite more mature foreign researchers and engineers to Japan for varying periods of time, the *Daily Industrial News* reports.

10.1.2 JAPANESE PROGRAMME FOR NON-LINEAR OPTICAL MATERIALS

The Japanese Ministry of International Trade and Industry (MITI), based in Tokyo, has launched a ten year research and development programme on non-linear optical materials as part of its 'Basic Technologies for Future Industries' project.

The programme is headed by H. Nakashini of the Molecular Engineering Laboratory, Research Institute for Polymers and Textiles, Tsukuba, and is divided into three sections:

*   Section One will last four years and will examine all the candidate materials;

*   Section Two will be three years long and will lead to the selection of the final materials; and

*   Section Three, taking three years to complete, will involve the manufacture of devices.

The programme will involve private companies, government research institutes, and universities.

10.1.3 NEW YORK STATE CREATES CERAMIC TECHNOLOGY CENTRES

New York State in the USA is providing a US$109 million budget to create two centres where companies wishing to exploit emerging technology in the field of advanced ceramics or glass ceramics can make a start.

The idea is to exploit the knowledge existing in the New York State College of Ceramics, at Alfred University, and at Corning Glass Works, according to *Ceramic Industry International*. The centres will be located near these two institutions.

Activities in the centres will range from the final stages of research through to prototype manufacturing, testing and product development. Overseas companies will also be eligable to apply for permission to use the centres.

10.1.4 AUS$1.5 MILLION FOR AUSTRALIAN CERAMIC RESEARCH

The Australian Government's Industry Research and Development Board has awarded three grants, totalling more than Aus$1.5 million, to collaborative projects aiming to develop ceramics.

The first is a three year, Aus$479 000 grant to a consortium which aims to develop high temperature superconductors for electrical power cables. The partners in the consortium are the Australian Nuclear Science and Technology Organization (ANSTO), Metal Manufacturers Ltd, the Electricity Commission of New South Wales, the Commonwealth Scientific and Industrial Organization (CSIRO), and the University of New South Wales.

ANSTO and CSIRO, in collaboration with Telectronics Pty Ltd, have also received a grant of Aus$514 000 for research into ceramic capacitors for high energy density applications. The aim of the project is to evaluate the breakdown characteristics of ceramics and apply processing techniques to develop advanced electroceramics. Such techniques, says the Board, would lead to the production of high energy density capacitors for applications in advanced microelectronics. Priority areas in this research are listed as high temperature superconductors and magneto-electronic materials and devices, as well as engineering ceramics.

The Board has also awarded a research grant to Melbourne's Monash University in commercial collaboration with ICI Australia Pty Ltd. The Aus$585 000 grant is for research into novel ceramic powders with enhanced separation characteristics. Its aim is to develop new ceramics with surfaces modified and clad with chemical or biological derivatives. These materials will allow for advanced high resolution separation applications in the chemical, pharmaceutical or biotechnology industries.

10.1.5 AUS$333 000 GRANT FOR ZIRCONIA RESEARCH

The *Daily Commercial News, Sydney* reports that a cooperative venture between the Australian Nuclear Science and Technology Organization (ANSTO), the Commonwealth Science and Industrial Research Organization (CSIRO), the Nilcra division of ICI Advanced Ceramics and Sydney University has been awarded a federal grant of Aus$333 000 over the next three years to further research into zirconia-based advanced ceramics.

Australia posseses about 50% of the world's supply of zircon sand, the primary raw material for the production of zirconia products. Currently around 90% of Australia's output is exported as the raw material; however, export earnings could be massively increased if value-added products were exported instead.

The current world market for zirconia powders and ceramic components is estimated to be Aus$70 million, and to be increasing by 15–25% each year.

10.1.6 AUSTRALIAN INSTITUTE FOR MICROWAVE PROCESSING

A Microwave and Materials Research Institute (MMRI) and associated Centre for Advanced Materials and Surface Engineering (CAMSE) have been created at the University of Wollongong in Australia.

The Institute's primary objective is to aid industry by assisting in the development of new products and processes by research, training and consultancy, according to *Metals and Materials*.

10.1.7 UK MICROWAVE PROCESSING GROUP CREATED

A Materials and Microwave Processing Group (MAMP) has been created at the University of Nottingham, UK under the guidance Dr Jon Binner.

The group currently has nearly 20 industrial, research institute and university members, with total funding in excess of £300 000 over the next three years. In addition, a grant application for a further £200 000 has been made to the UK Department of Trade and Industry. Membership is open to industry, research institutes and higher educational establishments throughout the European Economic Community (EEC) and the European Free Trading Association (EFTA).

The principal objectives of the group are:

*   to examine the potential offered by the use of microwave energy to the processing of materials, both in terms of improved microstructure control and the development of better properties and production economics; and

*   to design and fabricate superior materials for use in microwave components.

Materials considered by the group will initially focus on advanced ceramics, composites, and polymers. However, as the membership of the group increases it is anticipated that further materials will be considered. For example, a project has recently been submitted concerning the application of microwaves to biological systems.

10.1.8 INDUSTRIAL MATERIALS FOR THE FUTURE IN THE UK

The UK Department of Trade and Industry (DTI) and the Materials Science and Engineering Commission of the Science and Engineering Research Council (SERC), based in Swindon, UK, have completed a joint study of industrial materials for the future. The aim is to help identify the requirements of UK industry for new structural materials into the next century.

The Federation of Materials Institutes carried out the study under contract and interviewed top executives from a wide range of UK companies, representing users and producers as well as suppliers of materials. The study is intended to guide SERC in selecting strategic areas of generic materials research best suited to underpin industry's programme of applied research and development.

It is expected that the results will be published in the foreseable future.

10.1.9 ANGLO–SOVIET JOINT MEMORANDUM

The UK's Science and Engineering Research Council (SERC) and the USSR's State Committee for Science and Technology have signed a joint memorandum to facilitate collaboration between Soviet and UK scientists in certain areas of science and technology to which the State Committee attaches priority importance. Two of these areas are high temperature superconductivity and new materials.

10.1.10 UK RESEARCH COUNCIL'S INDUSTRIAL AFFAIRS UNIT

The Science and Engineering Research Council (SERC), based in Swindon, UK, has recently established a unit for Industrial Affairs to coordinate SERC's interactions with industry and provide a first point of contact.

The unit has absorbed what was the SERC LINK initiative. Two new leaflets entitled 'Industry' and 'SERC' are available from the SERC to provide a brief outline of the SERC's collaborative schemes and the role of the Industrial Affairs unit.

10.1.11 £1.65 MILLION FOR EVALUATING UK SUPERCONDUCTING DEVICES

The UK Government has awarded a £1.65 million contract to a consortium of UK companies to test and evaluate high temperature superconducting microwave devices, sensors and actuators, according to ICI Advanced Materials, Runcorn, one of the successful bidders.

The Department of Trade and Industry will provide 50% of the funding for the project, which also involves Plessey Research, Caswell; Lucas Automotive Systems, Solihull; and Birmingham University.

The research will focus on three areas:

*   the search for new compositions with high transition temperatures;

*   development of and fabrication of microwave and radio frequency devices; and

*   work on sensors and actuators.

ICI Advanced Materials will produce complex shapes of superconductors. Plessey will evaluate the microwave devices and Lucas Automotive the sensors and actuators. Birmingham University will be responsible for design and basic research and development.

11. INDUSTRY NEWS

The items in this section involve reports about companies using advanced ceramics. They span a broad range of interests, from the production of powders to sale of components and the manufacture of equipment. Because of the large number of reports, they have been approximately subdivided. First come the new companies that were created, then those which made significant expansions or investments. It is interesting to note that this latter group is by far the largest within this section. Reports covering joint ventures or agreements between companies are followed by news of takeovers or mergers, and then finally details of restructurings. It should be noted that the vast majority of items originate from Japan.

11.1 NEW COMPANIES

11.1.1 DIAMOND COATING COMPANY FORMED

Air Products and Chemicals Inc, based in Allentown, Pennsylvania, USA, has spun off its diamond coating unit into a venture capital-backed independent company, to be known as Diamonex Inc. It will be led by a group of former Air Products and Chemical employees.

The move is intended to facilitate rapid commercialization of products based upon recently developed technology for coating a variety of components with diamond films. Air Products will maintain an equity shareholder position in Diamonex, thereby keeping an interest in the technology base it has been developing in diamond films since 1985. Diamonex will establish research, manufacturing and administrative facilities near Allentown.

The Diamonex technology centres on a variety of plasma- and ion-assisted chemical vapour deposition techniques for producing polycrystalline and amorphous diamond coatings. It also includes an exclusive licence from the US National Aeronautics and Space Administration (NASA) for dual ion beam deposition of optically clear diamond-like carbon coatings.

Diamonex's specific product focus will be on the coating of eyewear for scratch resistance; on heat sinks, advanced packages and X-ray lithography windows for the electronics industry; and on speciality wear resistant products, including magnetic media, glass, ceramic and metal components and biomedical implants.

11.1.2 COMPANY ESTABLISHED FOR ZINC OXIDE WHISKER PRODUCTION

Matsushita Amtec Co Ltd of Osaka, Japan, a 100% owned subsidiary of Matsushita Industrial Equipment Co Ltd, has been set up to produce zinc oxide whiskers.

The new company will have a ¥200 million plant in Osaka with 3300 m^2 of floor space. It is intended to yield 200 t a month of the whiskers by 1993. However, production is already under way and by mid 1990 the plant was making 50 t a month. When running at maximum capacity, the company expects the plant to have sales of ¥6000 million.

The principal application for the ceramic whiskers is reinforcement for metals, plastics or rubbers, according to the *Japan Industrial Journal*.

11.1.3 KYOCERA FORMS SUBSIDIARY TO MARKET ADVANCED CERAMICS IN USA

Kyocera International Inc of San Diego, California, USA, has formed a subsidiary to market mechanical and industrial ceramics in North America.

The subsidiary, Kyocera Industrial Ceramics Corp, will also be based in San Diego. It was formerly a division of Kyocera America Inc. The new company will market industrial ceramics, automotive ceramics, ceramic substrates, electro-optical products, thermal and light emitting diode (LED) printheads, liquid crystal displays, contact image sensors and amorphous silicon drums.

Kyocera describes itself as one of the world's largest manufacturers of advanced ceramics for electronics, industrial, medical and consumer product applications, quoting worldwide sales exceeding US$3 billion.

11.2 EXPANSIONS/INVESTMENTS

11.2.1 JAPANESE PREPARE FOR SINGLE EUROPEAN MARKET

Several Japanese companies have revealed plans to build plants in Europe before the advent of the single European market on 31 December 1992.

According to the *Nikkei Industrial Daily*, the advanced ceramics companies include:

- Toshiba Tungaloy Co Ltd, Kawasaki, which aims to produce cutting tools;

- Mitsubishi Metals Corp, Tokyo, which intends to double production of Mitsubishi Metals Espanea by building a new cutting tool plant with a budget of ¥1.2 billion;

- Asahi Diamond Industrial Co Ltd, Tokyo, which has started production of diamond saws in the UK; and

- Minebea Co Ltd, Tokyo, which has already constructed a new plant in Lincoln, UK, for its subsidiary, Rose Bearing Ltd, to produce bearings for aircraft, with an annual production of 800 000 pieces. Miniature bearings are also to be produced, to compete with imports from Thailand.

Under the rules for the single European market, which are due to come into force at midnight on the 31 December 1992, if a foreign company has a European plant then its products will be treated as entirely European and, as a result, have free access within the European Community.

11.2.2 JAPANESE CEMENT COMPANIES EXPAND IN ADVANCED CERAMICS

Three Tokyo-based Japanese cement companies are currently expanding their advanced ceramics interests.

Nihon Cement Co Ltd is expanding its plant which makes ceramic substrates for electrical sensors. The work will have a budget of ¥300 million. In addition, the company is increasing its production of abrasion resistant and refractory ceramics, and is starting to commercialize its latest developments expects 1990 should have seen double the sales for advanced ceramics that it had in 1989 (¥1000 million).

Mitsubishi Mining and Cement Co Ltd has expanded its Yokose Plant which produces fine ceramics. Sales are now expected to be about ¥3000 million per annum, an increase of ¥700–800 million over 1989.

Chichibu Cement Co Ltd has established mass production facilities for very fine grained, high purity mullite powder in its Kumagaya plant and is expecting to produce 60 t of the material each year. If this business is successful, plans have been formed for further increases to take production up to 240-300 t per year.

11.2.3 MITSUI MINING MOVES INTO CERAMICS BUSINESS

April 1990 saw the commercialization of two ceramic products by Mitsui Mining Co Ltd, based in Tokyo, Japan.

'Macerite-Sp' is a machineable mica which is selling for only one third the price for which the prototype version has been available. 'Almatite' is an aluminium titanate which is also machineable and has a negligible thermal expansion coefficient, Mitsui Mining claims.

In addition to the sale of these two ceramics, which Mitsui Mining will manufacture itself, the company has also started importing barium titanate from the Ferro Corp in the USA. This is for wholesale in Japan, the *Daily Industrial News* reports.

11.2.4 NIPPON OIL MOVES INTO THE ADVANCED SPACE MATERIALS BUSINESS

Nippon Oil Co Ltd of Tokyo, Japan, has developed a carbon fibre/carbon matrix composite material and intends to develop it for application in the space construction material business, according to the *Daily Industrial News*. So far it has produced a gas jet nozzle of very complex shape and a 30 mm diameter cylinder which it claims can withstand a temperature drop of 1000°C across it. Further products are to follow.

11.2.5 GORHAM ADVANCED MATERIALS OPENS EUROPEAN SUBSIDIARY

Gorham Advanced Materials Institute (GAMI) of Maine, USA, has set up a subsidiary company, Gorham Advanced Materials (Europe) Ltd.

The company will provide direct access for GAMI's European clients to a range of services — such as contract research, development and consultancy — and will initiate single and multi-client studies within Europe.

11.2.6 ASTM EUROPEAN OFFICE OPENED IN THE UK

The American Society for Testing and Materials (ASTM), based in Philadelphia, USA, has opened an office in Hertfordshire, UK, in order to serve its European members better. The office will improve communication between ASTM's European members and ASTM headquarters in Philadelphia. UK staff will also answer questions about ASTM, coordinate symposia and standards technology training courses held in Europe, and provide rooms for standards development meetings.

11.2.7 ABAR GROUP TO ESTABLISH CENTRES FOR THERMAL PROCESSING

The Abar Group USA of Feasterville, Pennsylvania, consisting of Centorr Furnaces, Vacuum

Industries, and Abar Ipsen Industries, has announced plans to establish engineering centres for thermal processing excellence.

The restructuring will include combining the engineering resources of both Centorr and Vacuum Industries and redefining the role of Abar Ipsen engineering. However, the identity of the Centorr, Vacuum Industries and Abar Ipsen product lines will be maintained within the centres.

11.2.8 SHARP BUILDS RESEARCH LABORATORY IN UK

The Japanese-based company Sharp has established a laboratory in Oxford, UK, to conduct fundamental research in optoelectronics and information technology.

Sharp Laboratories of Europe Ltd will investigate electronic materials; optoelectronic systems to cater for the needs of information technology, using Sharp's developments with semiconductor lasers and liquid crystal displays; and applications of artificial intelligence, focusing on European languages for use in word processors, machine translations, electronic organizers, and personal computer equipment.

Sharp Corp (UK) Ltd will provide 75% of the initial capital of £0.5 million for the laboratory with the company's West German and Spanish subsidiaries sharing the rest of the costs.

Later this year, the enterprise will begin construction of a building at the Oxford Science Park, owned by the University's Magdalene College.

Sharp says it will recruit most of the employees from within Europe, particularly the UK. There will be 30 staff at first, with plans to expand this to 100 in the future.

Sharp Laboratories of Europe also aims to establish partnerships with UK and European research organizations.

11.2.9 ELECTRONIC CERAMIC PLANT FOR UBE INDUSTRIES

A dielectric ceramic production facility which cost ¥350 million has been built by UBE Industries Ltd, Tokyo, Japan. The plant is now complete and has begun production.

Operated by Mine Electronics Co Ltd, a subsidiary of UBE Industries located in Yamaguchi Prefecture, the principal products at present are parabolic antennas for satellite broadcast reception and component parts for car telephones. Total sales for the first year are predicted to be ¥600 million, rising to ¥1300 million by 1993, according to the *Japan Industrial Journal*.

11.2.10 ZIRCONIA OXYGEN METER PRODUCTION TO INCREASE

A five-fold increase in production of zirconia oxygen meters is planned by a joint venture company based in Kyoto, Japan.

Horiba-Westinghouse says it has recently developed a meter, called the 'World-Class 3000', which costs ¥650 000. The company is looking not only at the Japanese combustion control market, but also those of Korea and Taiwan, the *Daily Industrial News* reports.

Horiba-Westinghouse is a joint venture company owned by Westinghouse Electric Corp, Pittsburgh, Pennsylvania, USA, and Horiba Ltd, based in Kyoto, Japan.

11.2.11 OXYGEN SENSOR COMPANY STARTS PRODUCTION

A Japanese company which was formed in December 1989 to produce oxygen sensors for automobiles has now started production.

Ceramic Sensor was formed by NGK Spark Plug Co Ltd and NGK Insulators Ltd, both based in Nagoya City. The company has a staff of 226 and has a plant with 12 000 m^2 of floor space.

11.2.12 JAPANESE PLANT FOR OPTICAL AND ENGINEERING CERAMICS

Toto Ltd, Fukuoka Prefecture, Japan, has acquired a site in the city of Nakatsu in order to construct a plant to make ceramics components.

The site area is 54 713 m^2 and is located in the city's industrial park. Toto is to build a 10 000 m^2 facility for manufacturing optical and engineering ceramic components. The total investment (including the site) is ¥4000 million.

The *Japan Industrial Journal* reports that Toto expects total annual sales to have reached ¥120 000 million by 1993 - almost four times the company's current sales at existing facilities.

11.2.13 NIPPON SANSO TO SUPPLY SUPERCONDUCTOR RAW MATERIALS

Nippon Sanso KK of Tokyo, Japan, is to start to supply raw materials for making thin films of high critical temperature (Tc) superconductors using the metal organic chemical vapour deposition (MOCVD) method.

Nippon Sanso expects sales amounting to ¥20 million for the first year of business, rising to ¥100 million per year by the third year. Prices are quoted as ¥6000 per gram for barium, strontium, yttrium, copper and bismuth organic compounds, according to the *Nikkei Industrial Daily*.

11.2.14 TOYO ALUMINUM COMMERCIALIZES ALUMINIUM NITRIDE PRODUCTION

Tokyo Aluminum KK of Osaka, Japan, has announced that it has increased production of aluminium nitride ready for full industrialization of the process by the end of 1990. Production will amount to approximately 1 t per year and this will be used to produce integrated circuit substrates.

11.2.15 SILICON WAFER PRODUCTION PLANT OPENED

A silicon wafer production plant has opened in Wasserburg, West Germany, which will supply the European Community and the USA.

Wacker-Chemitronic GmbH, a subsidiary of Wacker-Chemie GmbH, Munchen, West Germany, opened the plant at the end of 1989. The company says it has equipped the buildings to produce epitaxy coated silicon wafers for use in the manufacture of semiconductor components. The facilities include class 10 clean rooms, i.e. areas where a maximum of 10 dust particles with a diameter greater than 0.5 μm are permitted in every 28 litres of air.

The company already has plans to expand the Wasserburg operation. At the moment about 100 workers are employed at the site, but this will rise to around 500 within a few years, according to Wacker-Chemitronic. Ultimately, the company hopes to employ 1500 people at Wasserburg.

Fairchild, a semiconductor component manufacturer, originally developed the site, selling it to the West German company in July 1988. Wacker-Chemitronic says it saw this as an opportunity to consilidate its position in the growing microelectronics market. Japan uses 58% of the silicon wafers produced in the world, and has five of the top seven suppliers. Wacker-Chemitronic describes itself as one of the two major non-Japanese manufacturers.

11.2.16 DU PONT TO BUILD TITANIA PLANT IN BRAZIL

A US-based company began construction work in September 1990 of a titania (TiO_2) plant in Uberaba in the South-Eastern state of Minas Gerais, Brazil.

E.I. Du Pont de Nemours and Co, Wilmington, Delaware, USA, is investing US$200 million in the plant which it expects to start operating at the end of 1991. The initial output will be 20 000 t a year; however, output is eventually expected to reach 80 000 t a year, with 50 000 t intended for export, mainly to Europe and Latin America. According to the *Gazeta Mercantil*, the company anticipates applications for the titania will cover the ceramics, paint and paper industries, among others.

11.2.17 FUMED SILICA PLANT TO BE CONSTRUCTED IN THE UK

A subsidiary of a US company is to build a plant in the UK for the manufacture of 'Cab-O-Sil (R)' fumed silica.

Cabot Carbon Ltd, part of Cabot Corp of Waltham, Massachusetts, USA, says the plant will produce 10 000 t a year. It will be located near Barry, Glamorgon, next to a Dow Corning plant, offering the potential for integration of feed, by-product and utility streams from each facility.

Fumed silica is an amorphous form of silicon dioxide. It is an inert substance used extensively as a reinforcing agent, viscosity modifier or flow improver for a range of products.

According to the company there are only four manufacturers of the material in the world.

11.2.18 FUMED SILICA PLANT FOR USA

A West German company plans to build a US$40 million plant in the USA to supply hydrophobic fumed silica.

Degussa Corp, part of Degussa AG of Frankfurt am Main, West Germany, will build the plant in Waterford, New York, where it aims to produce 6000 t a year. The site is adjacent to General Electric Co's silicone manufacturing facility.

When completed in 1992, the company intends the plant, which it claims is the first of its kind in the USA, to supply both GE's requirements and those of the general US market which is currently supplied by imported grades.

11.2.19 MASS PRODUCTION OF CERAMIC ENGINE PARTS

Kyocera Corp of Kyoto, Japan, has begun mass producing ceramic components for automotive engines at its Kokubu plant in Kagoshima Prefecture. The company plans to build more production lines at the site, to be operational early in 1991, according to the *Japan Economic Journal*.

11.2.20 CATALYTIC CONVERTER OUTPUT INCREASED

The decision by the European Commission (EC) to impose emission limits on cars with engine capacities below 1.4 litres has resulted in the West German company Degussa significantly increasing its production capacity for catalytic converters. The EC regulations, to come into force at the end of 1992, will mean that oxygen sensor-based emission control will become essential and will therefore require the increased use of catalytic converters. On hearing of the decision taken in

Brussels, Frankfurt-based Degussa immediately set about planning for an annual catalytic converter production capacity of 7 million units by the end of 1991. This represents an increase of 133% over the current capacity of 3 million units per year.

11.2.21 EXHAUST GAS PURIFICATION SYSTEM PRODUCTION TO INCREASE

A trial batch of 5000 ceramic exhaust gas purification systems for diesel engines to unnamed potential customers has recently been shipped by NGK Insulators Ltd, based in Nagoya City, Japan. With the steadily increasing regulations for exhaust emissions, NGK has already laid plans for increasing its production. The company estimates that it will be producing some 50 000 ceramic pieces annually by 1993.

11.2.22 NGK'S US PLANT COMPLETED

NGK Insulators Ltd, based in Nagoya, Japan, has finished construction of its subsidiary company's plant in Moorsville, North Carolina, USA.

NGK Ceramics USA of Wilmington, Delaware, USA, will produce 'Honey Ceram', a car exhaust gas purifier at the rate of 6 million pieces a year within the next three years, according to NGK.

11.2.23 DOWA MINING EXPANDS ZIRCONIA BUSINESS

Dowa Mining Co of Tokyo, Japan, has announced the construction of a furnace capable of producing 20 t of partially stabilized zirconia a month. The furnace will be used to produce electroceramic products, such as oxygen sensors and electrical and optical devices. The budget for this expansion to Dowa Mining's zirconia product line is 300 million, according to the *Nikkei Industrial Daily*.

11.2.24 PARTIALLY STABILIZED ZIRCONIA PRODUCT RANGE

A range of partially stabilised zirconia (PSZ) ceramics is now part of the product line of a US company.

Using its proprietary extrusion and injection moulding techniques, Orientation Inc, of Hudson, Massachusetts, says it can produce sub-micron precision parts, complex shapes and custom dimensions, as well as PSZ sheets with sizes ranging from 0.25–1.5 mm thick, up to 100 mm wide and 1300 mm long.

Typical properties of the material are:

- a bend strength of 1.2 GPa at room temperature, 300 MPa at 1000˚C;

- a Young's modulus of 200 GPa;

- a fracture toughness 7 $MN.m^{-3/2}$;

- a hardness of 13 GPa at room temperature, 4 GPa at 1000˚C;

- a thermal shock resistance of 250˚C;

- a thermal conductivity of 2.93 $W.m^{-1}.K^{-1}$; and

- a thermal expansion of 9 x 10^{-6}˚C^{-1} at room temperature, 11.5 x 10^{-6}˚C^{-1} at 1000˚C.

Current applications include cutting blades, abrasion resistant components, substrates, precision gauges, mechanical seals, bushings and pistons. Orientation Inc says it also offers a specialized hot isostatically processed zirconia for applications where great toughness is required.

11.2.25 CERAMIC DIE PRODUCTION PLANT

The *Daily Industrial News* reports that the Japanese company Hara Precise Dies Corp, based in Saitama Prefecture, has now completed construction of a manufacturing plant for making ceramic dies. The plant is located in Aomori Prefecture and was built with a budget of ¥400 million. The company is predicting sales amounting to no less than ¥1000 million yen by 1992-3.

11.2.26 ZIRCONIA GRINDING MEDIA

Dowa Chemicals Co Ltd from Tokyo, Japan, is to begin producing sintered zirconia spheres for use as a grinding media. The main product will be partially stabilized zirconia (PSZ) with radii in the range 0.5–10 mm. The move is part of the strategy of the parent company (Dowa Mining Co of Tokyo) to increase its ceramics business, according to the *Nikkei Industrial Daily*.

11.2.27 SEISHIN ENTERPRISE TO PRODUCE ALUMINA CRUCIBLES

Seishin Enterprise Co, based in Tokyo, Japan, a manufacturer of pulverizing machines, has announced that it has moved into the business of making high purity alumina crucibles. Currently it produces a range of crucible sizes, from 15 to 1200 cm^3, in 99.8% pure alumina. The price of these crucibles is quoted as being about the same as that for standard 95% pure alumina crucibles, i.e. 500 cm^3 crucible: ¥13 000; 150 cm^3 crucible: ¥5200.

11.2.28 EUROPEAN DEALERSHIPS FOR AGRICULTURAL CERAMICS

A UK company is allocating dealerships to European retailers for its range of agricultural products made from ceramics.

Stowmarket-based Agricultural Ceramics offers spring tine points, drill coulters, mole expanders, eradication tines, subsoiler shins, and nozzles for spraying, all fabricated from alumina by the UK firm Morgan Matroc. The products are more expensive than their traditional metal counterparts, but are much harder and so have a longer lifespan, according to the company.

Agricultural Ceramics believes that farmers are now ready to invest in the new technology. While exhibiting the products at the Smithfield (agricultural) Show, it says it received between 400 and 500 enquiries from interested parties. A similar high level of interest was shown at an event held in Denmark in January 1990.

Associated Farmers plc which owns three farms in Norfolk and Suffolk, UK, has just bought Agricultural Ceramics. The former says that its 2400 hectares of land will provide a testing ground for the technology.

11.2.29 LARGE GLASSY CARBON SHEETS

Hitachi Chemical Co Ltd of Tokyo, Japan, has begun manufacturing large sheets of glassy carbon (700 x 700 mm) at its Sakuragawa plant, according to the *Nikkei Industrial Daily*.

The company has constructed a ¥200 million production plant with a 2 t a month capacity. Hitachi Chemical expects annual sales to be about ¥500 million.

11.2.30 ION BEAM ASSISTED DEPOSITION FACILITY

The addition of a 'Nissin IVD' dual 2/200-40/40 unit with high capacity has expanded the ion beam assisted deposition (IBAD) facilities of Spire Corp in Bedford, Massachusetts, USA, the company has announced.

This coating system is customized to provide a range of corrosion resistant, electronic, optical, and/or tribological thin films on a variety of substrates, but with an emphasis on production service, Spire says. The facility will increase the coating capacity for aerospace, biomedical and industrial applications.

The system can provide multilayer thin films with precise composition and uniformity, according to Spire. Superior adhesion is accomplished through a graded interface layer which forms an integral mechanical bond between the substrate and the coating.

As well as pure metals, the IBAD system will also produce many ceramic compounds such as alumina (Al_2O_3), aluminium nitride(AlN), boron nitride (BN), titanium nitride (TiN), silicon nitride (Si_3N_4), and molybdenum nitride (MoN).

The facility joins Spire's existing services which include low temperature sputtering, ion beam mixing, ion implantation, and plasma assisted chemical vapour deposition.

11.2.31 ALUMINA FIBRES MANUFACTURING FACILITY

Rath Keramikfaser, based in Mönchengladbach, Germany, is to construct a facility capable of producing alumina fibre with a service temperature of 1800˚C. The fibre is intended for the European market and will have compositions in the range 80-95% alumina.

The facility will be in Mönchengladbach, next to the existing facility for fibres with a service temperature of 1400˚C. The first phase of plant construction will cost some DM13 million, *Ceramic Forum International* reports.

Applications for the fibres will include insulation for a number of industries and reinforcement for ceramic, metal and plastic matrices. The processing route has been licensed from the Japanese company Denka.

11.2.32 CARBON FIBRE PLANT

A plant for the production of carbon fibres has been built by Mitsubishi Gas Chemical Co Inc of Tokyo, Japan.

The plant, which has a budget of ¥500 million, will eventually produce 1000 t of carbon fibres each year, although initially it is only producing 300 t. Production costs are only one half those of conventional pitch-based carbon fibres, as a result of a processing method it has developed, the company says.

The method involves manufacturing the fibres from pitch using hydrogen fluoride and boron trifluoride as catalysts. Graphite fibres result with a yield of 80%.

Applications for the products include the aerospace and leisure industries, according to the *Nikkei Industrial Daily*.

11.2.33 US$9 MILLION CERAMIC FIBRE PLANT COMMISSIONED

A UK-based company has commissioned a US$9 million plant for manufacturing ceramic fibres and claims the site will double its production capacity in Europe.

Carborundum Resistant Materials has built the facility at Rainford, UK. Advanced furnace technology, combined with electronic process controls, enables a submerged electrode furnace to process increased volumes covering a range of chemistries — including alumina/silica/zirconia formulations, the company says.

The Rainford plant also converts high purity spun and blown fibre to paper, board, felt and fibre-based cements. In addition, the plant includes a laminating process for the production of ceramic fibre blankets backed with aluminium foil for applications such as fire protection.

Carborundum Resistant Materials says that, by virtue of its spinning technology, it can now offer a wider product range. Additional grades of 'Fiberfrax' ceramic fibre blanket and modules with new patented anchoring systems were marketed from April 1990. In addition, 'Durablanket S' is available in three standard densities: 64, 96 and 128 kg.mm^{-3}, in thicknesses from 19 to 50 mm.

11.2.34 TEXTRON DOUBLES CAPACITY OF 'AVCARB' CARBON FIBRE

Production capacity of 'Avcarb' carbon fibre will double to 34 000 kg a year following completion of a batch graphitizer furnace, according to Textron Speciality Materials of Lowell, Massachusetts, USA.

The furnace pyrolyses the carbon fibres at up to 1800°C for 12–24 hours, which Textron says reduces the levels of impurities such as alkaline metals to less than 50 parts per million (ppm).

The company has produced Avcarb carbon fibres for ten years. They are used in carbon–carbon composite brakes and rocket nozzles. Qualification has been obtained for their use in the composite brake systems of the McDonnell Douglas F-15 and F/A-18, Lockheed C-5A/B and Boeing 757 aircraft, Textron reports.

11.2.35 ZOLTEK CARBON FIBRE PLANT EXPANSION

Zoltek Corp of St Louis, Missouri, USA, has announced plans to double its current polyacrylonitrile (PAN)-based carbon fibre manufacturing capacity by building a facility in the University Research Park 25 miles west of St Louis.

Zoltek's existing carbon fibre plant is in Lowell, Massachusetts, USA, and was acquired from the Stackpole Corp in 1987. The combined capabilities of the two plants will allow the company to produce more than 450 000 kg of 'Pyron' oxidized PAN fibres and around 110 000 kg of 'Panex' PAN-based carbon fibres a year.

Facilities for carbon fibre manufacturing, batch carbonizing and carbon textile fabrication will be included at the new plant.

11.2.36 SUMITOMO TO DOUBLE PRODUCTION OF ALUMINA FIBRES

Sumitomo Chemical Co Ltd, based in Tokyo and Osaka, Japan, says it intends to double its production of long alumina fibres during the course of 1990.

According to the *Nikkei Industrial Daily* the increase in capacity will cost the company some ¥500 million, but it will allow them to drop the price from ¥)60 000 to ¥70 000 a kilogram to about ¥50 000 a kilogram.

11.2.37 QUARTZ FIBRES

Quartz and Silice of Nemours, France, a subsidiary of Saint Gobain, has now added a 14 μm diameter fibre to its range of Quartzel silica fibres. This complements the existing 9 μm fibres (and a 7 μm fibre which is available if requested).

The company says the fibres are all 99.99% pure silica. Yarns are available in 40–640 tex, 12–920 filaments per yarn with binders compatible with a range of polymer matrices. The existing 9 µm fibres are also produced as rovings and chopped strands.

11.2.38 FLOAT GLASS PLANT

After a long series of on-off negotiations it appears that a float glass plant is to be built near the town of Ferrol in Spain. The Spanish Government will put 60% towards the cost of the plant, estimated at US$149 million. However, the plant will be owned by Societa Italiana del Vetro (SIV), a subsidiary of the Italian firm, Efim. The plant will be SIV's second float glass plant in Spain and is expected to employ 400 people, according to the *American Glass Review*.

11.2.39 NIPPON SHEET GLASS CONSTRUCTING GLASS SUBSTRATES PLANT

Nippon Sheet Glass Co Ltd, based in Tokyo, Japan, started construction in May 1990 of a manufacturing plant for glass substrates for magnetic discs. When the plant is finished, in the middle of 1991, it will produce 100 000, 3.5 inch discs each month, the *Nikkei Industrial Daily* reports.

The plant is located in Yokkaichi City and has a budget of ¥3000 million.

11.2.40 HIGHER PRODUCTION OF FLUORESCENT GLASS

Production capacity for silver activated phosphate glass which fluoresces in the presence of radioactive contamination is to increase at Toshiba Glass Co Ltd in Tokyo, Japan. The company will now make 70 000 pieces a month, compared with 20 000 previously, exporting most of the extra glass, the *Nikkei Industrial Daily* reports.

11.2.41 JUMO EXPANDS

The German company M.K. Juchheim GmbH Co (Jumo), based in Fulda, has opened branch offices in the old East German towns of Erfurt, Leipzig and Magdeburg. The move is seen as intensifying Jumo's marketing activities in Germany. The company manufactures measuring and control instrumentation for temperature, pressure, moisture and other process variables. In 1989 Jumo achieved a total sales volume amounting to some DM 138 million.

11.2.42 LANXIDE REPORTS GROWTH

Lanxide Corp from Newark, Delaware, USA, has reported results showing growth in sales to US$905 000 during the third quarter of 1990. This is an increase of 470% on the same quarter last year. Total sales for the first nine months of this fiscal year amount to US$1.6 million, an increase of 437% on 1989.

Sales are primarily to customers for test and evaluation purposes, holding out the prospect for further substantial growth in the future, the company says. However, Lanxide recently completed the placement of US$30 million of three million shares of convertible preferred stock to private investors. This has resulted in a substantial increase in the firm and contingent resources that the company has committed to the development and commercialization of their ceramic and metal composite technology — now standing at a total of US$280 million.

11.2.43 PPG EARNINGS INCREASE

Second quarter 1990 earnings of US$141.0 million on sales of US$1.57 billion have been reported by PPG Industries, the US glass and coatings concern. Corresponding figures for 1989 were US$127.3 million earnings on sales of US$1.49 billion. The company says that this stronger performance is despite lower sales of its glass fibres during this period. The downturn in glass fibre sales has not dissuaded PPG from continuing its acquisition policy under which the company has recently gained complete ownership of Silenka BV of the Netherlands, following the purchase of shares from partners Akzo Chemicals International BV and NV Noordelijke Ontwikkelingsmattschappij.

11.2.44 CONSARC OPENS SALES OFFICE IN KOREA

Consarc UK and Consarc USA, manufacturers of vacuum furnaces, have combined forces in a joint venture to enlarge their network of sales offices by extending into Korea. The sales facility is located in Seoul and is focusing particularly on opportunities in the aerospace and automotive industries.

11.2.45 COTRONICS APPOINTS UK STOCKIST AND DISTRIBUTOR

The US company, Cotronics Corp, has appointed R.H. Symonds Ltd of Enfield, Middlesex, to be its UK stockist and distributor.

Cotronics manufactures such products as insulation, cements, castable and machineable ceramics, tapes and fabrics, sealants and binders and potting compounds.

11.3 JOINT VENTURES/AGREEMENTS

11.3.1 EUROPEAN CONSORTIUM FOR MATERIALS RESEARCH

The European Community's Joint Research Centre (JRC) has joined forces with three other institutes to develop and test advanced materials.

The JRC's Institute for Advanced Materials at Petten in The Netherlands will cooperate with Toegepast Natuurwetenschappelljk Onderzoek (TNO) from The Hague, The Netherlands, the National Engineering Laboratory (NEL) in East Kilbride, UK, and the Centro Informazioni Studi Esperlanze (CISE), based in Milan, Italy. The joint venture will be known as the European Materials Research Consortium (E-MARC). E-MARC says it will have about 1000 researchers and technicians offering a range of skills from materials innovation to engineered prototype. It has begun with initiatives to secure contracts in industrial sectors, such as the automotive industry, aerospace, off-shore engineering and energy generation.

11.3.2 STRUCTURAL CERAMICS DATABASE

The National Institute of Standards Technology, based in Gaithersburg, MD, USA, has announced a new database, the NIST Structural Ceramics Database (SCD). Version 1.0 contains thermal and mechanical properties of silicon carbide and silicon nitrides in a stand-alone, user-friendly database system that operates on DOS-based personal computers.

Searches of the data are conducted by means of the SCD's combination of menus, query-by-example technique and computer-assisted entries. Users may search for properties of a selected ceramic or find ceramics satisfying specified property values. NIST claims that this database is a representation of the current state-of-the-art of materials property data across all silicon carbides and silicon nitrides.

The PC system requirements are: DOS 2.1 or later, 640 kilobytes of memory, 3.7 megabytes of hard disk storage space, an EGA colour display card and a colour monitor. The cost of the database is US$495.

11.3.3 MATERIALS INFORMATION FOR DEVELOPING COUNTRIES

Materials Information of Ohio, USA, has signed an agreement with the United Nations Industrial Development Organization (UNIDO) to improve the dissemination of information about materials science and technology in developing countries.

Subject selections from the METADEX, Engineered Materials Abstracts, and Materials Business File Databases were made available on diskette as part of this agreement in mid-1990, along with a specially designed application of UNESCO's Micro-ISIS text retrieval software.

The diskettes being distributed, at no charge, to UNIDO's Industrial and Technological Information Bank (INTIB) and national focal points worldwide, to help in their provision of technical assistance to industry. Materials Information will also market the diskettes worldwide, offering a reduced price to developing countries. The cost per diskette subscription will depend on the quantity of data.

According to Materials Information, the scheme commenced without funding, but it is hoped that offers of help from industry and other organizations will make this project a success and enable it to develop into a second phase beginning in 1992.

11.3.4. MATERIALS INFORMATION AVAILABLE ON DISC

The 'Dialog OnDisc Metadex Collection' is a CD-ROM that provides international coverage on engineered materials, according to Materials Information. Contents include technical developments, business and economic news, and commercial reports from Materials Information's three databases: Metadex (1985–present), Engineered Materials Abstracts (1986–present) and Materials Business File (1985–present).

The information covers ceramics, metals and polymers, and a total of 350 000 documents can be accessed. These include abstracts, indexes and bibliographic citations. Subject emphasis is on materials, their properties, processes, products and forms.

Available now the Metadex Collection is being updated quarterly and an annual subscription is US$6950.

11.3.5 LICENCE TO MAKE ALUMINIUM NITRIDE POWDER IN USA

A Japanese company has granted a US manufacturer a licence to make and sell aluminium nitride (AlN) powder.

Tokuyama Soda Co Ltd of Tokyo will allow customers of Dow Chemical Company, Midland, Michigan, to fabricate and sell translucent AlN products made from the powder.

According to *High Tech Ceramic News*, Tokuyama Soda first developed its aluminium nitride powders over ten years ago and claims to have been the first company in the world to sinter translucent ceramic products at atmospheric pressure. Today, Tokuyama Soda has the capacity to produce 130 tons of AlN powder per year.

11.3.6 US–JAPANESE AGREEMENT TO DEVELOP ELECTRONIC MATERIALS

A Japanese company has signed an agreement with two subsidiaries of a US chemical firm to develop electronic materials.

Matsushita Electric Industrial Co Ltd of Osaka will cooperate with Du Pont Japan Ltd of Tokyo and Du Pont de Nemours International SA from Geneva, Switzerland. The companies say they aim to draw upon Matsushita's experience of electronics and Du Pont's expertise in materials technology. By exchanging information, they hope to develop speciality materials for electronics manufacturers.

The agreement stipulates the terms and conditions for the exchange and for the protection of proprietary knowledge. The companies say they will discuss each individual project on the basis of the contract, which will last three years with an option to renew.

A spokesperson for Du Pont in Geneva says the company's direct sales to Matsushita currently include thick films for hybrid circuits, materials for printed circuit boards, polyester films, connectors and interconnections, and synthetic resins and fibres.

11.3.7 DU PONT–LANXIDE JOINT VENTURE FOR MICROELECTRONIC COMPONENTS

A joint venture to commercialise high performance microelectronic components using the patented 'Primex' ceramic reinforced metal technology has been announced by Lanxide Corp of Newark, Delaware, USA, and Du Pont Co of Wilmington, also in Delaware.

The new company will be called Lanxide Electronic Components LP, and will offer hybrid circuit electronic packages, carrier plates, substrate chassis and support structures.

Products will be tailored to meet the needs of heat loss, material compatibility and structural demands. Initially, materials will be based on ceramic particle reinforced aluminium. Future products under development will use ceramic fibre reinforced metals.

A 3500 m^2 production facility at Newark is under construction for the venture, which is the third to be formed between Lanxide and Du Pont. The previous companies formed were Lanxide Armor Products Inc and Du Pont Lanxide Composites Inc, both of which were established in 1987.

11.3.8 JOINT VENTURE TO PRODUCE SHORT ALUMINA FIBRES

200 t a year of short alumina fibres will be produced at a site in West Germany, following an agreement between a Japanese company and its Austrian partner. Tokyo-based Denki Kagaku Kogyo KK and Rath Co expect construction to start shortly and to be complete around the end of 1990, says the *Daily Industrial News*.

11.3.9 AUSTRIAN–JAPANESE JOINT VENTURE FOR ALUMINA FIBRES

The Austrian company Rath Co, of Vienna, and Tokyo-based Denki Kagaku Kogyo KK, have formed a joint venture with the intention of producing and selling alumina short fibres, primarily in the USA.

A plant has been built in Austria by the two companies for the manufacture of alumina short fibres and Rath has created Rath Performance Fiber Co, based in Delaware, USA. The latter has already received its licence to sell alumina fibres in the USA. The final part of the package concerns Denki Kagaku Kogyo which is about to start a market survey for alumina short fibres in the USA.

11.3.10 KOREAN–FRENCH AGREEMENT TO MAKE GLASS FIBRES

A Korean company and a French company have agreed to a joint venture to produce glass fibres.

Hankuk Glass Industry and the French Vetrotex International, a subsidiary of Saint-Gobain, will construct a factory capable of producing 46 000 t per annum to the southwest of Seoul, Korea.

However, initially it will only produce about 20 000 t per annum. Vetrotex's share of the costs is US$150 million, according to the *American Glass Review*. Current glass fibre production amounts to about 9000 t per annum, insufficient to supply the demand which is estimated at 26 500 t, largely from the manufacture of printed circuit boards and the automotive industry. The plant should begin production in 1991.

11.3.11 JAPANESE–US COOPERATION WILL PUT US GLASS INTO JAPAN

An announcement has been made by the Pittsburg Plate Glass Co (PPG), based in Pittsburg, Pennsylvania, USA, that it has established a joint sales company in Japan in cooperation with the Japanese company C. Itoh and Co Ltd, based in Tokyo. The new company will eventually import heat reflecting and reinforced glass from the USA, in addition to other decorative sheet glass which will be imported from China.

11.3.12 POTASSIUM TITANATE WHISKER PRODUCTION UNDER WAY

A joint venture company based in Kanagawa Prefecture, Japan, has begun production of potassium titanate whiskers. Japan Whisker Co Ltd says capacity is 1200 t a year. The whiskers will cost ¥1500–2000/kg.

According to the *Nikkei Industrial Daily*, one of Japan Whisker's parent companies, Tokyo-based Toho Titanium Co Ltd, developed the manufacturing process. The other owners of the company are Mitsui and Co Ltd, Tokyo, and Otsuka Chemical Co Ltd, Osaka.

11.3.13 THREE-WAY COLLABORATION ON SUPERCONDUCTORS IN THE USA

With the aim of developing high temperature superconductors, three organizations have agreed to one of the largest collaborative research and development programmes between private industry and a US Government laboratory, according to a report in *European Chemical News*.

Under a three year US$11 million agreement, E.I. Du Pont de Nemours and Co (Wilmington, Delaware), Hewlett-Packard Co (Palo Alto, California) and the Department of Energy's Los Alamos National Laboratory (Los Alamos, New Mexico) are focussing on the development of thin film high temperature superconductors for electronic components. They will pool resources in basic and applied research, materials production and processing, and electronic applications.

11.3.14 SUPERCONDUCTIVITY PILOT CENTRE GETS FIRST RECRUIT

The General Electric Co's Research and Development Center, based in Schenectady, New York, USA, is the first US company to join the newly formed Superconductivity Pilot Center, at the US Government's Argonne National Laboratory, Illinois. The Pilot Center will examine the structure of bismuth-based high temperature superconductors, according to *Interceram*.

11.3.15 SEIKO INSTRUMENTS AND AIR PRODUCTS REACH AGREEMENT ON DIAMOND FILMS

Seiko Instruments Inc, Tokyo, Japan, has concluded a technical agreement on diamond thin film coatings with US gas manufacturer, Air Products and Chemicals Inc, Allentown, Pennsylvania, USA.

Seiko Instruments has been investigating the practical use of thin films made by microwave plasma-based chemical vapour deposition. Air Products and Chemicals has been looking at other techniques, such as ion beam production of diamond-like carbon films.

The agreement will allow Seiko Instruments to make progress towards developing its diamond film business, the *Japan Industrial Journal* reports.

11.3.16 SILICON CARBONITRIDE FIBRE LICENCE FOR HERCULES

An agreement between Rhone-Poulenc of Paris, France, and the US-based Hercules Inc provides the latter with exclusive rights to sell the French company's 'Fiberamic' in North America.

Fiberamic is a silicon carbonitride fibre, used primarily as a reinforcement in ceramic and metal matrices for high temperature applications.

Rhone-Poulenc currently produces Fiberamic at its French facility. However, under the terms of the agreement, if demand in North America were to rise significantly then the option of producing the fibre on that continent could be pursued. The agreement also allows both companies to work on further developments of ceramic materials.

11.3.17 US–ITALIAN AGREEMENT FOR ULTRA HIGH VACUUM EQUIPMENT

Elettrorava SpA of Torino, Italy, and the US company, MV-Systems Inc, based in Golden, Colorado, have announced an agreement relating to the worldwide marketing and production of ultra high vacuum (UHV) systems for the fabrication of thin film semiconductors and dielectrics. In particular, the UHV plasma enhanced chemical vapour deposition (PECVD), designed by MV-Systems Inc, will now be assembled under licence by Elettrorava.

11.3.18 BATTELLE LICENSES MILLER THERMAL FOR PLASMA-ARC SPRAY CONTROL SYSTEM

Miller Thermal Inc of Appleton, Wisconsin, USA, has signed an agreement to be the exclusive worldwide licensee to manufacture and distribute Battelle's plasma-arc spray control system.

Battelle claims that the system improves control and repeatability when applying coatings of a range of materials, including advanced ceramics and metal alloys. This is achieved by the use of sophisticated data acquisition hardware and software to monitor and record appropriate process data. This is then processed to obtain a value designated the 'net plasma energy', which when incorporated into the automated control system is the key factor in providing peak process performance and coating quality on a consistent basis.

Applications for the system include the production of alumina, chromia, tungsten carbide and nickel and cobalt alloy coatings for abradable seals, biomedical applications, corrosion protection, preparation of composite materials, spray forming of parts, thermal barriers and wear and erosion resistance.

11.3.19 GERMAN DISTRIBUTOR FOR LLOYD INSTRUMENTS

Lloyd Instruments plc, Fareham, UK, has announced the appointment of Elscolab as their distributor for materials testing systems in Germany. Elscolab, with operating companies in Belgium and Holland, will be opening a sales and service office in Cologne.

11.3.20 SHORT WAVE INFRA-RED HEATERS NOW AVAILABLE IN UK

A US company has appointed Astro Technology Ltd of Fareham, as UK distributor for its products which include short wave infra-red heaters.

Research Inc, based in Minneapolis, Minnesota, says that the heaters, such as the model '5070 Multi-Zone Load Test Heater', are designed for use in conjunction with mechanical testing equipment. They should fit between the jaws of most standard testing machines. For instance, the 5070 is capable of heating a 1.25 mm thick specimen to 1500°C in less than 90 s, according to Research Inc.

11.4 TAKEOVERS/MERGERS

11.4.1 ICI AUSTRALIA TO ACQUIRE US COMPANY

The US-based Fine Particle Technology (FPT) Corp has agreed in principle to sell its R&W Ceramics operation in Auburn, California, to ICI Australia.

R&W Ceramics employs approximately 100 people and manufactures high performance alumina ceramic components and metallized assemblies for a range of industrial applications, including speciality valves and pumps, medical and scientific instruments, power generation and wear resistant applications. R&W's customers are concentrated primarily in the USA, according to the company.

ICI says the acquisition will complement their Nilcra Ceramics business which supplies related ceramic products in partially stabilized zirconia (PSZ) from its plant in Melbourne, Australia. In addition, ICI wants to address new market opportunities with further product lines and develop a manufacturing base in the USA.

11.4.2 COORS CERAMICS BUYS GE CERAMICS

Coors Ceramics, which sold its substrate plant in Wales, UK, to Cookson Ceramics, has now purchased General Electric Ceramics Inc, based in Chattanooga, Tennessee, USA. The latter will be renamed Coors Electronic Package Co and will operate as a wholly owned subsidiary of Coors Ceramics.

11.4.3 FELDMÜHLE TAKES OVER CERAMX CORP

The German company, Feldmühle, based in Plochingen, have announced that they are taking over the Ceramx Corp, owned by Eagle Industries Inc of Chicago, USA. The move, which encompasses three different companies, Interceram (Middletown, New York), AlSiMag (Laurens, South Carolina) and Ceramaseal (New Lebanon, New York), is aimed at increasing Feldmühle's international activities, primarily in the electroceramics field. Ceramx Corp's activities are focussed principally on substrates and casings for electronic components such as thyristors and diodes, although they also manufacture structural ceramic components for telecommunication and computer applications. One of the principal advantages for Feldmühle is expected to be the receipt of Ceramx's extensive metal/ceramic technology.

11.4.4 UK RESEARCH INSTITUTES COMBINE

Two UK research institutes, Fulmer Research Institute (currently based in Stoke Poges) and BNF Metals Technology Centre of Wantage, are to combine to form a single organization working primarily on composite and metallic materials development and associated enabling technologies.

The grouping has arisen from the purchase of Fulmer Research by BNF Metals Technology Centre from the Institute of Physics. The latter retains control of the remaining parts of Fulmer Ltd, i.e. Fulmer Yarsley and Yarsley Quality Assurance. The new organization will trade under the name of Fulmer Materials Technology.

11.4.5 CRAY ADVANCED MATERIALS SOLD TO PRIVATE INVESTORS

Cray Electronics (Holdings) plc has sold the entire shareholding in its subsidiary company, Cray Advanced Materials Ltd, to private UK investors. The change in ownership is reflected by a change in name to Advanced Materials Systems (AMS).

The company, based in Yeovil, describes itself as one of the leading enterprises developing and exploiting metal matrix composite (MMC) technology. The particular expertise developed by AMS is in the liquid pressure forming route to net-shape MMC parts.

The technology is protected by patents held by the UK's Ministry of Defence and licensed to AMS, and consists of the infiltration of a ceramic fibre preform with molten metal. Following the sale, AMS plans to extend its operations into ceramic and polymer matrix composites and will launch its fibre and particulate preform technology early in 1991.

Structural analysis and design operations are to be developed; AMS offers itself as a source of finite element analysis for composites, and the company is currently recruiting in these areas.

11.4.6 UK MATERIALS INSTITUTES FAIL TO MERGE

On the 23 October 1990, Extraordinary General Meetings were held by all three of the main materials Institutes in the UK. The purpose of the meetings was to decide whether the Institutes of Ceramics, Metals, and Plastics & Rubber should merge and form a single Institute of Materials.

The Institute of Ceramics returned a vote of 58.8% in favour (38.4% poll), the Plastics & Rubber Institute 72.6% in favour (45% poll) and the Institute of Metals 96% in favour (only 90 members present). As the required percentage of 75% in favour was not achieved by all three Institutes the merger will not now be able to go ahead as proposed.

Subsequently, the Institute of Metals has indicated that it intends renaming itself the Institute of Materials, if necessary without the support of the other two bodies.

11.5 RESTRUCTURING

11.5.1 HITACHI UNIFIES AND EXPANDS CERAMICS OPERATIONS

Hitachi Metals Ltd from Tokyo, Japan, is unifying its electrical and structural ceramics operations, marketing all such goods under the single name of 'Iso-Ceram'. The company is also looking to increase its sales of advanced ceramics, according to the *Daily Industrial News*.

1990 total sales are predicted to be approximately ¥1.5 billion, a figure Hitachi Metals expects to double during 1991. A target of ¥7.5 billion, five times the 1990 figure, has been set for 1994.

11.5.2 PERKIN-ELMER INTEGRATES EUROPEAN OPERATIONS

Perkin-Elmer, Münich, Germany, has integrated the European operations of its Physical Electronic Division (PHI) with Atomika, which was acquired by the group in 1987.

The newly formed organization, Perkin-Elmer Surface Science Europe, is located in Oberschleissheim, West Germany, a suburb of Munich. The company will produce depth profiling, secondary ion mass spectrometry (SIMS) instrumentation, x-ray surface analysis (XSA), and ultrahigh vacuum products.

The facility also includes an analytical laboratory with the latest PHI and Atomika Auger, x-ray photoelectron spectroscopy (XPS), SIMS, XSA, scanning Auger microprobe (SAM) and multiple technique (XPS-SAM-SIMS) instrumentation.

11.5.3 ERA TECHNOLOGY REORGANIZES

ERA Technology Ltd, the Leatherhead, UK-based research and development company, has reorganized its materials activities within a new unit known as the Materials and Structural Integrity Centre.

The Centre comprises a Materials Applications Division, a Structural Integrity Division and a Materials Testing Department. The Structural Integrity Division has three Departments; Life Management Technology, Plant Assessment Services, and Mechanics and Dynamics. The Materials Applications Divisions' work includes: electronic materials, thin film technology, advanced materials and devices, and engineering polymers amongst others.

11.5.4 MITSUBISHI MINING TRANSFERS CAPACITOR PRODUCTION

Mitsubishi Mining and Cement Co Ltd, Tokyo, Japan, has transferred the production of multilayer capacitors to its subsidiary KCK Co Ltd, also based in Tokyo.

The move is a result of continuing problems with the business, which according to the *Japan Industrial Journal* has largely failed to mature. While the transfer takes place, the company will try to improve the production process in the expectation that the business can be made more profitable.

Mitsubishi Mining is also changing production at its Yokose factory to the manufacture of surge absorbers which, the company says, produce higher profits as a result of their higher added value.

11.5.5 DU PONT CEASES PRODUCTION OF PITCH-BASED CARBON FIBRES

Du Pont de Nemours and Co of Wilmington, Delaware, USA, has decided to cease producing its range of pitch-based carbon fibres.

Sources at the company say the decision was prompted by the large investment deemed necessary to sustain and develop activity in this area. The long term returns from the market for high and ultra-high modulus carbon fibres were not considered sufficient to justify further outlays - despite the considerable sums of money already invested.

Du Pont's pitch-based fibres have tensile moduli of 724 GPa ('E-105') and 894 GPa ('E-130'). E-105 and E-130 both maintain a strain to failure of 0.55% and have tensile strengths of 3.3 and 3.9 GPa, respectively.

According to the company, existing stocks of the fibres will be used to honour current commitments to customers and research laboratories but production has now ceased.

CHAPTER 2

COMPOSITES

11.　INDUSTRY NEWS

11.1　NEW COMPANIES

11.2　EXPANSIONS AND INVESTMENTS

11.3　JOINT VENTURES AND AGREEMENTS

2. EXECUTIVE SUMMARY

1990 was a year of consolidation and preparation for the advanced composites industry. It started with fear over the impact of defence cutbacks in the USA and Europe on the back of improved East-West relations and ended with the US and UK economies in recession, further uncertainty resulting from the possibility of a Gulf War and longer term worries over ever-tightening health and safety regulations and environmental concerns. Not the conditions to instil business confidence or to encourage cautious companies to switch to unfamiliar advanced materials.

In many respects the industry has stood up remarkably well. While many companies have cut back on staff, relocated, pulled out of projects and generally tightened their belts, others have continued to expand and invest for the future. The significant increases in world-wide carbon fibre production capacity for example, instigated in 1989, show no signs of a slowdown. There has also been a noticeable increase in activity in new material forms — existing fibres and matrices produced as preforms, hybrid yarns and novel fabrics, where the goal has been cost effective manufacturing rather than increased performance.

The uncertainties regarding military aircraft programmes have made the industry examine new markets with increasing interest — in rail and marine applications, general engineering and construction. Meanwhile the long cherished automotive market shows signs at last of embracing composites in a big way — even if it is still sheet moulding compound which is taking the largest share.

Traditionally it has been the military aerospace sector in the USA which has provided the biggest stimulus for advanced materials development — and paid most of the development costs.

The big military programmes have had an extraordinarily chequered career in 1990, with the ATF advanced tactical fighter, now known as the YF-22/YF-23, surviving just, the B-2 stealth bomber programme barely alive, and the A-12 navy attack plane finally cancelled. In Europe the EFA Eurofighter programme is intact — for the time being.

The industry has however found a number of alternative drivers from the civil sector. The Airbus family continues to exploit large quantities of composites but the project currently providing the greatest excitement is the new Boeing 777 where the specifications for tough polymer matrix composites issued by Boeing are proving to be demanding. The downturn in defence related composite procurement may ultimately be offset by the boom in composite-intensive civil aircraft widely predicted at the beginning of 1990, though the economic downturn may mitigate against this in the short term. Recession and fear of flying have dramatically hit the profitability of airlines throughout the world just as they were about to embark on the largest re-equipment exercise of all time. The result has been lapsed options, aircraft cancelled and deliveries deferred.

Raw materials suppliers who service the aerospace industry will undoubtedly be under pressure in the short term, with those more reliant on military contracts being the hardest hit. The long term demand for high performance aerospace prepregs must be good but given that there is a degree of over-capacity in the sector at the moment it would not be surprising to see some rationalisation during 1991, with some companies withdrawing from the sector altogether.

The increasing need for composites in aerospace to be competitive on capital costs as well operating costs is providing an even greater boost to the search for cost effective manufacturing systems, either through materials innovation or automated machinery

One of the categories of advanced composites that was expected to improve manufacturing costs was the high performance thermoplastics such as APC-2, carbon/polyetheretherketone. The widespread adoption of these materials has failed to materialize despite demonstrator programmes which have shown that technically complex aircraft structures can be produced. It seems that costs associated with processing and the difficulties with tooling have proved significant obstacles at the current time, and the additional benefits offered by thermoplastics (such as increased toughness) are being matched by the current crop of highly modified epoxy based thermosets. The future for advanced thermoplastic composites possibly lies in the various co-mingled and interdispersed thermoplastic polymer/reinforcing fibre fabrics beginning to emerge — a good example of a new material form solving the industry's problems without necessarily invovling completely new materials.

If the future of thermoplastics in aerospace looks uncertain, their position in the automotive industry may be more promising. Significant quantities of composites are beginning to appear on vehicles in a range of applications, from body panels to load bearing components. In terms of body panels, SMC seems to be most significant in tonnage terms at present, particularly in the USA where the General Motors' APV advanced van has a completely composite body and is produced in significant volumes. The general accepted wisdom in the industry is that these materials will eventually give way to composites produced from a resin transfer moulding or liquid composite moulding process. A nagging worry however is the increasing power of the environmental lobby and how composites will be regarded — environmentally friendly or otherwise. The answer to that question will depend on how broadly based an environmental audit is performed on the materials, but issues such as recycling are becoming an important priority for raw materials suppliers and users alike. Thermoplastics may have some advantage over thermosets here, given their theoretical potential for thermal re-processing into some form. The increasing range of stampable thermoplastic sheet composites derived from impregnated mats or via paper making technology is likely to be supplemented by high volume fraction aligned fibre products that are in many respects glass/polypropylene equivalents of the carbon/polyetheretherketone thermoplastics on offer to the aerospace sector.

The need for high volume production of polymer based composites for industries such as automotive industry and general engineering, coupled with an increasing need to produce large thick structures for marine and construction industries, has accelerated the drive into preform technology with a host of new techniques being developed to produce both high and low performance preforms utilising textile technology, 3D braiding, spray techniques and thermoforming.

Preform technology is also likely to be the key to the eventual profitability of metal and ceramic matrix composites. Particulate reinforced light alloys appear to have some future but metal matrix composites based on fibres are still struggling to find a market and one suspects that the manufacturing technology for these materials is in need of further improvement before the materials can be truly competitive. The success of companies such as Lanxide in the USA and AMS (formerly Cray Advanced Materials) in the UK, pioneering variations on liquid metal infiltration, is crucial in this respect.

The infiltration process for ceramics may be revolutionised by the exploitation of sol-gel techniques to complement existing chemical vapour infiltration methods. The market for advanced ceramic matrix composites looks extraordinarily promising, although there is a recognised need for fibres with higher operating temperature capabilities (1500°C) if the true potential for high temperature applications is to be realised. Significant research funding is likely to be directed towards high temperature fibres in the coming years. Cheaper ceramic fibre reinforcements are also required to push the materials into general engineering applications with more modest service temperatures and there has been some progress in this area, notably with aluminium borate whiskers from Japan.

Fibre development is also proceeding apace for polymer matrix applications. The properties of carbon fibres have been improved over recent years, with higher strain-to-failure rates and modulus values from both pitch and polyacrylonitrile based fibres. Currently, there is an awareness of the need to improve strength in compression, with the microstructure of the fibres appearing to be the key to obtaining a balance between stiffness and strength.

Some other approaches include the exploitation of non-circular cross-section fibres. It is known that many carbon fibre producers are experimenting with multi-lobal fibres, while tri-lobal and bi-lobal glass fibres are now available. Hollow fibres are also being made available in experimental quantities.

A fibre type that has become increasingly prominent during the year is the high modulus polyethylene fibres such as Allied's Spectra in the USA and DSM's Dyneema in Europe. There are many who feel that these fibres are just the beginning of a whole new class of high performance organic fibres that may come to rival aramids in their importance to the industry.

While composites and processes have received considerable attention during 1990, the infrastructure necessary for a successful composites industry has also been developed. Groupings such as SACMA in the USA have become more active. An international consensus seems to have emerged regarding the need for standardization on testing in order that the data bases can be amassed that allow composites to compete on equal terms with metals. The cost of testing is still considered by many to be a significant negative factor for composites, and there can be no justification for increasing these costs through unnecessary duplication. ASTM has established a sub-committee to work towards the harmonization of international standards which augers well for the future.

New materials, new materials forms, increased realism and a pressing need to explore new markets should culminate in some exciting progress for the composites industry in 1991 and beyond. High added value markets such as medical implants, and massive volume industries such as construction, look like offering excellent opportunities for composites. Markets as yet to be defined await the ultimate composite forms, namely multifunctional and smart materials.

3. MARKETS

With defence cutbacks causing concern throughout the composite industry in 1990, there was perhaps been even more interest than usual in business surveys looking at the prospects for existing and alternative markets for composites.

3.1 BOOM IN ADVANCED COMPOSITES

The Verbundwerk '90 meeting in September, 1990 held in Wiesbaden, Germany, led to reports in the German press that the worldwide market for composites (of all sorts) in 1988 had reached 2.83 million tonnes worth a total of US$3.17 billion, with predictions that this would rise at an annual rate of 25% to US$5 billion by 1995. These reports expected some 72% of the total composites market to depend on European consumption, with 26.7% going to Germany. It was not clear whether these optimistic figures took into account the developments in Eastern Europe which might further stimulate demand, particularly from the united Germany. At present the largest composites market, particularly for the higher performance advanced composites, is represented by the USA. During 1990, the Freedonia Group published a 123 page study 'Advanced composites and high performance fibers' which concentrated specifically on advanced composites in the US and predicted that the market for composites and fibres would reach 18 000 (tonnes) by 1994.

3.1.1 EUROPEAN MARKET FOR ADVANCED COMPOSITES

The European market was assessed in detail during the year by Frost and Sullivan Ltd of London, UK. They predicted that the European market for high performance composites would exceed US$1 billion by 1994, with European suppliers being capable of supplying their home demand during this time as well as making significant inroads into the US sector.

'The European market for high performance composites' says that aerospace will remain the major growth sector. It considers that costs are still too high for the high volume end of the automotive industry, although some further progress will be made into trucks and racing cars, particularly for springs, clutches and brakes.

Testing is identified as a significant element in the cost of advanced composites and the report states that this will also make it difficult for newcomers to break into the field. Material costs in Europe are predicted to fall as new capacity comes on stream, including production of carbon fibre in Germany, and polyethylene and aramid in The Netherlands.

The report predicts that the European carbon fibres market will expand at the expense of aramid, with carbon fibre consumption increasing from 1059 to 2922 t during the period 1989 to 1994, while aramid falls from 310 to 60 t. At the same time, epoxy consumption is predicted to rise from 680 to 1498 t with thermoplastics and particularly polyetheretherketone, also showing significant progress.

France is identified as the current European leader in composites based on high performance resins, with a consumption in 1989 of 706 t which will rise to 1563 t in 1994; Germany will be using 646 t in 1994 and the UK about 615 t.

The full report costs US$4200.

3.1.2 EUROPEAN MARKET FOR GLASS FIBRE COMPOSITES

In tonnage terms glass fibre composites continue to dominate the market and figures released by French glass fibre giant Vetrotex revealed that 1988 and 1989 were bumper years for the European glass-fibre composites industry. The 1980s were in fact revealed as a decade of consistent growth. In 1981 the total European market for glass reinforced plastics amounted to 576 000 t. This had risen to 858 000 t by 1986, with 1987 and 1988 figures amounting to 962 000 and 1,095 000 t respectively and, more importantly perhaps, representing the largest annual growth figures, 12.1% and 13.8%, for the decade.

Among the various countries in Europe, it is the German market that is the largest, with 1988 figures of almost 293 000 t and an astonishing growth rate of 26.7%. The French and Italian markets are very similar, totalling 192 000 t and 195 000 t respectively and showing similar growth rates of 17.6% and 17.9%. The UK comes fourth with 140 000 t and a growth of 12.8% and the Benelux countries combined amount to a market of 109 000 t and a growth rate of 10%.

1989 was an equally good year with figures indicating growth rates across Europe as a whole exceeding 10%.

According to Vetrotex, growth in glass fibre composites in Europe is now exceeding that of the USA where a slightly larger market (1,207 000 t) grew by only 5.1% in 1988. The fastest growth is taking place in Japan where the market, estimated at 533 000 t in 1988, represented a growth of 16.6% on the previous year.

3.1.3 WORLD GLASS FIBRE DEMAND REVIEWED

Glass fibre manufacturer PPG Industries in the USA sees demand for its products growing fastest in the far East. Annual growth rates of between 4% and 5% over the next five years are predicted in the comparitively mature markets of the USA and Europe, whereas the Far East averaged a healthier 5–9%. Domestic consumption in the USA was actually predicted to have fallen slightly in 1990 as a result of a downturn in the transportation, marine, construction and consumer applications markets. Good growth is predicted in corrosion resistant application areas as economic conditions improve.

The growth in the market for glass fibre is steady if not spectacular, and a large quantity of additional capacity is due to come on stream soon. PPG itself was busy in 1990, tripling the capacity of both its UK plant and a joint venture operation in Venezuela and starting up a new joint venture plant in Korea.

3.2 AEROSPACE MARKETS

Aerospace has been the traditional driving force behind the growth in advanced composite markets, although it has been military programmes that have feature most prominently in the past. This is likely to change if the revised estimates for worldwide civil aircraft demand issued by both Boeing and Airbus Industrie in 1990 are to prove correct. Boeing's predictions run to 9935 commercial aircraft between 1990 and 2005 with a combined value of US$626 billion. Airbus has settled for

12 200 aircraft between 1988 and 2008 with a total value of US$700 billion. Despite the variations these are staggering figures and would suggest that any downturn in the composites market that might have resulted from a reduction in military spending in the 1990s will be more than compensated for by civil developments.

The increased numbers of civil aircraft being built would be good news for the composites industry even if the levels of composites on new airframes were to remain at current levels. However, a survey published in 1990 by Business Comunications Co (see below) points to an increased penetration by composites into the commercial aircaft market.

3.2.1 COMPOSITES IN AIRFRAMES

A survey by Business Communications Co (BCC) looks at the market generally for advanced materials in airframe applications. The survey indicates that the market projected for 1990 for all materials will total US$3021 million, with a surprisingly high percentage of this total (US$1432 million) being made up from advanced composites. The value of the composites sector is somewhat out of proportion to its volume due to the high material costs. Growth of the total materials market for airframes up to 2000 is projected at 8% per annum, with advanced composites (of all matrix types) in the region of 12.4% per annum. According to their figures, the value of the composites market should be double that of metals by the turn of the century. The report, 'Conventional and advanced materials for airframes', costs US$2950.

3.2.2 ADVANCED SUPERSONIC TRANSPORTS

A number of consortia are now emerging that are looking at the future potential for advanced supersonic transports. These will inevitably require large tonnages of most types of advanced composites.

The only grouping to have actually built a successful commercial supersonic transport in the past, i.e. British Aerospace and Aerospatiale, has announced that it is to study both the market and technical viability of a successor to Concorde. The companies are at present spending about US$10 million on a study which envisages a larger plane, about 100 m long carrying 200–300 passengers. Preliminary estimates put the cost of such a project at US$10 billion, with a target in-service date of 2005.

Rumours abound that eventually the partnership involved in such a project would be larger, possibly involving other European concerns and American companies. Already Boeing and McDonnell Douglas have performed studies on such a plane, and NASA has begun a five year feasibility study (spending a reported US$284 million) on a Mach 2–3 supersonic transport. According to a *New York Times* report, Boeing and McDonnell Douglas have concluded that a market for such a plane would exist if a range of greater than 6000 miles for a payload of 250–300 passengers could be achieved, coupled with cruise speeds of 1300–2000 mph. A project of this magnitude would also be attractive for Japan where it is believed the government plans to study the possibilities of high speed commercial aircraft.

3.3 ARMOUR

3.3.1 MARKET FOR ADVANCED ARMOUR TO INCREASE

A market study carried out by Kline and Co of Fairfield, New Jersey, USA, addressed the potential for advanced armour systems.

The company says that in recent years improvements in munitions have rendered conventional steel plate armour systems vulnerable to penetration and as a result the market for advanced armour, based on ceramic and fibre reinforced polymers, had grown to an estimated US$200 million by the end of 1989. A defence turndown was allowed for in Kline's predictions of a US$400 million market for such materials by 2000, with a 10% annual growth rate. Kline's study covers all advanced armour materials and their usage in application areas ranging from personal armour to space armour, through structures, land vehicles, shipboard systems and aircraft.

According to Kline, the percentage of polymer composites in the advanced armour market has reached 45%, with ceramics set to increase their percentage substantially. Personal protection, land vehicles and shipboard applications represented the three major end-uses for the market in 1989.

3.4 METAL MATRIX COMPOSITES

Metal matrix composites is an area where the annual research bill is high but as yet the industrial applications are small. Although little in the way of hard information became available regarding the market potential for such materials during 1990, a number of new surveys had been commisioned. Several of these are likely to make rewarding reading.

3.4.1 METAL MATRIX COMPOSITE SURVEY

TechTrends published a new report concerned with the technology and industrial applications of metal matrix composites.

The report reviewed the nature of reinforcements available for metal matrix composites and the properties that may be obtained. Processing and fabrication aspects were considered with current limitations and recent developments in processing technology. Potential application areas were identified in aerospace, defence, automotive and sports equipment. Finally, the activities of a number of significant companies in the field were summarised.

This report completes a series of three reports by TechTrends on advanced composites, with similar publications available (for the same cost) on polymer matrix composites and ceramic matrix composites.

The 179 page report is entitled 'Metal matrix composites: technology and applications' and is available from Innovation 128 at US$500.

3.4.2 DEVELOPMENTS AND TRENDS IN METAL MATRIX COMPOSITES

A survey on current developments and future trends in metal matrix composites (MMCs) has been published by Elsevier Advanced Technology.

The report was co-authored by Dr Brian Terry from Imperial College, London and Dr Glyn Jones who is a director of Technical Communications (Publishing) Ltd.

The report is in four sections, covering concepts and classifications of MMCs, properties processing and potential applications of cemented carbides, aluminium, magnesium, titanium and iron based MMC, applications in the aerospace, automotive, leisure and other industries, and future prospects, market forecasts and potential for commercialisation.

'Metal matrix composites' costs £125 and is available from Elsevier Advanced Technology.

3.4.3 PERA TO STUDY POTENTIAL FOR METAL MATRIX COMPOSITES IN HIGH SPEED MACHINERY

PERA, the UK research establishment, has launched a new programme to investigate the potential for metal matrix composites in high speed machinery.

The PERA study is funded under the UK Government's Link programme where up to 50% of the funds are provided by the Department of Trade and Industry and the Science and Engineering Research Council. The balance of funding is provided by industry.

The aims of the programme are to generate a store of information on the design manufacture and performance of specific components suitable for high speed machinery; to quantify the benefits of using metal matrix composites in terms of improved mechanical properties and reduced mass and inertia; to determine the design methodology required for specifying high performance components as a function of materials selection and manufacturing routes; to examine a selected number of case studies, in collaboration with the industrial participants in the project.

3.5 PITCH-BASED CARBON FIBRES

3.5.1 GROWTH FOR PITCH-BASED CARBON FIBRES

Pitch-based carbon fibres will continue to make inroads into the total carbon fibre market, currently dominated by polyacrylonitride based fibres. The current market for all types of carbon fibres totals US$400 million and will rise to US$600 million by 1995 and possibly US$1000 million by 2000, with pitch-based materials representing about 5% of this total, according to a report produced by two Japanese companies.

The report, by Techno Co Ltd and Shinko Research Co Ltd, is available in an English edition through Omnia of Raleigh, North Carolina, USA. It says that the market share for pitch-based fibres could rise to 10% by the year 2000 if improved properties and lower production costs are realized.

The report looks at the background to pitch-based fibres, their manufacturing, current capacity, markets and applications, and compares competitive fibres. The current market is considered in detail to allow predictions up to the year 2000. Other issues, such as the international patent situation, and proposed methods for cost reduction, are also explored.

The report costs US$9700.

3.6 ALUMINA

3.6.1 ALUMINA INCREASINGLY IMPORTANT

Alumina is becoming increasingly important in the advanced composite field as a reinforcing fibre, as a constituent material in the production of ceramic fibres with more complex compositions, and as a matrix material, according to Mitchell Market Reports (MMR) of Monmouth, UK.

The second edition of MMR's report 'Alumina' is published in three volumes. The first considers the different forms of the material, world production and various end uses. Volume one also contains the executive summary and conclusions.

Volume two details alumina-based activities in some 30 countries covering over 750 companies and organizations, while volume three looks at the import and export records of aluminium hydroxide and non-metallurgical grades of alumina.

All three volumes may be purchased together for an inclusive price of £750.00 (US$1315) from Elsevier Advanced Technology.

3.7 THERMOSETS

3.7.1 POTENTIAL FOR THERMOSETS ASSESSED

Rapra Technology in the UK has published a major new study into the potential future market for thermosetting resins and compounds in the European market.

While thermosets are well established in many markets, with end uses such as matrices and adhesives for the composites industry, their position has been increasingly threatened by the emergence of engineering and high performance thermoplastic polymers. Rapra state that this has caused an apparent relative decline for thermosets. Despite this, growth in some sectors remains strong and increases of up to 10% per annum are forecast.

The survey features: a technical review of the competitive situation, taking into account new material developments (both thermoplastic and thermosets); market forecasts for the early 1990s; and identification of new business opportunities for polymer manufacturers, processors and users.

The total project final report is available for a fee of £5000 but contains specific modules which may be purchased separately.

4. MATERIALS

4.1 REINFORCEMENTS

1990 was not a particularly dramatic year in terms of new reinforcements coming onto the market. However, a number of the more pertinent issues in terms of fibre performance do seem to have been addressed in 1990 by some of the suppliers.

4.1.1 HIGH MODULUS, HIGH STRENGTH CARBON FIBRE

Toray Industries of Tokyo, Japan, has introduced a new carbon fibre which it states has an excellent combination of strength and stiffness.

The fibre, 'Torayca-M60J', is a polyacrylonitrile (PAN) based fibre with a quoted tensile modulus of 600 GPa, and a strength of 3.9 GPa. The modulus is exceptionally high for a PAN-based fibre, particularly with the attendant strength values. In contrast, an earlier Toray product, M-50, exhibits a stiffness of 490 GPa but a strength of only 2.45 GPa.

The heat treatments used for producing high strength or high stiffness are not always compatible. Toray claims that the unique properties of M-60J result from the special disordered structure developed during the heat treatment at temperatures between 2000 and 3000°C.

The search for better high temperature ceramic fibres for metal and ceramic matrix composites continued throughout 1990, with activity in the USA, Europe and Japan.

4.1.2 SILICON CARBONITRIDE FIBRE

More details emerged at the 1990 European Conference on Composite Materials (ECCM) regarding the new silicon carbonitride fibre, Fiberamic, being developed by Rhone Poulenc in France.

The fibres are being developed in an attempt to produce a higher temperature resistant fibre than the current silicon oxycarbide fibres (such as 'Nicalon' and 'Tyranno') that would match pure silicon carbide monofilament (where the high filament diameter of some 140 microns rules out most fabric production).

The composition of Fiberamic was stated by Perez and Caix from Rhone Poulenc Recherches as being 57 wt% Si, 22 wt% N, 13 wt% C and 8 wt% O. About 5–10% of the phases present are free carbon with the remainder $SiN_xC_yO_z$. The fibre structure is described as homogeneous, with the free carbon distributed as small clusters throughout the material.

Fibre diameter varies within a range 12–16 microns — 14 microns being the average. The elastic modulus is 200 GPa and average tensile strength is between 1500–2200 MPa.

The fibre retains some 80% of its room temperature strength when tested at 1400°C, and in an oxidizing atmosphere it retains 85% strength after 100 hours at 1000°C and 10 hours at 1200°C.

The fibre is at the pilot plant production stage of development at present but agreements have already been reached with Hercules for marketing and distribution (and possibly manufacture) in the USA.

4.1.3 'TYRANNO' FIBRE COMPOSITES STRONGER THAN 'NICALON'

Textron Speciality Materials claims that there is now conclusive evidence that ceramic composites produced with 'Tyranno' fibre reinforcement are stronger than composites produced using Nicalon fibre reinforcement.

The two competitive fibres are essentially ceramic yarns comprising mainly beta-silicon carbide with excess carbon and oxygen, and in the case of the Tyranno fibre a homogenous dispersion of a small quantity of titanium. The Nicalon fibres are produced by Nippon Carbon and Tyranno by Ube Industries, both of Japan. TSM represent and distribute the Tyranno fibre in the US and Canada and also hold an option to manufacture.

The claims that Tyranno fibres are superior are based on independent data published by the Oak Ridge National Laboratories early in 1990 (ORNL Report no TM-1154). The mechanical testing performed by ORNL appears to have consisted mainly of flexural tests performed at a range of temperature up to 1200°C on composites prepared by ONRL's forced flow thermal gradient chemical vapour infiltration process. The Tyranno fibres under these conditions exhibited a 10–20% higher strength than the Nicalon equivalent, according to TSM.

TSM also cite additional advantages of Tyranno fibres including a fibre diameter of 8 microns compared to 15 microns for Nicalon. This results in considerable benefits in weaving and braiding with these brittle fibres.

Many fibre suppliers are looking for novel routes to improved fibre performance, including non-circular cross-sections and hollow fibre constructions.

4.1.4 HOLLOW GLASS FIBRE

A hollow version of the successful 'S-2' glass fibre has been developed by Owens Corning Fibreglas Corp of Toledo, Ohio, USA. The company says that its objective was to produce a glass fibre with the properties of E-glass but with a substantial reduction in weight.

The hollow S-2 glass fibres are some 20-30% lighter than solid E and S-2 glass and most of their properties appear to be reduced on a pro-rata basis from that of solid S-2 glass due to the reduction in net section thickness. E glass has a tensile modulus of 6.5 MSI compared to the 8.5 MSI of S-2 glass. However, the hollow S-2 glass retains a modulus of 6.92 MSI. The strength of the hollow fibres is 210 KSI compared to 160 KSI for E-glass and 290 KSI for solid S-2.

The company says that particularly attractive properties include the low coefficient of thermal expansion of, 1.8×10^{-6} compared to 1×10^{-6} for solid S-2 glass and 3.8×10^{-6} for E-glass, and the dielectric constant of 3.2 (at 24°C and 10 GHz) compared to 4.7 for E-glass and 3.9 for S-2 glass. This is approaching the levels of quartz fibres and suggests that the hollow fibres might be suitable for radome applications.

The Owens Corning team says it is uncertain of the final applications areas and even the matrix systems that the fibres will be compatible with (i.e what sizes will be available). The design of the fibres (i.e. the diameter of the internal hole; the fibres themselves are 10 microns in diameter) has been chosen to restrict wicking of resin into the hole in the event of resin transfer moulding processing with low viscosity resins.

Companies that can see a potential market/application for such fibres are encouraged by Owens Corning to get in touch to discuss possible collaboration.

4.1.5 SHAPED FIBRES EVALUATED

Owens Corning has announced that it is evaluating fibres with bi-lobal and tri-lobal cross-sections. According to Owens Corning, the non-circular cross-section requires modifications to the fibre forming process in order to overcome high surface tension and viscosity effects. Fibres have so far been produced with effective diameters of 6.5 to 25 microns.

The potential value of non-circular cross section fibres lies in the greater surface area/mass ratio compared to equivalent circular fibres. This, it is hoped, will improve certain strength characteristics in composites, provide lower thermal conductivity in dry-laid mats or felts and vary packing characteristics. The end uses identified by the company include filtration applications as well as composite reinforcement.

At this stage the materials are not generally available and Owens Corning is working closely with selected customers to identify uses.

4.1.6 GLASS FIBRES COMPATIBLE WITH ACETAL

Manville Corp, Waterville, Ohio, USA, has introduced a new glass fibre reinforcement for use with acetal and acetal copolymers.

'Star Stran 751' is a chopped strand with a sizing consisting of a film forming ingredient and an aminosilane. The new product has been developed using technology acquired under an agreement between Manville and Nippon Electric Glass of Japan.

Reinforcing acetal polymers in the past has provided problems. Poor adhesion of fibres to the polymer has necessitated the use of additives which have deleterious effects such as introducing discolouration, restricting flow during moulding and causing gaseous emissions during processing.

The sizing on Star Stran 751 is claimed by Manville to provide good adhesion which eliminates the need for the additives to the polymer, and as such should encourage further applications for reinforced acetal systems.

4.1.7 HYBRID YARN FOR COMPOSITES REINFORCEMENT

A hybrid yarn from Japan is reported, in European Patent 0-326-409, to give a laminated material that is excellent not only in tensile strength, interlaminar shear strength and Charpy impact resistance, but also in compressive and flexural strengths. This, it is claimed, offers a way of creating a unidirectional prepreg which can give laminates with such properties.

It is being developed in the Hirakata Laboratory of Ube Industries Ltd, Osaka, and is a hybrid filament yarn combining carbon fibre and an inorganic fibre composed mainly of such elements as silicon, titanium or zirconium, carbon and oxygen. The ratio of the tensile modulus of the carbon fibre to tensile modulus of the inorganic fibres is in the range 0.6–1.4.

In an example, a piece of carbon fibre (Besfight HTA 6000, from Toho Rayon, 7 microns diameter, 1.77 specific gravity, tensile modulus 24 t/mm^2, 6000 fils) and an inorganic fibre composed of Si, Ti, C and O (Tyranno, from Ube Industries, 8.5 microns diameter, 2.33 specific gravity, tensile modulus 2.1 t/mm^2, 800 fils) were taken through pipes through which water was flowing and then directed into a water tank. These fibres were then spread, while being agitated mechanically, to bring the filaments of the two components into contact with each other.

This combined filament was taken through a 2 wt% concentrated epoxy emulsion tank and then dried and sized to produce the hybrid yarn. The size was based on 1 part to 100 parts fibre.

Observation of the resultant yarn with a scanning electron microscope showed that the carbon fibre filaments and the inorganic fibre filaments were uniformly combined.

4.1.8 FLAX FIBRES FOR COMPOSITE REINFORCEMENTS

According to the Institut Textile de France, Lyon, flax fibres have the potential to act as reinforcing materials in composites. ITF director Dr Michel Sotton, in an item carried by Techtextil-Telegram reports that flax fibres treated by a special process developed at ITF can obtain mechanical properties comparable to aramid fibres. Typical fibres are 10–15 microns in diameter and 10–60 mm long. They are suitable for incorporation in plastic and concrete matrices and may also find applications as asbestos substitutes in, for example, brake linings (where they would be competing with aramids).

4.1.9 SILICON CARBIDE WHISKERS FROM RICE STRAW AND SUGAR CANE LEAF

It is well known that silicon carbide whiskers can be produced by carbothermal reduction of rice husks. This is possible because rice husks contain both silica and carbon. A paper from workers in India ('Silicon carbide from sugar cane leaf and rice straw', M. Patel and P. Kumari, *Journal of Materials Science Letters,* Vol. 9, April 1990) reports on work aimed at producing such whiskers from alternative organic sources. In particular, the authors attempted to obtain silicon carbide from rice straw and sugar cane leaf. The initial results presented are encouraging, with evidence for silicon carbide whisker formation in small quantities. The addition of an external catalyst was not required.

4.1.10 NEW FIBRE TECHNOLOGY — CARBON FIBRES

A number of papers were published in 1990 that looked at a few important aspects of carbon fibres. A critical property nowadays for carbon fibres and their composites is compression strength. M.G. Dobb, J.J. Johnson and C.R. Park from the Textile Physics Laboratory of the University of Leeds, UK, have studied the compressive failure of carbon fibres 'Compressional Behaviour of Carbon Fibres', *Journal of Materials Science,* Vol. 24, February 1990). In particular, the authors consider the difference in compressive fracture mode between polyacrylonitrile-based fibres and fibres produced from mesophase pitch. A recoil test method was used to induce a compressive failure in individual fibres and compressive strengths and strain to failures are presented for a range of fibres. The polyacrylonitride based fibres exhibited the highest compressive strengths and this is linked to the existence of a two phase structure within the polyacrylonitride fibres, i.e. a layer-plane structure and a disordered carbon structure that allows some localized deformation. The pitch-based fibres contain few disordered sites and shear is easier between layers. Schematic diagrams illustrating the failure mechanisms in both classes of fibres are included in the papers as are some electron micrographs of failed fibres.

A related paper, 'High resolution scanning electron microscopy of PAN-based and pitch-based carbon fibre', by D.Z. Vezie and W.W. Adams, was published in *Journal of Materials Science Letters.* This paper is particularly interesting due to the attempts to correlate the structure of carbon fibres with their mechanical properties — and in particular, compressive strength. The paper presents some excellent scanning electron micrographs (SEM) of polyacrylonitrile fibres where a rough grainy texture was observed on fracture surfaces of AS-4 and T-300 and T-40 fibres, and pitch fibres where sheet-like structures were observed on P-25, P-75 and P-100 fibres. The higher modulus polyacrylonitride fibres such as GY-70 were found to possess structures that had assumed similar layered structures to the pitch fibres. The value of this paper is primarily that a reference set of structures are now available for further correlations between structure and properties from the authors or other workers in the field.

A further paper on this subject, 'Correlation of structure and compressive properties in pitch-based carbon fibres' was presented at the 5th American Society for Composites meeting in East Lansing, Michigan, by A.S. Crasto and D.P. Anderson of the Dayton Research Institute. Their micrographs included a number from the range of high modulus fibres from Toray (e.g. M-40J) which feature improved compressive strengths. These fibres were found to retain the nodular, grainy fracture surface of the lower modulus polyacrylonitrile fibres rather than the layered or sheet-like structures characteristic of pitch fibres. This observation poses the question: is it possible to raise the compressive strengths of very high modulus pitch fibres by structural modification without losing tensile stiffness?

Another important property area of carbon fibres that needs improvement is the stability of the fibre in oxidative environments at elevated temperatures. High temperature use of carbon fibres in carbon/carbon composites is of course only practicable if the fibres do not change into carbon dioxide too readily. The problem of fibre degradation is also perceived as increasingly important now that high temperature thermoplastic matrix composites are being used where processing at 400°C for an extended time may introduce a degree of property loss. Y. Deslandes and F.N. Sabir, from the Division of Chemistry of the National Research Council of Canada, Ottawa, report encouraging progress with a sol-gel coating technique to inhibit oxidation of carbon fibres ('Inhibition of oxidation of carbon fibres by sol-gel coating', *Journal of Materials Science Letters,* February 1990, Vol. 9). The technique they report involves the use of a very thin silicon dioxide coating applied from a sol of tetraethylorthosilicate in methanol. The coatings are of the order of 0.2 micron thick (NB: a thin coating is required to ensure that the film does not crack and affect the flexibility of the carbon fibres). Oxidation studies showed comparable Hysol E/IMS carbon fibres oxidising at 500°C uncoated and 630°C coated with the silicon dioxide SiO_2 film. The authors state that their process is far from optimized but are encouraged by these initial improvements.

Polyacrylonitrile-based carbon fibres are expensive 'largely because of the cost of the precursor' and this has been a main reason behind the search for a low cost pitch-based fibre. However, currently available low cost pitch fibres do not have adequate properties while higher performance pitch fibres are expensive. There has been interest expressed in the past in producing carbon fibres from cheap polymer fibre precursors, such as polyethylene. This route has been re-examined by workers at the University of Groningen, in The Netherlands, namely A.R. Postema, H. de Groot and A.J. *'Pennings whose paper 'Amorphous carbon fibres from linear low density polyethylene'* appeared late in 1990 in the *Journal of Materials Science*. The objective of the work of Professor Pennings' team is not to produce carbon fibres with good absolute properties, but to provide good properties for an economic cost. Low cost carbon fibres with adequate properties could compete with, say, glass fibres if the balance is right. The fibres produced from the polyethylene had a strength of 1.15 GPa, a Young's modulus of 60 GPa and a very high strain to break of 3%. Details of the cost of the process are not reported in the paper, but a conversion temperature of 900°C was required which seems reasonably low and therefore inexpensive.

'Processing of carbon fibre reinforced aluminium composite using K_2ZrF_6 treated carbon fibres: a degradation study', is by S.N. Patankar, from New Mexico Tech in the USA, and V. Gopinathan and P. Ramakrishnan, from the Indian Institute of Technology in Bombay.

This work examines the effects of fibre surface treatment by K_2ZrF_6 on the fibre degradation caused by melt infiltration of aluminium. The results indicate excellent wetting of the treated fibres by aluminium, which is difficult to achieve without a fibre treatment (or alternative addition of wetting agent to the matrix), but with the likelihood of fibre degradation unless temperatures and wet-out times are not closely monitored during the processing.

4.1.11 SURFACE TREATMENT OF POLYETHYLENE FIBRES FOR INCREASED BOND STRENGTH

One of the potential drawbacks of organic fibres and in particular polyethylene fibres has been the lack of good bonding to potential matrix systems. Various treatments can be applied to the fibres in order to increase this adhesion. Reports in *Advanced Composites Bulletin* in 1988 highlighted work on plasma treatment of polyethylene where significant property improvements where recorded. A paper in *Composites Science and Technology,* 1990, Vol. 38 No. 1, pp.1–21, by S. Holmes and P. Schwartz of Cornell University in the USA continued this theme. Their paper, 'Amination of ultra-high strength polyethylene using ammonia plasma' reveals that the wettability of Spectra fibres (supplied by Allied Signal, Petersburg, USA) increases significantly after treatment with an ammonia plasma. This increase in wettability translates into improved bond strengths as measured indirectly using a peel strength test on a fabric-epoxy laminate. Peel strengths rose from 6.88 N/cm to 9.65 N/cm at the optimum treatment level of 50 W for 5 minutes. Some interesting observations from the authors include the fact that the fabric showed no sign of change in texture after plasma treatment. This is in contrast to other studies that have shown significant degradation of the fibre surface, and indeed in many cases improvements in bonding have been attributed to mechanical keying effects due to surface roughening of the fibres by the discharge. A reduction in fibre strength of up to 10% was nevertheless recorded for fibres extracted from the fabrics after treatment.

4.2 MATRICES

Developments in the field of composite matrices have been driven by two requirements: the need for improvements in composite properties and the need for better processing.

4.2.1 ALUMINIUM–NICKEL ALLOYS FOR METAL MATRIX COMPOSITES

In the metal matrix field most interest continues to be centred on alumnium alloy systems. An aluminium alloy specifically for use as a metal matrix in conjunction with ceramic fibres has been developed by Ube Industries, of Yamaguchi, Japan.

The alloy, which is described in European Patent 0 335 692, comprises aluminium with 6-11 wt% nickel. The company says that its aim was to provide an alloy which is as compatible with ceramic fibres (such as its Si-Ti-C-O fibre) as pure aluminium, but with improved strength.

Although pure aluminium does not react to form brittle intermetallic crystals at the fibre interface, it has a relatively low strength. This means that the transverse strength of such a metal matrix composite is low. Alloying elements such as bismuth, antimony, cadmium and barium increase strength but reduce corrosion resistance.

Ube says that the addition of 8 wt% nickel increases transverse strength of a typical unidirectional Si-Ti-C-O fibre/aluminium alloy composite by a maximum of 80%. The longitudinal strength initially drops with addition of nickel, but then rises to a peak at about 8-9% Ni to show a small improvement of about 10% Ni.

In the polymer matrix field composites a major driver in terms of downstream composite properties has been the need to improve high temperature performance.

4.2.2 HIGH TEMPERATURE THERMOSETS

A family of high temperature thermosetting resins that are easy to process and have mechanical properties claimed to be equal or superior to current bismaleimides and polyimides have been developed by Allied Signal. According to *Materials and Processing Report*, the resins are phenolic triazine polymers, called PT resins, and retain the flame retardance of phenolics whilst overcoming the processing problems of polyimides.

Prepregs may be readily produced using solvent impregnation techniques (either methyl ethyl ketone or acetone). Composites produced from these prepregs have low void contents and, according to preliminary figures from Allied, exhibit similar properties to PMR-15 polyimide composites at high temperatures (330°C).

The cost of PT resins are also expected to be lower than current polyimides when the polymers are produced in commercial quantities.

4.2.3 IMPROVED THERMID POLYIMIDES

A paper in *SAMPE Journal*, September/October 1990, reports on success in improving the mechanical properties of Thermid matrix composites by the use of matrix modification through semi-interpenetrating networks.

Thermids are acetylene end-capped polyimides which are in principal easy to process thermosetting polyimide systems, cured via an additional polymerisation process. In practice the materials are very difficult to convert into composite components due to a narrow processing window and, in addition, they are extremely brittle. The goal for the research team at NASA-Langley led by Dr Ruth Pater, has been to improve both the mechanical properties and the toughness of these materials by creating semi-interpenetrating networks with inherently tougher thermoplastic polyimides.

The Thermid materials studied were those supplied by National Starch and Chemical Corp (namely Thermid, LR-600, AL-600, MC-600 and FA-700) and the thermoplastic polyimide was a commerically available system (NR-150B2) supplied by Du Pont.

The results presented in the SAMPE paper reveal significant improvements in neat resin toughness but, more importantly, reduced microcracking in composite laminates and improved processability. Dr Pater states that these novel semi-IPNs represent a significant advance in Thermid-based material technology and increase their potential as composite matrices.

4.2.4 CYANATE ESTER RESINS

Rhone Poulenc Speciality Resins are developing a new range of cyanate ester resins which are claimed to offer the toughness of thermoplastics with the lower temperature and lower pressure processing of thermosets.

A low viscocity cyanate ester, Arocy L-10, is capable of dissolving more than 25% of various amorphous high temperature thermoplastics which on curing forms phase-separated alloys to provide a high toughness matrix which may be processed by low cost methods such as resin transfer moulding, pultrusion and filament winding.

Another new product, cyanate ester RTX 366, allows large parts to be produced at a maximum cure temperature of 150°C, combined with the lowest moisture absorption and dielectric loss of any cyanate ester, it is claimed.

A third system, Arocy M-20, is stated to combine epoxy-like processing with superior hot-wet performance. Low viscocity allows RTM injection at moderate pot temperatures. The resin is easily filmed and unidirectional prepregs produced from the resin exhibit excellent tack and drape. Hot-wet performance indicates that service temperatures of up to 177°C should be permissible.

The polymer composite processing routes attracting most attention these days seems to be resin transfer moulding (RTM) and pultrusion. Consequently, resin systems tailored for these processes are emerging, either as varients of existing resins or completely new products.

4.2.5 PULTRUSION AND RESIN TRANSFER MOULDING RESINS

A range of resin systems tailored for different processes has been announced by Reichold Chemicals of White Plains, New York, USA.

Two of the systems have been designed for use in pultrusion processes. Reichold says that its vinyl ester, Dion 31-034-01, permits higher line speeds than competitive vinyl esters, coupled with smooth surface finishes and the ability to produce profiles up to 9.5 mm thick. The resin also has a high heat distortion temperature, making it suitable for high temperature applications, the company claims.

The second pultrusion resin is a phenolic system developed by the company's Canadian subsidiary, Reichold Ltd. No details are available at present, but phenolic systems for pultrusion are rare and this will be an important product for Reichold, given the ability of phenolics to satisfy fire, smoke and toxicity regulations. Good surface finish and high processing speed are claimed for this system.

A vinyl ester optimized for resin transfer moulding has also been announced. Polylite 31-512 cuts cycle times to less than five minutes compared to 20-30 minutes for equivalent parts using conventional epoxies, vinyl esters and polyesters, Reichold claims.

Reichold has also produced a troubleshooting guide for moulders to assist in RTM processing. The six page document, 'Resin transfer moulding troubleshooting guide', contains a cross reference guide to help eliminate many common problems that occur and a resin selection guide to help select the best resin (from Reichold) for a given application type.

4.2.6 RESIN TRANSFER MOULDING EPOXY

An epoxy resin system designed to meet the requirements of resin transfer moulding (RTM) operations has been introduced by Shell Chemical Co of Houston, Texas, USA. The system is distinctive in that it features an undiluted, non-crystallizing resin with a non-methylene dianiline, non-corrosive aromatic amine curing agent.

The product, Epon DPL-862 resin with Epon W curing agent, is said by Shell to provide a low room temperature viscosity and a long working life, making it ideal for RTM.

A number of practical concerns expressed by moulders have been addressed by this system, such as the need to eliminate methylene dianiline curing systems and the problem of crystallization of many epoxies. This can occur in feed lines to the mould and jam the process.

The cure of the resin can also be tailored with adequate green strengths claimed. After curing at 93°C a subsequent post cure is required for final properties.

In the automotive field and chemical process industries well established products, such as sheet moulding compounds (SMC) and vinyl esters, continue to be improved.

4.2.7 RESINS AND FIBRES FOR SHEET MOULDING COMPOUNDS

A number of new constituent materials for sheet moulding compound formulations have been announced recently.

PPG Industries of Pittsburgh, Pennsylvania, USA, has launched two glass rovings tailored for sheet moulding compounds. The first, type 5510, is a medium-hard (sized) roving designed for sheet moulding compound applications where the part is to be painted. The obvious applications area is automotive exterior body panels.

A further roving, Type 5538, has been developed specifically for structural applications where mechanical properties need to be maximized. This product is classed as highly soluble and has the added feature of being visibly whiter than previous grades, according to PPG. This results in superior fibre hiding characteristics in pigmented applications, PPG states.

Meanwhile, the Freeman Chemical Corp of Port Washington, Wisconsin, USA, has introduced a new version of its XXCEL unsaturated polyester resin for sheet moulding compounds which, it claims, can provide 30 s cycle times with superior as-moulded surface quality. When this resin is combined with a low profile additive such as Union Carbide's Neulon T Plus, the sheet mouldng compound provides Loria numbers of 56-66 (obtained using a Loria surface analyser).

4.2.8 VINYL ESTERS HAVE INCREASED ELONGATION

Three grades of Derakane 411 vinyl ester resin were recently introduced by Dow Chemical of Midland, Michigan, USA. According to the company all three grades exhibit an increase in elongation of 30% over the previous products (now up to 8%!).

Derakane 411-350 is a corrosion resistant grade suitable for acids, alkalis, solvents and bleaches; Derakane 411-700T is a thixotropic grade for use in lamination of vertical surfaces; and Derakane 411-700PAT is a version containing pre-measured promoter and accelerator for consistency of gel time.

Full data on these new systems is available in an updated brochure produced for the Derakane 411 series. Dow has also updated its bulletin on fabrication with Derakane resins.

4.3 PREPREGS

4.3.1 LOW CROSSLINK DENSITY EPOXY PREPREG

A new '500' series of low crosslink density epoxy resins has been launched in a variety of product forms by the 3M Co, St Paul, Minnesota, USA. This includes a new prepreg formulation, SP 500-2 and the PR-500 range of resins for resin transfer moulding and filament winding applications.

In Scotchply SP 500-2, the use of the low crosslink density epoxy resin is claimed to minimise moisture absorption, thereby providing good hot/wet properties, without sacrificing toughness.

Prepreg produced with IM7/12K carbon fibres is quoted as having a compression strength of 1279 MPa at 82 °C wet and 1377 MPa at 121 °C dry. Interlaminar toughness is reported as $G_{Ic}= 600$ J/m^2 and $G_{IIc} = 871$ J/m^2, while post impact compression strength is given as 302 MPa.

The prepreg has an equilibrium moisture content of only 0.45%, according to 3M, which is matched to a Tg of 194 °C, allowing a hot/wet service temperature in the order of 150 °C.

The resin system is also available with other fibre types, and its low outgassing (0.186%) and low volatile content (0.003%) could make it a suitable material for space applications.

Storage life is one year at −18°C and three months at 24°C.

The PR-500 resins will also result in composites with 150°C operating temperatures and good impact/toughness properties, but in this case the low viscosity (at injection temperatures of 160 °C) is an added advantage for processing. The resin is a paste at room temperature but at 160°C the viscosity falls to less than 45 cps which should assist good wet-out in RTM.

Short beam shear strength tests for woven fabric laminates (Celion 3K 8-harness satin) give results of 83–90 MPa at 24 °C and 45 MPa at 160 °C.

4.3.2 TOUGHENED PREPREGS

American Cyanamid, Wayne, New Jersey, USA, has announced three new toughened prepregs. Cycom 3135 is a toughened bismaleimide, while Cycom 1840 and Cycom 1845 are toughened epoxy systems.

Cycom 3135 is offered as a material with improved toughness and hot/wet compression strength (to 177°C, with laminates free of the microcracking that affects bismaleimides. Compression after impact is quoted as 248 MPa.

Both tape and fabric forms of Cycom 3135 are available and the materials are said to possess good drape with light tack. Cyanamid reports a shelf life of six months at −18°C (0°C) and a shop life of four weeks at ambient temperature.

Cycom 1840 and 1845 are complementary epoxy prepregs developed from the company's earlier Cycom 1827 system, to provide higher impact resistance and better hot/wet properties. Cycom 1840 is a single phase resin in which the toughening results from improved resin chemistry, while Cycom 1845 exploits a dual phase structure to provide its toughness.

The best toughness is provided by Cycom 1845 which has a compression after impact strength of 324 MPa, while Cycom 1840 provides a CAI figure of 283 MPa. The mechanical properties of the Cycom 1840 laminates are, however, slightly better in other respects such as stiffness and compression strength, according to the Cyanamid data sheets. Wet compression strength of Cycom 1840 is 1105 MPa at 82°C while that of Cycom 1845 is 1070 MPa at the same temperature. Both systems have similar cure cycles (177°C, minimum out-times (14 days at 27°C), and are said to possess medium tack and good drape.

4.3.3 THERMOPLASTIC MATRIX COMPOSITES

A range of thermoplastic matrix composites based on a polyphenylene sulphide has been introduced by Phillips 66 Co of Bartlesville, Oklahoma, USA. The new material, trademark Avtel, supplants the earlier PPS Ryton unidirectional thermoplastic prepregs produced by the company.

Phillips says that users of the earlier product experienced poor interfacial bonding between fibre and matrix, leading to disappointing mechanical properties in areas such as damage tolerance. As a solution to this problem, the Avtel system uses a new polyphenylene sulphide resin together with a proprietary sizing agent while exploiting the prepreging process developed by the company.

Currently, prepreg is being produced by Phillips based on continuous E-glass fibre reinforcement at fibre weight fraction levels of 68% (product LG40-60) or on continuous E-glass fibre swirl mat at fibre weight fractions of 40% (product LG20-40).

The improved interfacial bonding of these materials is at least partially responsible for the unusually large 500% increase in fracture toughness claimed by the company over previous polyphenylene sulphide composites. The transverse tensile strength is also said to have increased, by 350%.

In addition, Phillips says that Avtel materials are easier to process. The new resin provides a wider processing window and allows a reduction in processing time due to a rapid rate of crystallization. Degrees of crystallinity of the order of 45% are said to be attainable, providing the material with excellent chemical resistance. Typical processing temperatures are in the range 315-345°C. Phillips supplies the unidirectional material as pre-consolidated laminates rather than prepreg tape. The company will produce laminates with a given ply stacking sequence according to customers' requirements. This laminate may then be thermoformed to shape (pressures required are 689–1378 kPa) at a later stage (e.g. by diaphragm forming, press forming). The swirl mat is similarly available in consolidated laminates where hot stamping is the preferred fabrication route (13.78–41.34 MPa).

The full range of Avtel materials will ultimately include both glass and carbon fibre reinforced versions.

4.3.4 CARBON FIBRE REINFORCED POLYETHERETHERKETONE

In a move that parallels that of Phillips 66 with their Avtel product, ICI Fiberite Composite Materials of Tempe, Arizona, USA, is offering its 'APC-2', carbon fibre reinforced poly- etheretherketone thermoplastic composites, in a consolidated sheet form.

To date, APC-2 has only been available as 12 inch wide unidirectional prepreg tape and related forms, such as prepreg tow and slit tape which are suitable for tape laying/autoclave processing, winding or pultrusion type operations. However, the company says that part of the attraction of thermoplastics lies in their potential for rapid processing, and the availability of preconsolidated sheet will mean that matched metal stamping or hydro-rubber forming will be more attractive.

Fiberite will supply the consolidated sheet in various thicknesses and sizes up to 1 x 3.65 m, with lay-up sequence produced to customer requirements.

4.4 PREFORMS AND PRECURSORS

4.4.1 PAPER TECHNOLOGY USED FOR GLASS MAT THERMOPLASTICS

A novel route for the production of random in-plane glass fibre reinforced thermoplastics is described in a patent application from Du Pont of Wilmington, Delaware, USA (European Patent EP 0 341 977 A2). The material is produced using paper making technology, and features glass and staple fibres, and a thermoplastic matrix polymer.

A number of different approaches to producing a stampable or thermoformable sheet of thermoplastic polymer reinforced with long discontinuous fibres are now available. The earliest products were based on the impregnation of a glass mat by a molten polymer under high pressures. More recently approaches based on paper making have emerged where the reinforcing fibres are mixed with powdered matrix in an aqueous slurry before being allowed to settle on mesh to form a precursor sheet. This is then pressed to form a dense well dispersed composite (a notable pioneer in this area was Wiggins Teape in the UK).

According to Du Pont, some of the problems inherent in the latter route stem from the need to finely disperse the polymer powder in the aqueous suspension of the glass fibres. Various solutions to the problem have been studied, including radical adjustment of the pH of the solution, foaming agents, the addition of flocculants, polymer latexes and pulps, but all produce a detrimental change in the properties of the finished product.

The use of polymer fibres instead of powder is claimed to remove these difficulties, allowing dispersion in near-neutral pHs. Application of a wetting agent to the thermoplastic fibre surfaces further improves dispersion. The addition of a small percentage of a relatively low melting polymer in fibre form to the main volume of fibres of a higher temperature polymer enables a mat to be formed which is lightly bound together to allow preform integrity during processing.

Standard paper making machinery has been used with polyester, polyamide, polypropylene and polyethylene to successfully produce void free composite sheets which, Du Pont says, have substantially improved properties over state-of-the-art materials.

4.4.2 HYBRID POLYMER YARNS

The Japanese supplier, Nitto Boseki Co Ltd of Tokyo, has joined the ranks of companies producing polymer matrices in fibre form for the production of hybrid yarns.

Nitto Boseki is producing fibres from a range of engineering polymers such as polycarbonate, nylon 6, 6.6 and 12, and polyetherimide which can be combined with carbon or glass fibres to produce hybrid yarn prepregs. This form of prepreg is inherently easier to mould into complicated shapes than alternative fully wet-out forms of thermoplastic prepreg. The company's choice of polymer systems is interesting, in that, with the exception of the polyetherimide fibres, the polymers are relatively low temperature systems and therefore not competitive with existing thermoplastic products such as APC-2 from ICI or Radel C from Amoco. ICI has, however, recently developed a glass fibre/nylon or polypropylene continuous fibre thermoplastic prepreg, Plytron. These materials are obviously aimed at the non-aerospace industry in the main and if successful could contribute to a significant expansion of the composites market.

4.4.3 INTERLACED THERMOSETS AND ECONOMIC THERMOPLASTICS

A new thermoset version of its interlaced biaxial prepreg materials has been introduced by Quadrax Corp of Portsmouth, Rhode Island, USA. To date, biaxial tapes manufactured by the company have been produced exclusively from thermoplastic unidirectional tapes such as ICI's APC-2 (carbon fibre—polyetheretherketone.

The thermoplastic biaxial tapes are made by interlacing thin unidirectional tapes of pre-impregnated material, around 0.5 cm wide, to produce a product roughly equivalent to a fabric based on woven yarns. Various analogues of satin and twill weaves may be produced. The flat nature of the tapes means that compared to a woven fabric the effective crimp in the plies is very small; the company claims that the in-plane properties are equivalent to those of 0/90 laminates produced from unidirectional prepregs. The principal advantages of biaxial tape for thermoplastics are drapeability and the ability to produce very wide broadgoods - up to 3 m.

These advantages are also desirable for thermosetting prepregs where wide conventional fabrics are expensive to produce and suffer from property reductions compared to unidirectional-based laminates. However, interlacing thermosets have a degree of inherent tack. It is the lack of tack in the thermoplastic, normally a disadvantage with such systems for manufacturing, that allows the interlacing process to work easily.

The manufacturing route adopted by Quadrax involves coating the prepreg with a thin scrim of a polyamide flock. This reduces the tack of the tapes sufficiently for the material to be processed. During the final cure of the thermoset, the random fibre scrim becomes incorporated into the matrix of the composite.

The system is at a relatively early stage in its development. Quadrax admits to some property reductions as a result of the polyamide scrim, but is confident that this is mainly a result of an increase in effective matrix volume fraction and says that as the levels of scrim are optimized the properties will improve.

Limited shipment quantities of the optimized materials were made available during the latter part of 1990.

Quadrax is currently offering biaxial tape based on well qualified materials such as ICI Fiberite's 1337 AU which is recognized as being equivalent to Hercules' AS4/3501-6. A demonstration Lockheed C-130 Transport engine nacelle fairing has already been produced from this material.

It is hoped that the use of thermoset biaxial tape will make composite parts less expensive to manufacture. As part of a novel package on offer to customers, the company is offering to supply thermoset biaxial tape cut to shape, stacked, preformed and enclosed in a vacuum bag ready for moulding. All the customer needs is the tool and the means of curing the part. Tooling itself is an application area that Quadrax has high hopes for with this material.

As a spin-off from the research performed during the development of the thermoset biaxial tape, Quadrax says that it has realised cost savings on the manufacture of its existing thermoplastic biaxial tape range. A 20% reduction in price has now been announced. Typical prices (depending on quantity) for a biaxial tape based on a high temperature engineering thermoplastic matrix polymer were previously US$46–79. The reductions bring this down to US$37-63.

4.4.4 THREE-DIMENSIONAL PREFORM BRAIDING SYSTEM

Albany International of Mansfield, Massachusetts, USA, has developed a new three-dimensional (3D) preform braiding system. According to the company, the system will allow interlocked flat and tubular braids to be produced for composite reinforcement. The novel feature of the system is that the braids are interconnected at adjacent layers only, rather than through the entire thickness of the preform.

It is expected that properties such as interlaminar shear strength will be enhanced by the interlocking, with minimal reductions in in-plane properties. However, the company does not have experimental verification at present. The system also allows for the introduction of axial yarns which Albany claims is not a feature of most 3D braids

4.4.5 UNIDIRECTIONAL SHORT FIBRE FABRICS

A process for the unidirectional alignment of discontinuous fibres for subsequent processing into high performance composites has been developed by Technical Fibre Products of Kendal, UK.

The advantages of using a discontinuous fibre form in a unidirectional composite is that elongations which exceed the elongation of the fibres themselves are possible (given an appropriate matrix).

Technical Fibre Products says that the product, called Disco, is particularly suitable for use in small and complex shaped components where local deformation may be required that does not disrupt the overall arrangement of the material. Trials are reported to have been successful with moulding of gear box shells, shaft couplings for helicopters, underfloor struts, stiffeners and skins for airframes, intake ducts for aero-engines and rocket motor nozzles.

Disco is available as carbon fibres in a lightly bonded or prepreg form but other fibres will be available including glass, silicon carbide and ceramics. Areal weights envisaged at first are 100 to 220 g/m^2 in sheet form of 1 m x 2 m.

4.4.6 FORMING PREFORMED SILICON CARBIDE ALUMINIUM WIRE

An improved technique for producing preformed silicon carbide fibre aluminium wire is claimed by the Agency of Industrial Science and Technology (AIST), Tokyo, Japan.

Preformed wire consisting of reinforcing fibres, such as silicon carbide or carbon, impregnated with a metal matrix, such as aluminium, is considered as a useful interim material for the production of aligned continuous fibre-reinforced metal matrix composites. However, various problems are associated with the production of such wire and these lead to a poor translation of properties in the final part. If the wire is formed by blowing fine particles of metal or a metal vapour against a fibre bundle by plasma jetting or other techniques, it is difficult for the metal to penetrate into the bundles. Alternatively, the fibre bundles can be passed through a molten bath of metal. Here, ultrasonic vibration may be used to assist in wet-out but this leads to disruption of the ordering of fibres within the bundles and consequent strength loss. Furthermore, any excessive dwell at molten metal temperatures will promote interfacial reactions that embrittle the composite. Further refinements to the process include spreading and arranging the fibres from a bundle in some way before entry into the bath of metal, coupled with the ultrasonic vibration, which results in a reduced time at temperature and increased strength.

The refinement devised by AIST involves replacing the pure aluminium with a eutectic alloy composition comprising aluminium with 5-7% by weight of nickel heated to no more than 50°C above the melt temperature (European Patent Application EP 0 337 034 A1). This lowers the melt temperature (thereby reducing interfacial reactions) and, by virtue of a narrow temperature range for solidification, reduces internal defects in the wire.

4.4.7 PROCESS FOR DISPERSING SILICON CARBIDE WHISKERS

A process for improving the dispersion of silicon carbide whiskers in ceramic matrix composites that uses an aqueous silane dispersant has been developed by Hoechst Celanese of Somerville, New Jersey, USA.

Typical fabrication routes for ceramic matrix composites involve mixing whiskers and ceramic powders in a liquid medium using high shear mixers or elaborate sedimentation techniques. This tends to produce agglomerations of whiskers which form clumps or nests. A homogeneous dispersion of the whiskers in the final composite is not achieved and properties suffer as a result.

The Hoechst process (European Patent Application EP 0 339 904 A1) involves deagglomerating the whiskers in an aqueous dispersant fluid comprising a silane, a cationic lubricant and an acid to adjust the pH to within 3.5–6. The silane has a formula of $R_n\text{-}SiX_{4-n}$ where the R is a non-functional hydrocarbon group, x is a hydrolysable group and n is an integer from 1–3. The role of the silane is to react with silicon or silicon oxide or silicon oxycarbide to form a surface film on the whisker which acts like a protective colloid, limiting adhesion and keeping the whiskers apart.

Hoechst claims that this process results in better dispersion in the composite and allows higher volume fractions of whiskers to be more readily achieved.

4.5 COMPOSITE SYSTEMS

4.5.1 HIGH TEMPERATURE MOULDING MATERIALS

A range of high temperature polyamide-based moulding compounds has been introduced to the UK market by ICI Fiberite. The materials, known as the PI 1000 series and exclusively available in the UK from Vigilant Plastics of Carshalton, are designed for use up to 325°C. The various grades available include: PI 1010, a chopped carbon fibre bulk moulding compound with fibre type and length specified by the user (lengths of 6–50 mm), and intended for compression moulding; PI 1020, a pelletized chopped carbon fibre compound for long flow transfer moulding operations; and PI 1030, a chopped glass fibre bulk moulding material, equivalent to PI 1010. An S-glass version of this grade can be produced if required.

4.5.2 REINFORCED CERAMICS AND METALS

Preliminary information is now available from Lanxide Corp of Newark, Delaware, USA, on its NX 5201 reinforced metal and NX 1010 reinforced ceramic systems.

Lanxide NX 5201 is a particulate silicon carbide reinforced aluminium with a particle volume fraction of approximately 60%. It is suitable for applications where high thermal conductivity, low thermal expansion and light weight are important.

At present, values for the coefficient of thermal expansion of 8.5×10^{-6}/K coupled with densities of 2.95 g/cm^3 are being achieved. The coefficient of thermal expansion of aluminium (which has a similar density) is of the order of 23×10^{-6}/K. The thermal conductivities of the two materials are 160 W/m.K for NX 5201 and 220 W/m.K for aluminium.

NX 5201 is a silicon carbide particulate reinfored aluminium alloy. It is produced using the Primex pressureless metal infiltration process which yields net or near net shape parts. Lanxide claims that tolerances and surface finish are similar to those of precision cast metals, eliminating the need for grinding or finishing operations and improving the cost effectiveness of the material.

Lanxide NX1010 is an alumina reinforced aluminium titanate ceramic matrix composite that is claimed to possess high strain-to-failure and excellent thermal shock resistance, together with a low thermal conductivity and a low Young's modulus.

The structure of the composite is characterized by micropores in the region of 50–200 microns, with general microcracking. According to Lanxide, the material is machinable using conventional tooling and can survive encapsulation via casting in aluminum and cast iron. Potential applications include use in engines and other high temperature devices.

At present the material is produced in thin walled geometries, although it is expected that a wide range of shapes will be possible in the future.

4.5.3 MOULDING COMPOUNDS COMBINE THERMOSET AND THERMOPLASTIC RESINS

ICI Fiberite, Wilmington, Delaware, USA, has introduced a new range of moulding compounds that offers high strength and high toughness by combining thermoset and thermoplastic resin technologies.

The new TEM 9000 range is supplied as a granular free-flowing compound which may be compression, injection or transfer moulded, and comprises glass or carbon fibre reinforcement and mineral fillers in a toughened epoxy matrix.

The novel aspect of the compound lies in the matrix itself. Here, thermoplastic polymer segments are chemically reacted onto the epoxy resin backbone. The thermal properties, dimensional stability and creep resistance of epoxies are retained but with the additional toughness of thermoplastics being added.

The basic concept behind this material is the same as that exploited successfully by ICI Fiberite in its 977 range of thermoplastic-toughened thermosetting prepregs. Like the 977 materials, the resin technology incorporated in the new moulding compound was the result of a joint development programme between Fiberite and ICI's Materials Research Centre at Wilton in the UK. The resin system differs from that in the 977 range due to the need to exploit different base epoxy systems where the process and property characteristics are tailored for the needs of the press moulding industry.

The TEM 9001 grade contains about 25 wt.% short (^{1}/8th inch) glass fibres and has been developed with military aircraft connectors in mind. This application requires a combination of good mechanical and electrical properties with good shock resistance. Notable properties for the material include a maximum force of 1.46 kN, energy of failure of 4.47 J, total energy of 6.5 J, and a crack initiation energy of 0.61 J.

The moulding temperatures specified are between 160 and 190°C.

The TEM 9015 grade differs in that the reinforcing fibres are carbon. The compound combines chopped 60% wt.% of 12K IM7 carbon fibre rovings, resulting in a high strength (tensile strength quoted as 280 MPa), high stiffness (tensile modulus, 42 GPa) material with high toughness (notched Izod impact strength, 1.06 kJ/m).

4.5.4 GLASS FIBRE-EPOXY/ALUMINATE LAMINATE

The continuing interest in the Arall concept (Aramid fibre-epoxy laminated with aluminium) has prompted further developments along this theme with the Dutch group Akzo announcing Glare (glass fibre-epoxy/aluminium laminate).

The original Arall materials are being actively promoted by Akzo in Europe and Alcoa in North America and have been shown to possess many attractive properties compared to conventional aerospace grade aluminium alloys. Tension dominated fatigue properties are particularly good in these multilayer materials which also exhibit a 15–20% reduction in density compared to aluminium and (in the fibre direction) a 60% increase in strength. An application already announced for Arall is the skin of the cargo door for the C-17 transporter in the USA. Other feasibility studies are underway, most notably involving Fokker in The Netherlands.

The Glare laminates from Akzo come in four grades. Glare-1 and Glare-2, like the equivalent Arall-1 and Arall-2 materials, possess unidirectional fibre reinforcement in the composite layers and differ in the specific aluminium alloy used (7075-T6 for Glare/Arall-1 and 2024-T3 for Glare/Arall-2). In contrast, Glare 3 and 4 possess biaxial fibre reinforcement in a cross ply lay-up, with a 50:50 and 70:30 ratio in the two directions, respectively. According to Akzo, the strength in the fibre direction of the 50/50 cross ply Glare-3 is 550 MPa, compared with 717 MPa for the unidirectional Arall-2 and 440 MPa for the 2024-T3 alloy. Glare 3 has a modulus of 57.7 GPa compared to 72.4 for the metal alloy and 64.1 GPa for Arall-2. The major benefit, for the new grades, however, is said to lie in their high strain-to-failure which is generally double that of the Arall equivalents which means that formability is increased and notch sensitivity is reduced. The material is damage-tolerant and even the fatigue properties are superior to Aramid Arall.

4.5.5 ALUMINIUM BORATE WHISKER COMPOSITES

The Agency of Industrial Science and Technology of Japan, together with the Shikoku Chemicals Corp, has developed a general range of metal matrix composites based on aluminium borate whisker reinforcement (EP 0 394 056).

The material is aimed at high volume applications of metal matrix composites based on aluminium matrices for use in aerospace, automotive construction and general engineering, where cost is a prime consideration. While silicon carbide and alumina whiskers are excellent in terms of wettability and mechanical properties, they are expensive. The companies state that at present the only economical whisker reinforcement is potassium hexatitanate. These whiskers are cheap but possess drawbacks due to interfacial reactions with aluminium which result in the brittle intermetallic Ti_3Al forming.

The aluminium borate whiskers overcome these problems. The whiskers are produced from the liquid phase and thereby are low cost materials compared to most whiskers produced from vapour phase deposition. The whiskers are typically 0.5 to 5 microns diameter and 2 to 500 microns in length. They are rendered more readily compatible with aluminium by modifying the surface by contact with lithium hydroxide (applied from solution). Processing of the fibres into a metal matrix composite involves mixing the whiskers with aluminium or aluminium alloy powder and subsequent hot pressing. Volume fractions within the range 5 to 40% are recommended. Particle size of the powders are preferably in the order of 20 microns. An alternative route would be melt infiltration of a whisker preform.

The mechanical properties of composites produced in this way compare favourably with alternative ceramic fibre reinforced melt systems. For a whisker fraction of 20%, quoted data shows moduli of 10.1 tons/mm^2 for the aluminium borate systems compared to 9.0 for alumina short fibres, 9.3 for silicon carbide, 9.3 for potassium hexatitanate, 10.0 for silicon carbide and 9.8 for silicon nitride whiskers. The corresponding strengths are 45, 31, 31, 44, and 42 kg/mm^2 respectively.

The first published details of this composite appeared in the June issue of *Journal of Materials Science Letters* (Vol. 9, 1990) 'Aluminium composite reinforced with a new aluminium borate whisker' by K. Suganuma, T. Fujita, N. Suzuki and K. Niihara. The fibre was revealed to have a composition of $(Al_2O_3)_9(B_2O_3)_2$ with a rhombohedral crystal structure.

Composites prepared by squeeze casting a random fibre preform of the whiskers with aluminium 6061 alloy gave modulus values of about 100 GPa and strength values of about 430 MPa, which are similar to those obtained from silicon carbide whisker composites with similar volume fractions (20%). The properties of potassium titanate whisker reinforced metal matrix composites are not quite so good, with strength data being only in the region of 300 MPa.

The sliding abrasion properties of the aluminium borate composites also looked more attractive than the other materials tested, with both low abrasion of the metal matrix composite and low wear of the mating steel surface being measured.

More details will need to emerge regarding the processability, fracture toughness and durability of aluminium borate whisker composites before their ultimate potential can be assessed, but at this time, given that only potassium titanate whisker metal matrix composites are competitive on price, they would appear to have a very promising future.

4.5.6 NOVEL CERAMIC COMPOSITE PRODUCTION ROUTE

A group working at ICI Advanced Materials have announced in *Nature* (Vol. 347, pp. 455–457) a method of manufacturing a novel tough ceramic composite. The most successful ceramic composite production method so far has been to infiltrate a ceramic fibre preform with a gaseous precursor. However, this is a time consuming process and requires several machining steps, making the process very expensive. The new technique adapts existing technology used for making multilayer capacitors, in which ceramic powders are formed into a sheet, coated to provide suitable interfacial properties, pressed into shape and sintered without pressure.

Clegg and co-workers took silicon carbide powder doped with boron in a solution of polyvinylalcohol acetate, pressed this into a 2 mm thick sheet and further rolled this into flat sheets 200 micrometers thick, later cut into squares 50 x 50 mm. The cut sheets were then coated with graphite, stacked and pressed to form a plaque 2 mm thick, heated at 1°C/min to 450°C under argon, and then sintered at 2040°C under argon for a further 30 mins. A monolithic silicon carbide slab was also made at the same time for comparison purposes.

The results of tests showed that the two materials had an identical flexural modulus. Bend strength in the composite was 633 MPa, compared to 500 MPa in the monolithic silicon carbide. Fracture toughness measured on samples notched perpendicular to the surface yielded values of 3.6 MPa/m for the monolithic material and 15 MPa/m for the composite.

4.5.7 HEAVY THERMOPLASTIC COMPOSITES

It is unusual for a composite to be designed to provide increased weight but a new grade of injection mouldable polybutylene terephthalate (PBT) from GE has increased weight as one of its features.

Application targets are consumer applications where weight is often important for customer acceptance and is synonymous with quality.

The PBT resin is filled with 65% of a particulate mineral filler (not identified as yet by GE) which results in a specific gravity of 2.4, compared to 1.3 for the unfilled PBT and 2.5 for glass.

The resin allows heavy parts to be produced with the processing advantages of plastics but without the need to introduce inserts, incorporate thicker walls or introduce some secondary process, all of which increase cost.

The negative side of the very heavy filler loading is the increased brittleness of the system. However, in most cases the materials that the filled PBT is aimed at supplanting (e.g. glass) are brittle in their own right and this will not be a significant problem.

Some companies are already evaluating the material for applications ranging from plumbing fixtures to cosmetics closures and GE is working on applications such as ceramic-like bowls and telephone housings.

4.5.8 STRUCTURAL FOAMS FOR BUOYANCY

The Insitut Francais du Petrole (IFP) has developed micro-cellular composite structural foams to provide buoyancy in underwater vehicles.

The need for underwater exploration at depths of 5000 m has produced a need for buoyant materials that do not collapse under the pressures involved. The materials developed by IFP consist of a syntactic foam, where the foam is created by the introduction of hollow spheres into a matrix. In this case, the spheres are glass and of reinforced resin. A new resin matrix formulation — Butalip — was developed to provide reduced water absorption compared to conventional epoxy resins. The resin system also features low viscocity and low density.

Current research at IFP is now centred on modelling the kinetics of thermal reticulation for better control of large volume production of the foams

4.5.9 MELT SPUN METAL MATRIX COMPOSITES

Sulzer Innotec, Winterhur, Switzerland, have developed a process for producing a composite based on amorphous metal matrices with hard ceramic particle fillers.

The process involves melt spinning metals into a foil form at very high rates in order to develop an amorphous structure in the metals. Hard ceramic particles are added to the melt and this results in a foil with either embedded or protruding particles. Where the particles are embedded, the foils are exceptionally wear resistant and as a result are suitable for application in highly stressed components such as yarn guides in weaving machines.

Where the particles protrude from the glassy metal, then the foil is extremely abrasive and is suitable for friction applications as well as cutting, grinding and polishing tools.

4.6 ADHESIVES, COATINGS AND FILMS

4.6.1 FIRE BARRIER ADDITIVE

An additive material which acts as a fire barrier for polymer-based materials has been launched by ICI's Chemical and Polymers Group. The system, called Ceepree, is a ceramic powder which is applied to the surface of a plastic-based article. It apparently melts progressively as temperatures rise from 350–900°C, creating a strong integral crystalline barrier which will protect the host material at temperatures up to 1100°C.

The white glassy powder can be formulated with a wide variety of plastics, including thermosets and thermoplastics, and is potentially of great interest for composite applications. ICI's corporate fire laboratory at Manchester in the UK has been made available to undertake trial work with potential customers' materials to all major existing fire standards.

4.6.2 HIGH TEMPERATURE ADHESIVE FOR USE AT 370°C

American Cyanamid, Wayne, New Jersey, USA, has introduced a new structural film adhesive which the company claims is capable of operating at over 370°C.

The new adhesive, FM 680, is an aluminium-filled condensation polyimide supported on a fibre glass cloth. The company recommend the adhesive for continuous operating temperatures of 250°C, intermediate operation (1000 hours plus) at 316°C and for up to 100 hour exposures at 370°C. Good bonding to a variety of metallic substrates, such as stainless steel and titanium, as well as to composite laminates, is reported. A primer, BR 680, is available for use with metallic substrates. The adhesive can also be obtained in foam and paste forms, and a grade without aluminium filler is available for radome-type applications where radar transparency is required.

The high temperature capabilities of the adhesive enable adhesive bonding to be exploited in high temperature operating areas, such as aircraft engines, and also allow co-curing and consolidation with composites requiring high temperatures cures, such as Du Pont's Avimid N and other polyimides. The film has a shop life of 15 days at 24°C at 50% RH. Processing can be achieved at temperatures as low as 316°C and free standing post cures at 391°C, are possible.

Much of the chemistry behind the FM 680 adhesive was developed by Du Pont (NR-150 B2) and acquired by Cyanamid through a joint development agreement.

4.6.3 SURFACING FILMS

Increasing interest in providing composite structures with surfacing films has prompted the release of new products by both 3M, St Paul, Minnesota, USA (Scotchply AF films), and Hysol Aerospace Products, Pittsburg, California, USA (SynSkin).

The surface skins are designed to improve surface finish and increase the environmental resistance of the composite. Surface imperfections such as honeycomb core imprints and porosity are eliminated, it is stated. Hysol claims that its SynSkin film also reduces crushing of any core structure by evening pressure distribution across a surface during cure.

Additional benefits include providing a barrier to chemical reagents such as paint strippers and hydraulic fuel, while there is evidence that impact resistance of laminates is also improved by the presence of surface films. This latter feature has been cited by 3M as a further advantage in situations where bead blasting techniques are used as an alternative to paint strippers; a specific product, AF 32, which is a nitrile rubber toughened film, is recommended in such cases. Another of 3M's products, AF 3113-5, is a structural adhesive film for surfacing carbon fibre laminates where operating temperatures up to 177°C are encountered. Hysol's SynSkin films have also been tested under extreme hot/wet operating conditions where, according to the company, they perform well.

4.6.4 FIRE-SMOKE-TOXICITY RESISTANT COMPOSITE

Ciba-Geigy Composite Materials, Anaheim, California, USA, has developed a low fire-smoke-toxicity composite system for aerospace and commercial applications.

The product, called Cibabarrier, is a sandwich consisting of honeycomb core faced with composite skins. The critical features of the system are a measured Ohio State University heat release value of less than 10/10 (compared with the new Federal Aviation Authority regulations stating a maximum of 60/60), a Ds value of under 50, and toxicity within the ATS1000 and domestic specifications. The company claims that variants on the product have withstood burn-through after one hour exposure to a 1093°C direct flame.

Ciba has applied for patent protection for the system, which is now ready for customer applications.

4.6.5 NOVEL HONEYCOMB STRUCTURES

Supracor of Sunnyvale, California, USA, has adapted its process for producing flexible honeycomb systems to the production of aerospace honeycombs reinforced with fibres.

The Supracor honeycombs have been under development since 1982. They are flexible systems usually based on thermoplastic elastomers that use a modified hexagonal structure which allows significant resistance to shearing, recovery after impact and little fatigue degradation. Antisotropy in the honeycomb can be introduced and exploited to provide specific shock absorption characteristics in applications such as sports shoes. The recent innovation concerns the adaption of the process to allow the honeycomb itself to be produced from fibre (and fabric) reinforced materials. The fibres (carbon, Kevlar, glass) are impregnated with thermoplastic elastomers prior to fabrication into the honeycomb using a proprietary fusion bonding process.

Applications envisaged include anti-ballistic apparel and boat hulls where high compression resistance coupled with excellent recovery properties is required.

4.6.6 POLYCARBONATE HONEYCOMB

In a further development from Supracor, a glass filled polycarbonate has been developed for use in honeycomb panels which need high compression strength and dimensional stability in combination with high shear and tensile strengths.

The company makes the material using a polycarbonate glass filled film called 'Makrofol', produced by A.G. Bayer in Germany.

Developments in which the polycarbonate is enhanced by fibre reinforcement (ceramic, carbon or glass), and where the honeycomb is fusion bonded to composite facings, are also contemplated.

4.6.7 POLYPHENYLSULPHONE FOR AIRCRAFT INTERIOR

Injection moulding grades of Amoco's Radel amorphous polyphenylsulphone alloy have just been certified by Airbus for use in aircraft interior applications.

Two grades of the materials, Radel R7000, an unreinforced polymer, and R7110, which is reinforced by glass fibre, have been approved and satisfy the ATS 1000.001 and the FAA's 65/65 fire regulations on heat release. The relevant German Aerospace standards DAN 1291 and 1292 have also been met.

Parts are now in production for various Airbus models by a range of component manufacturers, including Montaplast and Weber in Germany.

This is an important development for Amoco who can now compete in a market previously dominated by GE. The market for plastics based aircraft interior components is currently booming. A recent survey by the Business Communications Co, based in Norwalk, Connecticut, USA, states that the total plastics and plastics composites usage for civil aircraft interiors will reach 15.7 million pounds/annum (valued at US$427 million) by the year 2000. BCC cites the need to pass heat and smoke emission standards as the driver for increased use of higher specification theromplastic polymer systems like Amoco's Radel, and this will in turn lead to an increase in the cost of aircraft cabins (from US$16.5/ft^2 in 1989 to US$35.00/ft^2 by the mid- nineties.

Radel R 7000/7110 is not as yet qualified by Boeing, but it is understood that this may not be long in coming as the materials meet the stringent solvent resistance requirements specified by Boeing (but not by Airbus).

5. APPLICATIONS

5.1 AEROSPACE

5.1.1 COMPOSITE FIGHTER

A cheap combat aircraft for close support/anti-helicopter duties has been unveiled by Scaled Composites Inc of California, USA. The model 151 Agile Response Effective Support (ARES), which makes extensive use of composites, was designed by Burt Rutan of Scaled Composites Inc. Rutan achieved notable success by exploiting composites in the Voyager craft that circumnavigated the world non-stop on one tank of fuel. The new combat aircraft is comparatively slow, maximum speed around 350 knots, but the company claims it is very agile and is a stable weapons platform.

Each plane will cost in the region of US$4 million, according to Flight International. However, despite the attractive price and the potential of the plane for close support activities, there is some scepticism as to the likelihood of US orders for the plane. Business Week says that too many vested interests are busy protecting their own turf at present, in the climate of inevitable defence budget cuts, to support such a novel system. International sales prospects could, however, be far more promising.

5.1.2 COMPOSITE CARGO LININGS FOR CIVIL AIRCRAFT

Permaglass WFT/4, a woven glass fabric-phenolic laminate cargo-liner which its manufacturer, Permali Gloucester Ltd, UK, claims offers improved fire resistance, low smoke and toxic fume emissions, together with excellent impact resistance, has been approved by Boeing to its materials specification BMS-8-223 and subsequently installed in British Airways' fleet of Boeing 757 aircraft.

Shortly afterwards, the liners were also approved to the DMS 2226 specification of McDonnell Douglas of Long Beach, California, USA.

Existing Permali products are already widely specified by aerospace companies such as BAe, Fokker, and Airbus, and met the standards set by the FAA and CAA regulatory bodies.

5.1.3 BOEING FLOORING CONTRACT

Boeing Commercial Airplanes has awarded Ciba Geigy's Composite Materials unit based in Anaheim, California, a long term sole source contract for aircraft flooring material.

The award of this contract, worth over US$50 million over the next four years, signals a change of policy for Boeing who had previously insisted on dual sourcing. A key factor in its decision was apparently the statistical process control procedures now adopted by Ciba Geigy which cover the critical operating parameters in the manufacture of prepregs, honeycomb core and adhesives. These procedures effectively allow Boeing to reduce their own costs by performing little or no inspection on incoming materials. A new state of the art finished floor panel facility is soon to be completed by Boeing in Spokane, Washington State, USA.

5.1.4 THERMOPLASTIC COMPOSITE FLOOR PANELS FOR HELICOPTER

Floor panels made from carbon fibre reinforced APC-2 thermoplastic are to be used on the EH101 helicopter. Cost and weight reduction are said to be the reasons for the adoption of carbon fibre thermoplastic floor panels, which replace carbon fibre epoxy floor panels used in the original design.

The EH101 helicopter is being developed jointly by Westland in the UK and Augusta in Italy. Westland, through its Helicopter Division at Yeovil, is responsible for a number of composite items on the aircraft, including the blades, cockpit structure fuselage cowlings and the floor panels.

The floor panels are to be made from an ICI's APC-2, with AS4 carbon fibre reinforcement, as skins over an aluminium honeycomb. The aircraft is designed such that these panels add to the structural integrity as well as supplying payload support.

Westland Helicopters says that much of its composites expertise has been devoted to developing thermoplastic composites technology (including a proprietary hot-head filament winding system). This complements the activities of Westland Aerospace which specializes in thermosetting composites.

1990 was a good year for the Westland Group as a whole. Westland Aerospace won contracts worth US$120 million from McDonnell Douglas for the No 2 engine inlet duct for the new MD-11 airliner which is one of the largest composite structures designed to date for a civil aircraft (diameter 3.17 m, length 3.86 m). Westland now claims to be amongst the largest suppliers of composite engine nacelles.

5.1.5 APC-2 PART ON BOEING V-22

The Boeing V22 'Osprey' tilt rotor aircraft is to feature a front landing gear door produced from ICI's APC-2, a carbon fibre reinforced polyetheretherketone thermoplastic prepreg.

Originally the door of the V-22, which features a considerable amount of composite in its structure, was to be produced from an epoxy prepreg. The thermoplastic was substituted due to its improved impact performance.

The landing gear door is quite large, 4.5 x 1.5 ft, and is produced by Boeing using a diaphragm forming process from unidirectional prepreg tape. The curved outer skin and corrugated inner skins are produced separately and joined by a fusion bonding process. A 35% reduction in weight is claimed for the APC-2 part compared to the earlier designs.

5.1.6 BRAIDED HYBRID COMPOSITE PROPELLER BLADES

Dowty Rotol, Gloucester, UK, a manufacturer of aerospace propellers, has improved manufacturing output rates and propeller quality by exploiting braiding technology in combination with resin transfer moulding processing.

Dowty Rotol produced some 10 000 composite blades in the 1980s for aircraft such as the Fokker 50 and Saab SF340. The blades typically comprise carbon fibre blade spars supported by a foamed polyurethane core and enclosed by a glass fibre aerofoil envelope. The manufacturing process involves a complex series of operations performed manually, culminating in a final resin injection stage to impregnate fibres and fabrics preformed into the skin and spar assemblies. The most time consuming operations were identified as those involved with the production of the skins. Alternative mechanized processes for skin production were considered by Dowty (with the assistance of the National Engineering Laboratory and the Shirley Institute, both in the UK,) and a braiding process adopted as an alternative route. A preformed core/spar assembly is passed through the braiding machine to produce the skins in place. The whole assembly is then resin injected as before. The blade design has, according to Dowty, shown superior structural integrity, together with improved quality and reproducibility, by eliminating various stages of manual operation. The blades are to be used on the Allison T406 engine which has been selected for the Saab 2000 feeder airliner.

5.2 AUTOMOTIVE

5.2.1 ENVIRONMENTAL ISSUES — CALIFORNIAN LEGISLATION ON EXHAUST EMISSIONS

New and far reaching legislation has been passed in California which may have a significant impact on future automotive design and the use of new materials in future vehicles.

In an attempt to cut smog-producing emissions and also CO_2, there will be a requirement that a minimum of 2% of new cars by the year 1998 will have no harmful emissions whatsoever.

This implicitly means that the power source must be electric. The restrictions on the motor industry will become even tougher by the year 2003 when a total of 10% of vehicles will have to satisfy the zero emission rule. The desired result will be a 70% reduction in emission from burning hydrocarbons over this 13 year period.

Reports in the auto press suggest that the US Industry is sceptical of meeting the timescales involved but the Californian market is one of the largest in the world and cannot be ignored. A total of 2 million cars are sold in the State each year, suggesting a minimum market for electric vehicles of 200 000/year in 2003. General Motors in the USA had already taken the decision to proceed to manufacture the Impact electric cars, previewed to the motor industry in early 1990, and other manufacturers are bound to follow suit shortly.

The direct impact on the use of composites in vehicles is difficult to assess, but the use of lightweight materials in an electric vehicle is likely to be more critical to the performance of the vehicle than is the case with conventionally powered cars, and this therefore could provide a welcome stimulus to the industry.

5.2.2 FORD TESTS COMPOSITE CAR

Ford Motor Co of Dearborn, Michigan, USA, has produced a glass fibre composite bodied version of its US production car, the Taurus. The objectives of the project, which is not part of the Automotive Consortium project with Chrysler and General Motors, are to test the reinforced plastic's ability to handle torsional and suspension loads. There are no plans to produce such a vehicle at present and indeed the manufacturing techniques used to produce the test vehicle were labour intensive and not indicative of viable production routes.

Some of the cost savings that might result if high speed manufacturing problems can be overcome are illustrated by the fact that some 400 steel stampings were replaced by five components in the composite car.

5.2.3 US SPECIALIST CAR MAKER ADOPTS ADVANCED COMPOSITES

Specialist US car maker, Avanti, of Youngstown, Ohio, is using both carbon and Kevlar reinforced plastic body panels for its 1990 four-door sedan.

The carbon composite materials, which are supplied by Hexcel, are used in the roof, roof reinforcements, roof pillars and door beams. Kevlar uses include floor pan and bumpers. According to *Automotive News*, Avanti plans to produce its entire range of cars, including coupes and convertibles, with all-composite body structures.

5.2.4 COMPOSITE CHASSIS

The new Bugatti supercar is reported to feature an advanced composite chassis. The company enlisted the assistance of Aerospatiale on the project; Aerospatiale's expertise with composite, and in particular hybrid technology, enabled the company to develop a carbon-Kevlar hybrid chassis for the vehicle, said to be equivalent in size to the Porsche 911.

5.2.5 FROM SPRAY TO RESIN TRANSFER MOULDING

The Canadian manufacturer René Ltd of St Éphrem has switched from producing truck front ends by a spray-up process to one of resin transfer moulding.

The hood shell is moulded in one piece and uses the new ICI Modar modified acrylic resins and a thermoformable glass fibre preform mat from Vetrotex in France. The 35 kg shell requires stiffening which is provided by structural reaction injection moulded parts, produced from Dow urethane resins and glass fibre mats, which are then adhesively bonded to the shell. The 70 kg assembly is produced for the Mack Truck DM 600 model.

5.2.6 GKN BREAKS INTO JAPANESE MARKET

The UK-based engineering group GKN has announced a contract to supply its lightweight composite leaf springs to the Japanese truck market. The leaf springs will be manufactured at GKN Composites' Telford, UK, plant and then assembled into steel leaves in Japan, forming hybrid springs for the Mitsubishi Motors Corp.

The contract for the springs, which are to be used in the front and rear suspension units of Mitsubishi's eight tonne 'Fighter' truck, was won by Translite KK, a joint venture company set up by GKN and Mitsubishi Steel Manufacturing Co in 1985, specifically to develop the Japanese market. Mitsubishi Steel will assemble the spring units in Japan.

A weight saving of some 40 kg per truck is claimed by GKN through the use of the hybrid springs which are interchangeable with existing multi-leaf steel springs. Additional benefits are reported, including improved ride and handling, together with reduced interior noise.

GKN Composites operates within the GKN Automotive Group, and has already supplied composite leaf springs to Leyland Daf, Iveco, Mercedes Benz and London Taxi International. The springs are also being tested by a number of other vehicle manufacturers

5.2.7 CONTINUOUS CARBON FIBRE POLYETHERETHERKETONE FOR FORMULA 1 GEAR SELECTOR

A gear selector fork for use in Formula 1 racing cars has been produced from continuous carbon fibre polyetheretherketone (PEEK) (ICI's APC-2) as a result of collaboration between Williams Grand Prix Engineering of Didcot, and ICI's Wilton Materials Research Centre both in the UK.

ICI says that APC-2's ability to withstand the hot oil environment at 180°C, good wear resistance and a 40% weight saving over metal forks made it ideal for the application. Williams first used the new forks during the 1990 Grand Prix season.

5.2.8 LEYLAND-DAF CONTRACT FOR BTR PERMALI REINFORCED PLASTIC

BTR Permali RP, Gloucester, UK has won an initial £600 000 contract for hot press moulded glass reinforced plastic (GRP) components for a Leyland DAF four tonne military truck.

An order of 5350 of these trucks has been placed by the UK's Ministry of Defence and the truck goes into production in 1991. BTR Permali RP will be supplying a variety of interior components for the truck cab.

5.2.9 COMPOSITE FOR AUTOMOTIVE ENGINES

Composites are continuing to gain footholds in the automotive engine market.

In the USA, the Detroit Diesel Corp announced that it is to use a sheet moulding compound (SMC) based on Dow Chemical's Derekane vinyl ester resin for the rocker cover and oil pan for its new Series 60 11.1 and 12.7 litre diesel engines. In Europe, a consortium of companies funded under the European Community BRITE initiative has produced a prototype petrol engine where all parts are plastic and plastic composites, apart from the combustion chamber, cylinder walls and moving mechanical parts.

The European consortium, which consists of Ford Motor Co, DSM Resins, Vetrotex St Gobain, Galvanoform, GKN, Nottingham University and NEL, found that weight savings in its 1.0 litre single overhead camshaft four cylinder engine were disappointing but production costs were lower than expected. By passing a standard 200 hour durability cycle specified by Ford, the project has illustrated the potential for composites, in this case predominantly glass reinforced epoxy resins, in many new automotive engine and transmission applications.

5.2.10 COMPOSITE INLET MANIFOLD

A composite air intake has been adopted for BMW's 520i and 525i cars, according to the German manufacturer of the thermoplastic which is used. The manifolds are produced by Mann and Hummel by injection moulding with BASF's polyamide, 'Ultramid', with 35% glass fibre reinforcement.

BASF says the material can easily withstand the maximum operating temperatures experienced in the manifold, about 130°C, and is unaffected by exposure to fluids such as oil, fuel and cleaning solvents. The reinforced plastic part is about 50% lighter than the conventional aluminium intake which it replaces and is cheaper to produce.

The manufacturing process involves injection moulding into a cavity containing a low melting temperature metal core. The core, which is a precision casting from a tin-bismuth alloy, is subsequently melted out in a 150°C bath to leave a hollow component.

The low thermal conductivity of the polymer, coupled with the high conductivity of the metal alloy, ensures that the core does not fuse on contact with the injected hot polymer. Instead, the initial polymer in contact with the metal freezes, with the heat transported into the bulk of the casting and the frozen polymer acting as a thermal barrier, insulating the core from the remainder of the molten polymer in the cavity.

In the past, this particular moulding process has been successfully exploited for mass production items by Dunlop-Slazenger (for tennis and squash rackets) and by Ford (who pioneered such a process for the production of inlet manifolds for their small diesel engines.)

The BMW manifold differs from the Ford equivalent in that engineering thermoplastics are used instead of a thermosetting, glass-reinforced, polyester moulding compound.

5.2.11 PHENOLIC THROTTLE BODY FUEL INJECTOR

Detroit Advanced Systems and Technology Corp, USA, the research and development arm of Siemens/Bendix, has developed a glass reinforced phenolic automotive throttle body fuel injector.

The composite part, which is produced from Fiberite FM4056J, replaces an aluminium component and, the company claims, has reduced production and material costs.

Prototype throttle body fuel injectors have been moulded by Kurz Kasch Co of Newcomerstown, Ohio, and subjected to intensive testing including over 5000 miles of road tests.

The indications are that the production of the throttle body fuel injector from the glass phenolic moulding compound is viable, with some of the improved economics provided by a significant reduction in tooling costs: the number of tools was reduced from two to one compared with the aluminium unit. A further two years of testing is anticipated before the composite component goes into mass production as part of an electronic control unit for automotive engines.

5.3 RAIL

5.3.1 LOW SMOKE COMPOSITE FOR ROLLING STOCK

A high impact resistant, low smoke grade of glass fibre composite for use in railway rolling stock has been introduced by the UK company Permali Gloucester Ltd. The material, which is available in a number of grades, is a self-supporting phenolic resin matrix with a decorative melamine finish. Permali says that the Permaglass 22MFM/1 materials meet the relevant safety standards such as BS 6853:1987 Category 1, which relates to the design and construction of railway stock in the UK, and the UK Ministry of Defence's Naval Engineering Standard 713 for smoke and toxicity.

5.3.2 KEVLAR TABLES IN TRAINS

A French company is now manufacturing table tops based on Kevlar for the TGV Atlantique High Speed Train.

Trioplast of Bethune is manufacturing the tough lightweight table tops from Kevlar/glass hybrid fabric resin impregnated skins (Aramet-75K from Chomarat SA), supported by polypropylene foamed cores.

The table tops, which can withstand loads of up to 350 kg, need to be very strong and frequently suffer abuse not just from luggage loading but also from passengers standing on top of them to retrieve belongings from overhead racks. They replace aluminium tables which weigh twice as much.

5.3.3 MOBILE MAINTENANCE PLATFORM FOR LONDON UNDERGROUND

A mobile maintenance platform for London Underground is to be manufactured using fire resistant composites from Insulation Equipments Limited, Oswestry, UK.

Fire resistance is an important consideration in underground railway applications and London Underground severely restrict the materials that may be used to minimise the potential risks.

The mobile maintenance platform consists of a lightweight trolley which acts as both a work bench and tool store, and is moved along the track manually. The deck of the trolley is produced in a one piece moulding from Melaform, a phenolic glass reinforced composite which conforms to British

Standard BS 6853 governing the use of fire safe materials on rolling stock, and satisfies the impact resistance and durability required for this application.

5.4 CHEMICAL AND PROCESS PLANT

5.4.1 MARKET FOR FLUE GAS DESULPHURIZATION PLANT

A large market for composites in the construction of flue gas desulphurization plants for power stations in the next decade is predicted in a report by McIlvaine Co of Northbrook, Illinois, USA.

Glass fibre reinforced thermosets are proving the ideal materials for construction of flue gas desulphurization plants associated with coal fired power stations. Corrosion resistant metals are not readily suitable (except at great cost) for such applications where sulphuric acid at its dew point is encountered. Vinyl ester resins with acid resistant glass fibres have been successfully used in many installations to date (e.g. at BASF's plant at Ludwigshaven, West Germany).

McIlvaine has studied the potential market for flue gas desulphurization plants worldwide up to 1996. It concludes that some 50% of world orders by 1996 will come from US utility companies, driven by strong acid rain legislation and accelerated building of power plants. The need for new power stations has arisen from a recent underestimate of US power requirements by the utility companies. McIlvaine predicts that by 1996 the requirement for new flue gas desulphurization systems will exceed those for retrofit applications in existing power stations.

Substantial markets are also foreseen in the UK, Canada and Italy where retrofit flue gas desulphurization systems are being installed; while significant contributions from Asian and other European countries may also be expected by 1996.

The total value of the flue gas desulphurization market in 1989 was around US$2 billion which, according to the report, will rise to US$7 billion by 1996. The exact share of this market that will be secured by composites cannot be predicted at present, but the potential is reportedly considerable.

5.4.2 GLASS FIBRE COMPOSITES FOR FLUE GAS SYSTEMS

Dow Europe has disclosed two major applications for glass fibre composites based on its Derakane epoxy-novolac, vinyl ester resins for use in flue gas cleaning systems.

Snamprogetti, Italy's largest state engineering company is using Derakane 470 vinyl ester to construct flue gas scrubbers serving two waste incineration plants at Vercelli and Schio in Italy. The choice of Derakane follows experience with an earlier scrubber situated in Bolzano where toxic gases from a 200 ton/day throughput of waste had not resulted in any obvious corrosion after a year's service. Various parts of the scrubber operate at elevated temperatures under acidic conditions (with hydrochloric and hydrofluoric acid) or basic conditions, with soda added to absorb sulphur dioxide. The scrubbers are of the order of 10 m high and 2.5 m in diameter.

The second example comes from Dow Europe itself where a complete flue gas cleaning system (quencher, absorber system, fan, stack, ionising wet scrubber and connecting ducts) has been produced from Derakane-based composites, primarily to exploit the corrosion resistance of the material. The flue cleaning system was produced by MBT-Ceilcote GmbH of Germany and is installed at the company's Stade plant in Germany which handles 30 000 t/year of by products from liquid, solid and viscous materials. The glass fibre reinforcement used for these structures was ECR-glas, an acid resistant fibre produced by Owens Corning. Like the Italian system, Derakane

470 was used in most components, although the fire retardant grade 510 was used for the ionizing wet scrubber.

5.4.3 REINFORCED PLASTIC SCRUBBERS ENSURE SAFE WATER SUPPLY

The community of Delray Beach, Florida, USA, now has a safe water supply thanks to four large glass reinforced plastic (GRP) scrubbing towers. The towers were filament wound by Industrial Plastics Systems of Lakeland, Florida, using Type 1000 series multi-end glass fibre roving from PPG Industries.

The 3 m diameter, 17 m high towers are used to strip out volatile organic compounds (VOCs) that have seeped into the community's wells. The VOCs include petrol, paint thinners and solvents that may have leaked from landfills, industrial disposal sites or faulty underground metal storage tanks.

Glass reinforced plastics was specified for the tanks because of VOCs' corrosive nature and the towers' exposure to coastal salt air. Strength was also a major consideration; the towers must withstand winds of at least 160 km/hour. Industrial Plastics says that the wind loading was a major design factor for the towers. The company was able to select a winding angle which increased tensile strength in the axial direction to make better use of the high strength properties of the glass reinforcement.

5.4.4 CARBON–POLYETHERETHERKETONE IMPELLERS FOR CHEMICAL PLANT PUMPS

Reinforced Victrex polpolyetheretherketone provided distinct advantages over current materials during a pump development programme for the chemical industry, according to its producer, ICI Advanced Materials of Welwyn Garden City, UK.

The project involved ICI's Rozenburg site in The Netherlands, together with pump manufacturer Begemann, ICI Advanced Materials and the Dutch government.

In an early programme, valves lined with unfilled grades of Victrex polyetheretherketone were found to provide surprisingly good performance at high temperatures in a corrosive medium, being in excellent condition after six months service. This led to the development of other pump parts, e.g. those in contact with the product, pressure covers, housing and impellers, being manufactured from polyetheretherketone.

Impellers were injection moulded from both glass and carbon fibre filled polyetheretherketone and installed in the polyurethane plant at Rozenburg. Both impellers performed well, although, according to ICI, the carbon fibre impeller showed the best performance and has now given satisfactory continuous service for 18 months. This compares with a typical lifetime of two months for glass fibre epoxy impellers previously used in the plant.

5.4.5 FILAMENT WOUND TUBES FOR CRYOGENIC CONTAINERS

Glass reinforced plastic filament wound tubes, produced by Permali Gloucester Ltd in the UK, are being used as part of a range of cryogenic containers and pressure vessels, the company reports.

Statebourne Cryogenics of Tyne and Wear, UK, uses the vessels as neck tubes in its lightweight refrigerators and dewars. The tubes, which form the inner core of the refrigerators, are attached at the neck only, but support the full weight of the fully loaded vessel.

The composite tubes have to operate from room temperature to -196°C and withstand a vacuum of 4×10^{-10} mbar.

5.4.6 COMPOSITES FOR UNDERGROUND PETROLEUM STORAGE TANKS

The problem of corrosion in service station underground storage tanks for petrol is likely to result in a dramatic increase in the use of glass reinforced plastics for such applications in Europe, according to Amoco Chemicals.

In Europe, steel is conventionally used for underground forecourt storage but corrosion may result from both water contained in the fuel (a problem likely to be exacerbated by the introduction of lead-free petrol where the octane boosters have a large quantity of associated water) and from external attack from the soil.

In order to improve safety and minimize environmental problems major oil companies in Europe are now beginning to specify glass reinforced plastic, particularly in The Netherlands where there is an ever present problem with the high water table.

The European trend follows the pattern established in the USA where glass reinforced plastic has been used for some time, to the extent that now almost all forecourt storage tanks are produced from glass reinforced plastic. The experience generated in the USA over a 25 year period has been documented by Amoco.

5.5 MARINE, OFFSHORE AND CONSTRUCTION

5.5.1 OFFSHORE COMPOSITES PROGRAMME

A £1 million university research programme in the UK studying the 'Cost effective use of fibre reinforced composites offshore' has led to a second, two year phase, which began in October 1990.

The research programme is under the control of a steering body, Marinetech North West, established by a group of five universities in the North West of England. Funding for the programme came initially from the UK's Science and Engineering Research Council, the Ministry of Defence, the Department of Energy, and a consortium of industrial organizations, including multinational oil companies and chemical firms.

The next phase of the study will continue to place emphasis on the areas of fire resistance, degradation due to water ingress, blast, and hot gas flame and impact resistance.

5.5.2 GLASS REINFORCED PLASTIC STANCHIONS FOR NAVAL SUPPORT VESSEL

Permali Gloucester Ltd is to supply glass reinforced plastic stanchion sets for the Royal Fleet Auxiliary vessel, the Fort George, currently under construction in the UK for the Royal Navy.

The stanchions act as outriggers for the flight deck safety nets which are dropped out when helicopters land or take off. A radar transparent glass reinforced plastic has been used in this application to help reduce the vessels signature. The more conventional steel or aluminium stanchions would generate readily recognisable signals.

The Fort class RFA vessels are relatively large crafts whose job is to rendezvous with and replenish the stores of naval vessels. This task is becoming more important as military vessels devote more of their on-board space to munitions and less to storage of supplies.

5.5.3 PREPREGS IN MARINE APPLICATIONS

SP Systems, based on the Isle of Wight, UK, says it has demonstrated the value of aerospace style prepregs in the marine industry by the success of a number of craft produced with its systems.

Ampreg 75 is a prepreg material developed by SP Systems. It requires a relatively low cure temperature of about five hours at 75°C, without the need for high pressures. This allows boat constructors to eliminate the messy and inconsistent manufacturing methods associated with wet resin construction of composite boats, exploit a greater working time with the materials, and yet avoid the need for expensive ovens and autoclaves.

5.5.4 RESIN GAINS APPROVAL FOR MARINE STRUCTURES

The modified acrylic resin, Modar 836S, produced by ICI Acrylics of Darwen, UK, has obtained a Certificate of Approval from Lloyds Register of Shipping for its use in the resin transfer moulding of marine structures.

Modar 836S is a fast reacting system tailored for resin transfer applications, ICI says. Test results obtained for Lloyds approval have indicated that the resin offers superior hydrolysis resistance compared to isophthalic polyesters, it is stated, and formulations have been developed to provide effective low smoke fire performance.

5.5.5 PRE-STRESSING ELEMENT FOR CONCRETE

A pre-stressing element for concrete, based on Twaron Aramid fibres, has been launched by Akzo Fibres and Polymers Division of Arnhem, The Netherlands. According to Techtextil Telegram, the product is the result of five years study with the Dutch Beton-Groep. Arapree consists of Twaron HM fibres pultruded into continuous circular or rectangular sections within an epoxy resin matrix. Advantages over conventional steel pre-stressing elements include corrosion resistance, (an expected strength loss of only 14% is forecast after 100 years service), low weight (one fifth of a comparable steel element), and an improved fatigue life. The elements are also non-magnetizable and electrically non-conducting which may be an advantage in certain situations.

Likely applications include lightweight concrete components in corrosive environments, for structures under severe stress from fluctuating loads, such as railway sleepers, and where impacts are a problem.

5.6 GENERAL ENGINEERING

5.6.1 DOCTOR BLADES FOR PAPER INDUSTRY MADE FROM PULTRUDED THERMOPLASTICS

Pultruded doctor blades made from continuous glass fibre reinforced polyphenylene sulphide are providing considerable cost benefits to two major US paper making companies, according to their manufacturer.

The Fiberflex Plus doctor blades are made by Thermo Electron Web Systems of Auburn, Massachusetts, USA, which reports that one of its customers ran a machine continuously for six to ten months before the rollers needed resurfacing.

It is in the reduction of wear on the rollers in paper making machines that the pultruded composite doctor blade is said to be particularly impressive. Each doctor blade in a paper making machine is placed along the entire 9 m length of a roller (there may be up to 140 doctor blades in a single machine). The job itself is not a simple one, with typical rollers being 1.2 m wide and produced from solid granite. In order to eliminate resurfacing, the traditional steel blades need to be replaced every few days, whereas according to Thermo Electron, synthetic alternatives already tried, such as epoxy resin matrix composite blades, chip and consequently fail as doctoring systems.

Apart from improved performance, the pultruded blades represent an advance in manufacturing technique compared to laminated epoxy blades, it is stated, and come in longer stock lengths, eliminating much of the waste in cutting to size. The blades are supplied as 30.5 m pultrusions by Phillips 66 Advanced Composites Center, Bartsville, Oklahoma, USA. These are cut to length by Thermo Electron and then bevelled into 76 mm wide strips, 1.6 mm thick.

The pultruded blade offers additional advantages over traditional blades in possessing an ideal balance of flexibilities in various directions; this provides a better scraping ability and conformability to the roller profile. Another advantage is resistance to the corrosive chemicals used to bleach paper. The market potential for such applications can be gauged by the fact that Thermo Electron now sells some 150 000–200 000 doctor blades of all types annually to some 400 paper mills.

5.6.2 COMPOSITE PANELS-CLAD NOVEL FLIGHT SIMULATOR

The external shell of a new modular flight simulation system launched recently by Rediffusion Simulation Ltd, in the UK is produced from glass fibre composite sandwich panels produced by Polymeric Composites, Bristol, UK.

The novel simulator, Concept '90 represents a new approach to flight simulation where previously units were tailored for specific aircraft. During the switch to a modular system, the opportunity was taken to change the materials of construction for the shell, traditionally aluminium, and thereby improve both performance, access and appearance. The reduced mass of the composite shell reduces vibration, is lighter and stiffer and provides better sound attenuation. Polymeric Composites was involved in the design process from the outset and subsequently produced the panels for the prototype which has an overall size of 22 x 11 x 5 ft.

5.6.3 CARBON–CARBON NUTS AND BOLTS

In two separate developments, Nippon Steel Chemical Corp together with Nippon Steel Corp and Shikibo Ltd and its subsidiary company Shikishima Canvas Co have announced that they are to produce carbon–carbon nuts and bolts.

In both cases the nuts and blots are manufactured from blocks of carbon-carbon produced from tightly woven three dimensional fibre preforms.

According to reports in the *Japan Economic Journal,* the Shikibo/Shikishima bolts are claimed to withstand temperatures of 2500°C and to be two or three times more heat resistant than existing products.

Nippon Steel's products are meanwhile claimed to be 1.5 times as strong in shear and three times as strong in tension as conventional products. In both cases the conventional materials referred to are carbon-carbon laminates produced from 2D fabrics where low interply properties results in inferior absolute properties and further reductions above 150°C.

Applications for the nuts and bolts are expected in industrial machinery and, in particular, in the production of ceramics and nuclear fusion reactor insulators.

The Shikibo/Shikishima nuts and bolts are the result of technology developed by the two companies aimed at producing 3D fabrics for shapes such as T and L-shaped sections, cylinders and cones.

Shikibo now intends to establish a research and development institute in Yakkaichi, Shiga Prefecture to explore further the commercial applications of its products.

Another Japanese consortium of Kawasaki Heavy Industries and Kukuvi Chemical have announced plans to develop carbon fibre reinforced plastic bolts for use in aerospace.

The companies cite the tendency in the industry to use metal fasteners with carbon fibre reinforced plastics as a possible cause of galvanic corrosion in the airframe. This, the company states, which will be overcome by the use of the composite bolts.

5.6.4 ULTRA THIN CARBON–CARBON MEMBRANES

Le Carbone-Lorraine, a subsidiary of the French Pechiney organisation, has developed asymmetric ultra thin membranes from a carbon-carbon composite.

The membranes are tubular in shape and couple high strength and corrosion resistance with high temperature performance (up to 400°C). They consist of a microporous filtering layer, which is only a few micrometres thick, supported by a substrate layer which itself is less than 1 mm thick. For a given pressure, the flow through the carbon-carbon membrane is greater than conventional membranes where the thicknesses cannot be reduced without unduly compromising the mechanical properties. A brief periodic unclogging of the permeate return for a few seconds is claimed to be all that is required to maintain a stable flow through the membrane to avoid irreversible clogging. The membranes are also claimed to be biocompatible, water repellent and inert to virtually all chemicals except strong oxidising agents, and they are good conductors of both heat and electricity.

5.6.5 HOUSEHOLD APPLICATION FOR REINFORCED THERMOPLASTIC

The French company Bezault has selected 'Verton RF700-12EM', a 60% glass fibre reinforced nylon grade from ICI Advanced Materials of Welwyn Garden City, UK, for the reinforcing bars of its 'Riviera' door plaques.

According to ICI, the main reason for the use of Verton was a tripling of the price of the existing material, a malleable soft alloy, 'Zamac', which itself had only been used for two years. Verton is claimed to have many advantages over Zamac, including a reduction in the number of manufacturing operations. It is also stated to have advantages over alternative short fibre thermoplastics, in terms of surface finish and mechanical properties.

Bezault's commitment to Verton is such that the company has built a 800 m^2 moulding shop and is planning to install an injection moulding machine dedicated to Verton processing.

ICI says Verton is an injection moulding material produced from engineering thermoplastics, with a fibre reinforcement that is considerably longer than that found in conventional injection moulding granules of reinforced plastics, as a result of the compounding route.

5.6.6 COMPOSITE AIR DUCT ON 'M1' BATTLE TANK

A one piece composite airbox/plenum which forms part of the American M1 main battle tank air intake system has been developed by General Dynamics of Sterling Heights, Michigan, USA. The company says weight reduction was the primary aim of the project, with cost efficiency, part conformity and part consolidation cited as additional advantages.

The airbox is a precleaning unit containing filters, with the plenum diverting clean air into the vehicle engine. The part was originally made from welded aluminium but leakage due to poor part conformation was a problem at the welds. A weight saving of 53 kg was achieved using an E-glass fibre/epoxy resin composite.

General Dynamics says that the part was produced in one piece using a resin transfer moulding process with E-glass fibre wrapped around foamed cores and balsa wood. Tactix 123 epoxy resin from Dow Chemical was used for the prototype parts.

The parts have undergone qualification testing which involved their exposure to DS-2 decontaminant materials — highly corrosive systems used to decontaminate the vehicle after nuclear, chemical or biological attack.

5.7 MEDICAL

5.7.1 COMPOSITE HIP READY FOR HUMAN TRIALS

Biomet Inc of Warsaw, Indiana, USA, commenced human trials of its composite hip implants during 1990. A total of 500–600 people are expected to be given the composite hip as part of studies before the prosthesis acquires approval for general use by the US Food and Drug Administration (FDA).

The market for artificial hip joints is enormous. In the US it is estimated that some 200 000–270 000 hips are implanted every year, costing US$2500 to US$3000 each.

There are fears in the US that metal implants used to date, typically cobalt chrome or titanium alloys, may be linked to the onset of cancer in certain cases. The link is particularly tenuous at the moment, even for the medical field, and the drive for composite implants is based more on the design possibilities inherent with composites which allow the implants to be more mechanically compatible with the body.

The Biomet hip prosthesis is based on Amoco's polysulphone Udel resins and Hercules carbon fibres. Other US companies with similar products on the way include Osteonics Corp of Allendale New Jersey; Howmedica Inc, Rutherford, New Jersey; Zimmer Inc, Warsaw, Indiana; and Richards Medical from Memphis, Tennessee.

5.8 SPORTS

5.8.1 TENNIS RACKET USES POLYETHYLENE FIBRES

A tennis racket produced by Ellipse Sport of Altamonte Springs, Florida, USA, is a recent application for Allied Signal's high modulus polyethylene fibre, trade named Spectra. The racket frame has been produced from a blend of Spectra and carbon fibres, and a non-woven fabric of Spectra has been used to cushion the joints in the throat where the yoke meets the frame. The design results in a highly effective vibration trap, reducing vibration and shock to the player, Allied claims.

According to Dr Huy Nguyen, senior composite specialist for Allied Signal's High Performance Fibers Unit, the 'Fusion' racket results in about 25% less vibration transmitted to the player's arm on contact with the ball. The resistance of the racket to damage propagation is also increased, it is stated.

5.8.2 EPOXY PREPREGS FOR HANDCRAFTED SKIS

Evolution USA, the specialist ski manufacturer based at Salt Lake City, USA, is now using ICI Fiberite prepregs to produce its high performance handcrafted skis.

Conventional aerospace grades of woven fabric, carbon and glass fibre epoxy prepregs are used by the company for its low volume, specialist products. The 'Pearl', general purpose ski uses glass fibre prepregs while the 'Diamond' racing ski is of carbon fibre construction.

Mass produced skis often exploit composites, but use wet-processing where volume fractions of about 50% are achieved. The higher volume fractions offered by the prepregs provides better properties and reduced variability and the use of prepregs has the added benefit of improving working conditions.

5.8.3 KEVLAR SLEDS TO THE POLE (ALMOST)

The abortive attempt by explorers Sir Rannulph Fiennes and Dr Mike Stroud to trek on foot to the North Pole, did at least feature some successful technology. The sleds pulled by the two men were produced from a Kevlar fabric composite manufactured by Fothergill Engineered Fabrics of the UK.

The ultralight sleds were designed using computer modelling to allow optimum performance on hard ice and soft snow, with good impact resistance and minimum weight. The sled construction is the fifth generation design produced by designer Steve Holland for the Polar expeditions and involved just six plies of Fothergill's D0267 Kevlar fabric — down from ten in previous designs — laminated with a two part room temperature cure epoxy. The empty sled weighed only 65 lb but was capable of supporting the fully loaded weight of 300 lbs at the start of the expedition.

5.9 HIGH TEMPERATURE: STRUCTURES

5.9.1 THE ROLE OF CONTINUOUS FIBRE REINFORCED CERAMIC MATRIX COMPOSITES

A very upbeat article on the potential for continuous fibre reinforced ceramic matrix composites appeared in ASTM's Standardization News.

The author, Scott Richlen is a programme manager with the Office of Industrial Technologies at the US Department of Energy. Some of the figures quoted by Richlen are quite staggering and the potential market for continuous fibre reinforced ceramic matrix composites is on this basis, enormous.

While the potential of continuous fibre reinforced ceramic matrix composites has been appreciated for some time in the aerospace and defence industries, the role of the materials in industry in general has not been so clearly understood.

The high temperature capabilities of the materials, coupled with their excellent toughness compared to monolithic ceramics and whisker and particulate reinforced ceramics, makes possible industrial applications that could improve energy efficiency and reduce emissions. Surveys by the Department of Energy have indicated several industrial applications that could result in a cumulative annual energy saving in the US of 0.7 quadrillion Btu (approx 0.7×10^{15} J). A further 0.4 quadrillion Btu could be saved if continuous fibre reinforced ceramic matrix composites were to be used in power generation where nitrogen oxide emission could be cut by 600 000 tons/year.

Further environmental benefits could accrue from the use of continuous fibre reinforced ceramic matrix composites in high temperature incineration plants where the high temperature resistance under extremely corrosive environments would be an important advantage. Significant solid waste disposal, where a reduction in volume by 90% is typical, together with decreases in airborne pollution through a better breakdown of molecules, such as dioxin and PCBs, and fuel displacement by energy recovery, would all result from such applications.

The figures are again enormous. The total US municipal solid waste burden is forecast to reach 785 000 tons per day by the year 2000. At present only some 6% of the total waste is disposed of in waste to energy facilities. This could rise to a figure of 30%, with a market for incineration plant estimated at US$1.5 billion/year.

Legislation may well be the prime motivation for many environmentally beneficial applications of continuous fibre reinforced ceramic matrix composites. 300 million tons of liquid organics and sludges are generated each year in the US but only about 2% of this total is incinerated — a figure that must rise if strict legislation is brought in. Advanced burners and combusters, exploiting the high temperature properties of continuous fibre reinforced ceramic matrix composites may be developed to meet requirements on emissions of gases such as nitrogen oxide. Research by the Electric Power and Gas Research Institutes is investigating the use of continuous fibre reinforced ceramic matrix composites for catathermal combusters for land turbine power generation, where catalytic substrates result in *in-situ* reduction of nitrogen oxide. The market here should be judged on the basis of a total of 42 million trucks and buses currently registered in the USA alone.

The value of using reinforced ceramics in engines to allow more efficient power generation by higher temperature operation is more well known. The figures suggest that a four percent increase in thermodynamic efficiency would be achieved by using continuous fibre reinforced ceramic matrix composites in a typical open cycle gas turbine, equivalent to a 13% reduction in fuel consumption. Department of Energy/Energy Information Administration data, allowing for certain assumptions regarding fuel prices and capacity, suggests total fuel savings in the USA could amount to US$12.7 billion in the period 2000 to 2014 as a result of exploitation of ceramic composites in gas turbines.

By necessity, projections of this sort make many assumptions, but the pressures to maintain the environment are likely to increase rather than decrease over the next few years while assumptions made regarding fuel prices are likely to have been optimistic. In general, the future of continuous fibre reinforced ceramics looks good if the materials can be commercially developed in time.

5.9.2 FUSION REACTOR FIRST WALLS MADE FROM CARBON–CARBON

A potential application for carbon–carbon composites is the production of tiles for the inner walls of improved nuclear fusion reactors.

In 'conventional' fusion reactors, the first wall of the reactor i.e. that wall in greatest proximity to the plasma, is usually produced from graphite tiles mechanically fastened to a metal substrate. The very high temperatures experienced by the tiles requires an adequate heat flow through the mechanical fastener to be removed via the metal support plates. In current generation systems this process works, but it is anticipated that future reactors will subject the first walls to a more demanding heat flux, raising the surface temperatures to 2800°C and causing a loss in thickness of the tile from sublimation in the order of tens of microns/s. This would result in an unacceptable contamination of the plasma with impurities, and the loss in the tile thickness would be such that the effective life of a tile would be too short.

The solution considered by the Mitsubishi Kasei Corp, Tokyo, Japan, detailed in European Patent EP 0339 606, is to produce a carbon-carbon composite to replace the graphite, with the fibres predominantly oriented in the thickness direction of the tile which is bonded to a metal substrate. The carbon fibres employed in such a material could be of the existing types but the highly thermally conductive pitch-based carbon fibres are preferred. A thermal conductivity of at least 3 W/cm °C is required in the thickness direction. Numerous examples are quoted in the patent application of carbon-carbon composite constructions with fibres variously arranged in woven fabrics, non-woven fabric and short fibres formed into webs.

5.9.3 FIRST WALLS FROM CARBON–BORON COMPOSITES

Toyo Tanso Co Ltd, of Tokyo, Japan, has announced that it has developed a composite based on carbon microbeads and boron, for use in the first wall of a fusion reactor. This material was developed in collaboration with a research team at the University of California, Los Angeles, USA. It is claimed to be readily processable and economical compared to alternatives such as boron coatings.

5.10 HIGH TEMPERATURE: ENGINES

5.10.1 NON-CARBON COMPONENT FOR SPACE SHUTTLE'S ADVANCED ROCKET MOTOR

An integral throat entrance for an advanced solid rocket motor for the Space Shuttle is to be made from carbon-carbon composites by Textron Specialty Materials of Lowell, Massachusetts, USA. The company was picked by the Thiokol Corp of Brigham City, Utah, to make the part.

The integral throat entrance is a large three dimensional, ring-shaped structure, about 2.4 m in diameter, which is used to line the inside of the rocket nozzle where exhaust gases can reach 2800°C (5000°F) when exiting the motor. The company claims that this represents the largest carbon-carbon structure of its kind ever made.

The unit will be fabricated from a 3D woven preform constructed using Textron's, 'Autoweave', a proprietary computer automated weaving system.

Textron describes itself as a producer of carbon–carbon parts, with experience of smaller integral throat entrance units for various ballistic and other missile systems. Its annual production of carbon-carbon components currently amounts to 31 700 kg.

6. PROCESSING

6.1 CASE STUDY

6.1.1 THERMOPLASTIC COMPOSITE FUSELAGE

The experiences of a team from Lockheed's Composite Development Center in the USA involved in developing a thermoplastic composite fighter forward fuselage are detailed in a paper ('Thermoplastic fighter forward fuselage') in SAMPE Quarterly (Vol. 21, No. 1). The article illustrates the wide range of manufacturing options now available for the expanding family of novel composite materials forms on offer to the advanced composite industry.

The project examined a range of available thermoplastic composites, including ICI's APC-2 and HTA, Amoco's Radel C, Cyanamid's Cypac 7005 and Phillip's PAS-2. Unfortunately there is no discussion regarding the relative merits of these systems.

The paper details the forming techniques for the items that made up the fighter forward fuselage and the tooling systems that were adopted.

The high temperatures involved in processing the thermoplastics (from 230–360°C) requires specific tooling materials and favoured materials are castable ceramics such as thermosil 120, filled castable ceramics such as Comtek XS (from OxyChem) and integrally heated laminated ceramic tools such as those available from Comtool. The integrally heated tool system was particularly well received by the research team at Lockheed who confirmed the claims of Comtool that, in combination with an insulating blanket, low temperature bagging materials may be used for thermoplastics.

Consolidation techniques used ranged from autoclave consolidation to double diaphragm forming and press forming (using a rubber punch).

Assembly operations exploited both co-consolidation and thermoplastic adhesive joining. Co-consolidation of top hat stiffened lower skin and access doors involved supporting the preformed stiffeners during the secondary processing stage using a wash out tooling system, Caremold, developed by Composite Horizons. It is reported that a barrier film between the wash out tool and the thermoplastic is necessary to prevent the plastic penetrating the porous surface. The thermoplastic adhesive used for certain operations was a polyetherimide film, and parts were successfully produced from APC-2 and HTA using this route.

The fact that a complicated structure such as the fighter forward fuselage has been produced with thermoplastics must encourage other teams working with these materials. The technology is evidently available to produce components and structures. The relevant questions that now need answering concern the economics of the processes relative both to thermosets and improving light metal alloys.

6.2 PREFORMING/PREPREGGING

6.2.1 PREPREGGING PROCESS

Electrostatic Technology Inc of Branford, Connecticut, USA, has developed a novel prepregging process.

The process uses a proprietory dry powder applicator to apply polymers to the reinforcing carbon or glass fibres. By exploiting electrostatic powder application technology, the prepregging process can be much faster than using conventional solvent or hot melt processes, according to the company. Electrostatic has installed a prepreg production line at its Branford plant and long term plans include the sale of both prepreg and prepregging equipment.

6.2.2 PREFORMING PROCESS

An automated preforming process that is claimed to be energy efficient, simple, and effective at enhancing materials usage with the minimum waste, has been introduced by Freeman Chemical Corp of Port Washington, Wisconsin, USA.

The process, called Compform, involves the application of a binder resin to precut mats. Freeman says that new technology is used to heat the binder (with only minimal heating of the reinforcement or mould), mould the preform, and attach structural additions such as ribs and cores.

The attachment method, which allows encapsulation of metal, wood and foams, eliminates many engineering obstacles encountered with existing preforming methods, the manufacturer reports. These include the difficulty of engineering preforms with various types, thicknesses and configurations of supports in high stress areas of the part without adding to the overall thickness and weight of the preform.

6.2.3 NEW ROUTE TO PREFORMS

Shell International have developed a new method for producing fibre preforms for use with resin transfer moulding (EP 0 393 767A2)

Conventional preforming processes involve either the spray up or directed fibre placement techniques or the manipulation of fabrics. The first technique leads to variable fibre deposition and restrictions on the radius of curvature that may be introduced into the preform. The second approach utilising fabrics can involve extensive trimming operations, and may lead to fabric damage during shaping.

The approach developed by Shell follows the fabric route but has many novel features. The fabric to be preformed is placed between two elastomeric sheets and a vacuum is applied to the laminate so formed. The laminate assembly is then formed to the desired shape by a combination of vacuum forming and matched press moulding. The initial forming takes place by a vacuum forming process in which the elastomeric sheet maintains the preform fabric parallel to the mould surface and eliminates wrinkling. This stage in itself may be sufficient for gently curved parts, but for more severely contoured parts the final forming involves the action of a matched tool. Under normal circumstance, the excessive deformation might be expected to lead to tearing of the preform due to frictional forces on the mould surface. In Shell's process, however, the surface of the tool-laminate interface is lubricated by a fluid, which is air in most cases.

The elastomeric sheet materials, which are typically natural latex, may be re-used and by careful protection may be expected to last from 200–300 cycles.

Any fabric type can be used in this process, and if thermal binding systems are employed then the laminate assembly will be heated during or prior to the forming cycle and subsequently cooled to retain the shape. Certain fabrics will possess the ability to retain the moulded shape without such aids.

The system has been proven in laboratory trials on Shell's behalf by Doefer Inc, of Cedar Falls, Iowa, USA, and is now being used for the production of preforms for commercial parts.

6.3 MATERIALS HANDLING

6.3.1 WATER JET CUTTING TRIALS OFFERED

Ingersoll Rand, Bolton, UK, a manufacturer of specialized water-jet cutting equipment is now offering a fast cutting service from its operation in the Netherlands.

Straight line cutting trials are normally conducted free of charge while complex shapes will incur charges from £60-125/hour depending on the level of programming required for the computer controlled cutting equipment.

6.3.2 APPARATUS FOR SEAM WELDING THERMOPLASTIC TAPES

The Boeing Co, Seattle Washington, has patented an apparatus and method for joining together thermoplastic tapes to form broad goods width continuous sheet (EP 0 392 431 A2).

Most unidirectional carbon fibre reinforced thermoplastic tapes come in relatively narrow widths (on average 350 mm) and for the manufacture of large surface area parts it is advantageous to combine these tapes to form broad sheets of the material.

The Boeing equipment allows for multiple seaming with control of the seaming parameters and features a pneumatic control of the pressure at each seaming head. The axial spacing between rolls of tape and the tension in each tape are adjustable.

The equipment allows the automated conversion of tape into broad goods under controlled conditions with no apparent process limitation on size of the ultimate sheet.

6.3.3 PICK-UP ROBOT

A pick-up robot which is stated to permit economic prepreg processing has been developed by Arato Engineering of Buochs, Switzerland. The robot allows dry or tacky preforms, non-woven fabrics or mats to be picked up, transported and assembled without damage to the fibre arrangement, according to a report in *Techtextil Telegram*. A special version of the Arato gripper is used which allows the gripping depth to be reproducibly set. The new robot is said to be suitable for automated production of sandwich panels and similar structures.

6.4 PROCESS MONITORING

6.4.1 PULTRUSION PROCESS MONITORING EQUIPMENT

Two complementary pieces of equipment have recently been introduced by Pultrusion Specialties of Alum Bank, Pennsylvania, and Pultrusion Technology, Twinsburg, Ohio, USA.

Pultrusion Specialties is producing the CPAN thermal analyser, which may be used to perform a variety of on and off-line functions. The system can be used off-line to study resin standards and measure cure behaviour, exotherm, etc., and follow temperature profiles and exotherms during the process itself.

Pultrusion Technology, a division of Morrison Moulded Fiberglass, has released its Trendstar process monitoring equipment which continually monitors and records pulling loads and processing speeds over a range of selectable timetables up to 480 minutes maximum. A preset alarm function alerts operators to process deviations in sufficient time to make corrections.

6.5 CURE PROCESSES

6.5.1 CURING BY IONIZATION POLYMERIZATION

Aerospatiale has developed a novel method of effecting cure in thermosetting resin composites that does not require heat as is usual in most curing systems.

The process involves the use of an electron beam which c-scans the composite part to be cured. This results in an immediate reaction between the radiation and the resin, and polymerization takes place without any rise in temperature, according to reports in the *French Technology Survey*.

The principle is based on molecular chemistry where the energy of electron or X-ray beams break chemical bonds thereby generating radicals which may trigger a chemical reaction.

One of the major advantages of this curing route is claimed to be the time saved. A filament wound structure for a jet engine is cited as an example where a thermal cure might require a cycle time measured in days. The corresponding cure time for ionisation polymerisation would only be a few hours due to the immediate action of the radiation. The lack of a temperature rise is also significant in terms of differing thermal expansions for metal inserts and tooling and residual stresses in the part.

Aerospatiale is to construct a specific facility to exploit the technology, to be known as UNIPOLIS (Unite de Polymerisation par Ionisation des Structures Composites). The facility will comprise a 10 MeV (20 kW) electron accelerator capable of polymerising carbon composites up to thicknesses of 32 cm, a compartment measuring 22 x 12 m for polymerisation of parts up to 4 m diameter and 10 m long, and a programmable automatic system to pass the entire surface of the components in front of the accelerator in an appropriate manner.

6.5.2 CHEMORHEOLOGICAL MODEL FOR THERMOSET PROCESSING

'A model of the thermal and chemorheological behaviour of thermoset processing: (II) unsaturated polyester based composites' is the title of a paper in *Composites Science and Technology* (Vol. 38, 1990, No. 4, pp 339-358) by Kenny, Maffezzoli, Nicolais and Mazzola, from the University of Naples and Molding Systems SpA of Italy.

The authors have developed a computer based model, using kinetic and heat transfer models, to calculate temperatures and reaction profiles for composites during cure that will predict the corresponding viscosities according to an empirical model.

This has been applied to a real situation of resin transfer moulding where the real temperature at different positions within the mould could be measured. The necessary kinetic and thermodynamic data for the calculations was obtained by performing differential scanning calorimetry and dynamic viscosity measurements on the resin.

The experimental results show good agreement with the calculated values and the authors state that the numerical simulation of processing should allow optimum processing conditions to be predicted.

6.6 EQUIPMENT/MACHINERY

6.6.1 SYSTEM FOR INJECTION MOULDING POLYMER COMPOSITES

The French machinery manufacturer Billion says that it has introduced a new version of its ZMC system for injection moulding glass fibre reinforced thermosets.

Billion's ZMC machines are designed to allow the processing of a dough moulding compound/bulk moulding compound type material where a relatively long fibre length is retained in the final part. The improvements incorporated in the new Billion ZMC 300 system include a modified screw design and changes to the feed zone and injection chamber. A further reduction in fibre attrition during processing is claimed, which results in fibre lengths of the order of 5 cm being retained in the part rather than the 2.5 cm achieved with earlier systems. Impact properties and surface finish of the moulded parts are said to be improved.

The ZMC systems are supplied in two sizes to fit the range of Billion injection machines. According to *Plastics and Rubber Weekly*, the smaller ZMC 300 unit, the H2500, intended for use with Billion's 300 and 450 t injection machines, will have an injection volume of 2251 cm^3 at 1053 bar injection pressure. The larger unit, the H13800, has an injection volume of 10 650 cm^3 at 1250 Bar and may be used with the 900, 1150, 1800 and 2300 t machines.

6.6.2 LARGE RESIN TRANSFER MOULDING MACHINE

A resin transfer moulding machine capable of delivering 45 kg of matrix a minute has been designed by Liquid Control Corp of North Canton, Ohio, USA.

The variable ratio, hydraulic driven Multiflow-CMFH model is suitable for a variety of resin systems, including epoxies, polyesters, urethanes and methacrylates, the company says. The equipment is built with volumetric positive displacement Posiload metering pumps and can pump methylethylketoneperoxide and BPO catalyst emulsions undiluted at concentrations down to 0.5%. Ratios from 1:1 to 200:1 and flow rates from 2.3 to 45 kg/minute can be accommodated, Liquid Control says.

The machine is capable of handling a variety of fillers, including calcium carbonate, aluminium trihydrate and hollow glass microspheres, and the metering accuracy is not affected by back pressure from closed moulds. According to the manufacturer, a high flow over/under injection block at the mixer inlet ensures proper introduction of catalyst to resin. These materials are thoroughly blended in a mixer prior to injection into the mould.

An optional feature is a pneumatic cycle counter to automatically cycle the machine a preset number of strokes for precise mould filling.

6.6.3 50 TONNE ISOSTATIC PRESS

A new 50 tonne isostatic press has been launched by Simac Ltd, Rugby, UK.

The company claims that the new range of Densomatic 50 presses have been designed with flexibility and accessibility in mind, with space behind the tooling, an easy to reach intensifier, and pumps located near the motors, separate from the tank.

Automatic load and unload are features on the new presses which are capable of producing a wide range of shapes, including balls, rods, cones, nozzles and tubes, from materials such as free flowing ceramic powders.

A maximum work rate of four stokes per minute is possible with a total of seven different machine options available. Pressure capabilities range from 700 bar to 2100 bar, with fill diameters from 30 mm to up to 70 mm.

6.6.4 RESIN CARTRIDGE DISPENSER

A US equipment supplier says it has developed a composite pneumatic dispenser for its coaxial resin cartridges.

Liquid Control Corp, North Canton, Ohio, supplies equipment for the resin transfer moulding industry. It says the dispenser is compatible with its 'Supermix II' cartridges and has been produced from high strength injection moulded and aluminium extruded components.

Coaxial resin cartridges are securely held in place with a positive snap-lock cartridge retainer, reversible for left or right handed manual cartridge valve operation. Cartridge contents can be visually monitored with an external indicator and the entire dispenser is disassembled for cleaning and inspection by removing three fasteners.

The inlet air line is inside the air cylinder to guard against damage, and air pressure is relieved through an internal exhaust which is muffled and directed away from the operator.

6.6.5. 100 TONNE COMPOSITE PRESS

A 100 tonne compression press designed primarily for advanced composites has been introduced by Hull Corp of Sparta, New Jersey, USA.

The CP-100 is a downward acting press with a heavy duty design. Two 18 cm bore, 10 cm diameter rods with 81 cm stroke rams give a claimed platen deflection of less than 0.0008 mm/cm.

The heating system consists of six forward and six rear 500 W heater cartridges acting on the 91 x 91 cm square platens. A rate of heating of up to 2.8˚C/minute and temperature operation within the 82–982˚C is available, according to Hull.

Closed loop heat control is provided by a computer terminal and software developed by Hull.

6.6.6 COMPUTER NUMERICAL CONTROL SYSTEM

Cincinnati Milacron of Cincinnati, Ohio, USA, has developed a new computer numerical control system, the Acramatic 975 'F' series, for use with its tape laying technology.

According to the company's *Composite Comment*, it is the advent of Milacron's FPX 7-axis Fibre Placement Processing Centre that has required a new control system to be developed to accommodate the demands for more programmed points and greater programming power.

The 'F' series is said to combine the versatility and familiarity of standard NC programming with robotic Tool Centre Point programming. Other notable features include two 32 bit microprocessors which provide the additional computing power.

6.6.7 FILAMENT WINDING TENSIONERS

The Electroid Company of Springfield, New Jersey, USA, has unveiled its '2000' series of filament winding tension control modules.

There are two models in the range, the TCM-2000-B — a fully digital microprocessor-based passive system which allows limited takeback (up to 3 m), and the model TCM-2000-C — an equivalent active system which the company says will allow virtually unlimited takeback.

Both units can be interfaced with an IBM compatible personal computer for data acquisition. They contain their own power supplies and are modular in design. System expansion will allow up to 255 individual tensioners to be controlled by a single network manager system.

According to Electroid, both systems are held in stock for off the shelf delivery.

6.7 MANUFACTURING PROCESSES

6.7.1 EFFECTS OF DIAPHRAGM STIFFNESS ON THERMOPLASTIC FORMING OF COMPOSITES

The effects of diaphragm stiffness on the quality of parts produced by diaphragm forming of advanced thermoplastic composites is discussed in a paper in SAMPE Quarterly (July 1990).

The work was performed as a result of the collaboration between the University of Galway in Eire and the Centre for Composite Materials at the University of Delaware, USA.

Diaphragm forming is developed from vacuum forming of thermoplastic sheet. The laminate is placed between two deformable diaphragms which when clamped around the edges maintain biaxial tension on the laminate during deformation, thereby restricting wrinkling and buckling.

Various polymeric diaphragms were studied including 0.125 mm Upilex-R, Upilex-S and Kapton-H, together with 1.00 mm superplastic aluminium (Supral).

The studies showed that for constant pressurization rates the times to completely form a given article differed for each diaphragm material, with times of 10 minutes for Supral, 1 minute for Upilex-R and 6 minutes for Upilex-S and Kapton. However, the quality of the part increased with the stiffness of the diaphragm such that the Upilex-S and Supral formed parts with a better surface finish.

6.7.2 CO-CONSOLIDATED PARTS

The processing versatility of co-mingled and powder prepreg thermoplastics compared to equivalent fully wet-out thermoplastic tape is well illustrated by demonstrator parts produced by or on behalf of BASF during the year.

Co-mingled fabrics consist of reinforcing fibres such as carbon intimately mingled with fibres spun from the eventual matrix polymer which in the case of BASF's products may be either polyetheretherketone (PEEK) or PEKEKK (Ultrapek). Powder prepregs are an alternative form developed for use with polymer matrices that cannot be spun into fibres. Current products include a new thermoplastic polyimide/IM8 material. Both systems have been developed in order to retain the advantages of thermoplastic composites while overcoming the processing and handling difficulties experienced with stiff, boardy thermoplastic tape, by exploiting the inherent drapeabilty and formability of fabric systems.

Steve Olson of BASF Structural Materials, based in Charlotte, North Carolina, USA, summarises the various processing routes available for thermoplastic fabrics (both co-mingled and powder prepregs) in a paper in the September/October 1990 *SAMPE Journal*. Press forming and vacuum bag/autoclave consolidation are covered, as is a variation on a relatively new technique for forming thermoplastics, namely diaphragm forming. It is interesting to note that BASF recommend the use of monolithic graphite tools (5 cm thick) for the production of flat plates in a press forming operation, with Upilex or Kapton films for mould release, in order to improve surface quality and in particular to eliminate any surface fibre buckling.

The diaphragm forming technique has been developed by BASF for use with co-mingled and powder prepreg fabrics and differs from conventional diaphragm forming in that only one diaphragm is required. This means that complex parts such as hat stiffened panels may be formed in one operation, where the base panel and hat stiffening sections are co-consolidated.

The BASF diaphragm forming process has been demonstated using a Upilex re-usable diaphragm and a universal miniclave heated in a press to 400°C, with the processing pressure introduced from a nitrogen gas bottle. Various tooling materials were explored as mandrels for the hat section in the part, including steel, aluminium and wash-out tooling. The aluminium provides the easiest system to remove after processing but deforms under the temperature/pressure regime and ages rapidly. Wash-out tooling has many advantages but may be difficult to remove. Steel is re-usable but can also be difficult to remove.

Autoclave consolidation with steel tooling has been used in a separate demonstration exercise by British Aerospace to produce a tailpane unit for its Hawk fighter. This consisted of skin with integral J-spars, again from BASF co-mingled fabrics. In this case, the matrix was PEEK 150 and the reinforcing fibre was IM6. Once again the part was produced as a single piece. The thermal mismatch between the steel tool sections and the fabric used for the J-spars was exploited to increase local consolidation pressures.

The maximum process temperature, applied pressure and dwell times at temperature are all important variables for processing co-mingled and powder prepregs, as the polymer has not only to flow to allow consolidation to assume the final part shape, but also to wet-out each individual fibre — a process which has been demonstrated in the more conventional unidirectional tapes such as ICI's APC-2 and Phillips Avtel.

Another important consideration in processing such materials is the appropriate time to apply the consolidating pressure. This depends on whether fibre movement is required to form the part. If it is then the pressure should be applied after the melt temperature of the polymer has been reached. If fibre movement is not desirable then the pressure should be applied at least partially before the melt temperature of the polymer is achieved.

6.7.3 AUTOMATIC COMPRESSION MOULDING

A range of machines for the automated compression moulding of thermosetting materials has been designed by Battenfeld Duroplasttechnik of Meinerzhagen, Germany.

The 'BPA' series of machines can provide clamping forces in the range 1000–4000 kN. The clamping force is numerically adjustable, a prerequisite for the mass production of precision parts such as electrical goods. A hydraulic system permits the pressure and stroke-dependent control of all three phases of the compression moulding operation: breathing, preliminary pressure and moulding pressure. A material feed system, designed by Battenfeld, operates fully automatically by means of a filling gauge.

All functions are controlled by a 'Unilog 4000' closed loop multiprocessor control system, which needs to know the pressures, speeds, strokes, times, and temperatures specified for the process.

6.7.4 CONTROLLING FIBRE ORIENTATION

The processing of glass mat thermoplastics such that desired orientations of the reinforcement can be introduced during the manufacturing process is discussed in a paper in *Kunststoffe*.

The paper reports on studies of the orientation behaviour in glass mat thermoplastics and describes the concepts applied to the practical situation of an automobile bumper mounting.

Further details in 'Method for producing orientations in long fibre reinforced composites,' by H. Dittmar, L. Wahl, K. Bentrup, and C. Molitor, *Kunststoffe*, Vol. 80 (1990), pp. 563–571.

6.7.5 DEVELOPMENTS IN PULTRUSION TECHNOLOGY

American Composite Technology (ACT) Inc of Boston, Massachusetts, USA, is currently developing pultrusion technology in a number of areas.

The company claims as its objective the updating of pultrusion from its traditional form with attendant limitations on materials, geometry and quality to a more flexible process widely applicable to a broad range of products. Central to much of ACT's developments would appear to be the use of direct injection of resin into the special chambers in the pultrusion die rather than the more conventional impregnation of the reinforcement in an on-line bath. The direct resin injection into the die results in a short time between mixing and impregnation, allowing resin to be preheated — thereby minimising viscosity. Pot life problems are of course eliminated and rapid curing systems can be used. An additional benefit is the total containment of the resin system, thereby minimizing environmental problems.

These concepts have been successfully exploited by ACT to develop a phenolic pultrusion process. Pultruded phenolic sections have considerable potential in many industries, where fire, smoke and toxicity problems are concerns for composites based on polyester and epoxy, but the material has not proved particularly amenable to the process in the past. The ACT process employs a catalyst free phenolic specially formulated by Plastics Engineering Co.

In addition to material developments, ACT is exploring pull-winding and pull-pressing/pull-forming. The use of direct injection is said by the company to move the process towards a from of continuous resin transfer moulding while pull-pressing resembles a continuous compression moulding process. Pull-pressing involves a conventional pultrusion process except that when the resin emerges from the die it is only b-staged and the first puller of the pultrusion machine is replaced by a moveable heated press which both forms and pulls the composites at the same time. In many respects these ideas are similar to the pull-forming techniques already exploited by Alcoa-Goldsworthy of Torrance, California, and the sequential moulding process demonstrated by Harwell Laboratories in the UK.

A further area of innovation from ACT concerns the development of novel pressure transducers that may be used to measure the pressure at any position within the pultrusion die. The sensors are relatively cheap and come with IBM or AppleMac compatible data acquisition hardware/software.

6.7.6 SUPERPLASTICITY IN SILICON NITRIDE–ALUMINIUM COMPOSITES

Superplastic forming has been reported in many ceramic fibre reinforced metal matrix composites, but in most cases there are drawbacks to the technique as thermo-mechanical processing is required which raises costs. The rate of deformation at which superplasticity is observed is usually very slow — of the order of 10^{-4} s^{-1} which is unsatisfactory for mass production, and the temperature range is above the solidus line of the metal matrix, resulting in coexistence of solid and liquid and the inevitable formation of voids.

In a paper in the July 1990 issue of *Journal of Materials Science Letters* ('Superplasticity of silicon nitride whisker reinforced 6061 aluminium at high strain rate', M. Mabuchi and T. Mai, Vol. 9, 1990) workers in Japan report on a system that does not exhibit such problems.

The composite studied was a commercial 6061 aluminium with 20 vol% silicon nitride whisker reinforcement, prepared by a powder metallurgy route. The composite was extruded and given a standard heat treatment-ageing cycle (T6). Specimens rods of 3 mm diameter and 10 mm gauge length were then tested at strain rates between 0.16 and 0.83 s^{-1}, and temperatures between 515 and 565°C. An elongation of over 200% was recorded at the $1.6 \times 10^{-1} s^{-1}$ strain rate at 525°C which represented optimum conditions. At this temperature the alloy is below its solidus line and voiding would therefore not be a problem.

6.7.7 LOW PRESSURE PRODUCTION OF METAL MATRIX COMPOSITES

A process for the production of short fibre reinforced metal matrix composites that eliminates the need for high pressures is claimed by Toyota Jidosha Kabushiki Kaisha of Aichi-ken, Japan in *European Patent Application* (EP 0-340-957 A2).

The production of short fibre reinforced metal matrix composites usually involves some form of squeeze casting operation in which high pressures are required to infiltrate a molten metal into a fibre preform. This process is costly, requiring expensive plant, and may result in attrition in the fibre preform.

Various methods have been tried to ease the infiltration of metal into a preform, including alloy additions to the base metal in order to increase the affinity between fibres and metal, heating the fibre preforms and, in the case of carbon fibre reinforcement, pretreating the fibres with a chemical agent containing fluorine. All of these modifications to the process, though successful, are of limited applicability and usually they incur a cost penalty. The basic need for high pressures remains and in addition the process usually involves a large component of wasted metal which must be removed at the junction between the preform and the molten metal reservoir.

The Toyota process involves producing a preform that contains not only some combination of reinforcement (fibres, whiskers and/or particles) but also a third constituent metal or metal oxide, chosen so as to improve the affinity of the molten metal and the preform. For aluminium metal and alloys the third constituent may be selected from a wide group of metals, including Ni, Fe, Co, Cr, Mn, Cu, Ag, Si, Mg, Zn, Sn, Ti. For magnesium metal and alloys the list includes Ni, Cr, Ag, Al, Zn, Sn, and Pb. Oxides of W, Mo, Pb, Bi, V, Cu, Ni, Co, Sn, Mn, B, Cr, Mg, and Al may also be used with both metal systems.

Apparently, a relatively large volumetric fraction of the preform must consist of the additional metal (in powder or fibre form) typically between 5 and 80% of the preform total, and, therefore, the resultant properties of the composite will not be those of a simple composite based on the base metal matrix alone. However, Toyota claims that infiltration of molten metal into such preforms occurs so readily that there is no need for any pressure to be applied. A composite can be produced by simply dipping the surface of the preform into a bath of molten metal.

A composite can therefore be made roughly to net-shape without high pressures from an accurate preform, and without excessive solidified metal, apart from the reinforcement.

Toyota notes that, although not necessary, the process is further improved by heating the preform (though not to levels necessary for conventional preforms) and by the addition of certain alloying additions to the base matrix metal (e.g. more than 0.5% of Mg in an Al alloy).

6.7.8 LASER-PRODUCED METAL MATRIX COMPOSITES

Laser consolidation of preformed wire to form sheets of unidirectional metal matrix composites is discussed in an article *SAMPE Quarterly*, July 1990.

The article reports on work performed in Japan by the research team of Mitsuhiro Okumura, Shoji Murakami, Shin Utsunomiya, Takeshi Morita and Seigo Hiramoto of Mitsubishi Electric Corp. They took silicon carbide ('Nicalon') or carbon (unspecified fibre from Torayca) fibres preformed into a metal matrix composite wire by liquid infiltration of molten aluminium and consolidated these into sheet using laser irradiation and rolling.

According to the report, the process was successful in producing good quality metal matrix composite sheet, but, in the case of the carbon fibre metal matrix composite, strength reductions were observed if the power of the laser exceeded 600 W, due to excess heating and the formation of aluminium carbides at the interface. The technique is best suited to aluminium/silicon carbide composites.

6.7.9 SILICON CARBIDE–SILICON CARBIDE COMPOSITES BY FORCED CHEMICAL VAPOUR INFILTRATION

Composites consisting of silicon carbide matrices reinforced with continuous ceramic fibres, whiskers or platelets have been fabricated at the Oak Ridge National Laboratory in the USA, using a forced chemical vapour infiltration technique. The researchers say that this thermal and pressure gradient technique can produce thick composites of relatively simple shape.

The processing of composites reinforced with Nicalon (silicon carbide) fibre by forced chemical vapour infiltration has been optimized at the Laboratory after many years of study. The material is reported to exhibit useful flexural strength and high fracture toughness. Models have been developed which improve the understanding of the infiltration process and will be used to control the process more thoroughly.

Using the improved understanding developed during the modelling process, the researchers produced a number of composites from different reinforcements, including silicon carbide and mullite whiskers, and silicon carbide platelets. Composites with continuous fibres were also made using alumina cloth, and Nextel and Tyranno fibres. Preliminary assessments of their mechanical properties are said to indicate lower strengths and less fracture toughness than composites reinforced with Nicalon fibres.

6.7.10 IMPREGNATION OF CARBON–CARBON WITH CERAMICS

The development of a method for coating and impregnating carbon-carbon composites with ceramics is detailed in a report on work performed at the Israeli Institute of Metals. The objective is to make the materials suitable for use in oxidizing atmospheres at very high temperatures.

Electrophoresis, a technique applied extensively for biological systems, was explored as a method of displacing ceramic particles and depositing them on carbon–carbon substrates. The scope of the work was later extended to include electro-reduction as a depositing technique. This latter method involved synthesising ceramic oxides from an ionic aqueous solution following a chemical reaction.

The electrophoretic deposition was studied using colloidal and fused silica, silicon carbide, and silicon nitride. These ceramics acquire a charge and may be deposited under an electric field. The induction of particles into porous substrates was also demonstrated during the programme.

The 84 page report details the effects of deposition voltage, solvent properties, and particle concentration on the penetration of silica into the carbon–carbon composite and the deposition by electrochemical reduction of cerium dioxide, zirconium dioxide and alumina.

6.8 TOOLING

6.8.1 COMPOSITE TOOLING

A meeting on composite tooling, sponsored by *Advanced Composites Bulletin* was held in Amsterdam, The Netherlands, in November 1990. Many useful pointers to current practices, problems and future developments in the field were provided by speakers representing industry on both sides of the Atlantic.

Tooling Materials for Thermosetting Composites

A number of presentations were given by companies involved in low temperature curing tooling prepregs, including Advanced Composites Group, Cyanamid, Advanced Polymer Industries and Fiberite. These revealed that great improvements have been made in the critical areas of prepreg out-times and ultimate-use temperatures. A typical example is Cyform 22 from Cyanamid which can be cured at any temperature between 20 and 55°C, with a three day out-time at 20°C and an ultimate use temperature of 185°C after a full post cure.

Most suppliers offer a range of systems and cure temperatures, with curing at 60°C favoured for a combination of long out-times and low temperature curing. Some problems with this temperature range were highlighted by Chris Belk of Cyanamid. Sealing materials are used with all low temperature low cost master making materials, including acid-cured furane, urea formaldehydes, etc. and can cause problems at cure temperatures over 45°C by reacting with the prepreg resins. The volatiles emitted result in surface porosity and inhibition of the prepreg cure. If problems are observed, such as the tool face surface changing or neutralizing the master, sealant should be considered.

An interesting development is the move towards twill weave fabrics by many of the tooling suppliers. Advanced Composites Group (for their LTM prepreg range) and Cyanamid use twill weaves as they simplify the lay-up process via a through thickness and warp to weft symmetry (thereby eliminating the need to be continually aware during tool lay-up of the need to turn and rotate fabric layers in order to achieve a balanced and distortion free tool) while maintaining good drapeability and minimum fibre kinking.

Tool Design

Some general ideas regarding the philosophy of tool design for composites emerged from a number of contributions, notably those of Wilf Bishop (Aero Consultants) and Bill Lajoy (Airtech International), primarily aimed at autoclave manufacture. Whatever the material selected for the tool face, the structure is generally supported by some form of egg box construction to both retain its shape during heating and cooling and also to allow transport into and out of the autoclave. The design and construction of these support structures is as critical to the successful operation of the tool as the correct choice of tool face material, but according to Bishop and Lajoy some fundamental design faults have been incorporated into most traditional structures. For accurate control of the cure of a part, the part should be heated uniformly and from the tool face rather than directly from the hot gas stream. The 'old technology' egg box construction restricts the flow of hot gas through the tool structure and severely impairs the ability of the tool face to heat-up uniformly. This may result in the top surface of the prepreg stack curing first, due to hot gas heating.

The remedy proposed by Airtech International involves producing the support structure itself from tubular framework composite sections. The tool supports are lighter, and are less likely to impose thermal stresses on the tool face, simplifying attachment and resulting in cost reductions. Figures quoted by Bishop and Lajoy for a tool produced for CASA in Spain for the manufacture of undercarriage doors for the A320 Airbus (toolface area 6.9 m^2) indicate a reduction in cost by adopting a composite approach to the tool of over 58% compared to a traditional method, equal to US\$80 000/tool, with an accompanying reduction in weight of 62% (337 kg). The influence of backing structure on the heat throughput from the tool face was confirmed by Barry Wainwright (Brookhouse Patterns) who used thermograms to demonstrate uneven heating in tools with poorly designed backing structures.

Bill Lajoy provided some final tips regarding tool construction, including taking steps to ensure that a tool is not written off by the failure of a few surface plies. The tool face is generally produced as a monolithic structure. Lack of vacuum integrity or damage to the surface may result in the entire item being scrapped at great expense. Prior consideration of the pitfalls could lead to slight modifications in production that eliminate some of these difficulties. Why not, for example, produce the tool in such a way that the bulk of the tool does not need to be vacuum tight, instead, the final surface layers are produced almost as a slip sheet on the main part of the tool, with bagging materials in between.

High Temperature Tooling

At present there is probably only one freely available tooling material for complex parts made from high temperature curing polyimides and thermoplastic composites — monolithic graphite. This is an excellent material for tooling for carbon fibre composites in that the coefficient of thermal expansion is closely matched to that of the composites. Peter Roupp (Great Lakes Carbon, USA) pointed out that monolithic graphite is a readily machinable material that can be used within integrated CAD/CAM operations, eliminating the need for master materials. The difficulties encountered with the material come from the limitations on block size, which may require butt-joining of blocks using a proprietary adhesive to make large tools, together with porosity and lack of vacuum integrity, which may be overcome by tool design. Cost is a further factor to take into consideration. Successful thermoplastic parts have already been made with monolithic graphite tooling for the ATF fighter programme in the USA, the V-22 tilt rotor aircraft and large quantities of parts for the Boeing Condor pilotless aircraft project which uses 80% thermoplastic composites in the airframe. The tools are soft and great care and respect is required from operators during use in the moulding shop.

While monolithic graphite is a good option there are other materials on the horizon. All of the reasons for using composite tooling are as valid for high temperature cure as for lower temperatures so it is not surprising that a high temperature composite tooling system is now being developed. Du Pont has introduced a composite system, Avimid N/Graphite, which is based on polyamide resins. Jurgen Krey, responsible for the European development of this product, described the properties of Avimid N which appears to be very promising, being ideal for polyimides but probably not quite up to dealing with thermoplastics just yet.

A class of materials that is definitely about to assume a strong market position is that of the chemically bonded ceramics. These are slurries of ceramics where the bonding is chemical in nature — no sintering or other thermally assisted diffusion processes are required to form a monolithic structure. Sean Wise (Occidental Chemical) described progress with such systems. Chemically bonded ceramics may be used for both high and low temperature processing, indeed the material looks promising at all curing ranges. The chemically bonded ceramics are cast initially at room temperature; gel formation takes about 24 hours and then the tool may be demoulded from the master. Subsequent air drying is performed slightly above the expected use temperature. The Comtek materials from Occidental for use at temperatures below 200°C (Comtek 66, 100) are said to be excellent, giving a superb surface finish, according to Wise, and featuring good matching of coefficient of thermal expansions and low thermal mass, together with rapid heat-up cool-down rates. The grades manufactured so far for use with thermoplastics (Comtek XS) still require improvement as the company is still not satisfied with both the shrinkage of the part on forming and the coefficient of thermal expansion. Vacuum integrity does not seem to be a problem to date.

If chemically bonded ceramics are on their way in then coming along behind would appear to be reinforced chemically bonded ceramics. This very interesting class of tooling prepregs was discussed by Dale Woolum (Comtool Technology). The Ceracom carbon silica tooling system developed by Comtool consists of a prepreg of carbon fibres (or Nextel ceramic fibres) in a slurry of silica. According to Woolum, a lightweight, ultra high temperature service tool may be constructed from these materials (capable of use far above 1000°C). The unique feature is that the tools may be built with integral heating elements at the mid point of the tool thickness. The ceramic nature of the prepreg allows temperatures to be sustained within the tool that are higher locally than those required at the tool face. A further advantage of this system according to Woolum is that it is possible to process thermoplastics and achieve acceptable consolidation levels using vacuum bagging alone, with no autoclave. The heating is controlled and supplied uniformly at the tool face where initial melting occurs. The use of an insulation mat on the top of the prepreg stack increases the control of the heating process and further allows the use of low temperature nylon bagging materials with polyetheretherketone (PEEK) type composites. This system is initially cured at 120–176°C and is claimed to have excellent thermal shock resistance, a coefficient of thermal expansion of 0.4 x 10^{-6}/°C and to be capable of producing some 200 parts out of one tool.

6.8.2 ALUMINIUM–BRONZE TOOLING

The potential of cast to size aluminium-bronze alloys as tooling materials for advanced composites is considered in an article in Aerospace Composites and Materials (Vol. 1, No. 6).

Advantages of aluminium-bronze include durability - the material is hard and will withstand prepregs being trimmed on the tool, — together with a high thermal conductivity and good mechanical properties at elevated temperatures, making it suitable for thermoplastics.

However, the coefficient of thermal expansion of the material is relatively high, greater in fact than steel, which can give rise to problems with carbon fibre reinforced plastic materials for many part shapes. There is less of a problem with glass reinforced plastics where the coefficient of thermal expansion of the composite is higher and similar to steel.

Much of the development work on aluminium-bronze tooling has been undertaken in the USA where the International Copper Association has been involved with Sikorsky. Tools have been produced for the tail rotor shaft cover of the Sikorsky Blackhawk helicopter while other parts produced out of aluminium-bronze tools include items for the B-2 stealth bomber and snow deflector shields for helicopters.

6.8.3 HIGH TEMPERATURE PATTERN AND MODELLING BLOCK

The Advanced Composites Group of Heanor, UK, has announced a new tooling block material, ACG TB 650, which has been designed for use as a model and pattern for the production of prepreg composites tools and can also double as a mould material for vacuum forming operations.

The company claims that the block has excellent machineability, dimensional stability, low coefficient of thermal expansion and chemical compatability with tooling prepregs, offering benefits in the production of prepreg tools directly from the tooling block by vacuum forming or autoclave moulding.

The material is currently available in sizes up to 1000 x 1000x 125 mm and a complementary range of fillers and sealers is also offered.

6.8.4 SEMIPERMANENT MOULD RELEASE AGENT

A semipermanent multiple use mould release agent has been developed by Chem-Trend Inc of Howell, Michigan, USA. The company claims that Mono-Coat E179 has high temperature stability, high abrasion resistance and minimal transfer to parts. It is suitable for all types of composite parts and metallic or plastic-based tools. According to a report in *Plastics Engineering* the release agent is a clear solvent-based fluid (viscosity less than 10 cp at 21˚C) that can be applied by conventional wiping or spraying techniques to hot or cold tools and cured at room temperature.

6.8.5 MASTER MODELLING MATERIAL

Ciba Geigy is now in full scale production of its new master modelling material, Cibatool BM 5126. This new system is claimed by the company to be capable of being used at higher temperatures than any other carvable block medium and can apparently withstand autoclave conditions of 125˚C and 100 psi.

For use in autoclave manufacture, the material must be completely non-porous and vacuum tight and capable of consistently producing a high quality surface. According to *Resin Aspects*, carbon fibre reinforced tools can now be produced directly from the model and for short production runs, Cibatool BM 5126 can even be used as a tool making material in its own right.

The material has a hardness similar to mahogany and a coefficient of thermal expansion similar to aluminium. The density of the material is only 40% of aluminium.

6.8.6 NOVEL MANDREL CONSTRUCTIONS

Hercules Inc of Wilmington, Delaware, USA, has devised a new mandrel construction for use in filament winding or other fibre placement processes for thermosetting resin composites in which the mandrel needs to be disassembled after cure of the part. (European Patent EP 0 394 934 A1).

The production of hollow section components via a winding or placement process gives rise to problems in that rigid mandrels are required for support during cure under autoclave pressures but these need to be dismantled and removed after final cure. In addition, the mandrel is subject to all the normal consideration for tooling materials for composites and attention must be paid to the coefficients of thermal expansion of the mandrel relative to the composite, providing an additional constrain on mandrel design.

The construction developed by Hercules consists of an assembly of appropriately contoured quarter sections of the mandrel, produced from carbon fibre tooling prepregs mounted on T-stiffeners, which are secured to a bulkhead mounted on a central steel shaft. The seams of the sectional construction are filled with a putty. The construction provides compatibility of the tool with the coefficient of thermal expansion of carbon fibre composites and also eliminates leakage paths between the sectional elements of the mandrel.

6.9 ASSEMBLY

6.9.1 BONDING AUTOMOTIVE COMPOSITES

A novel method for assembly of automotive composite structures using adhesive bonding has been developed by Budd Co of Troy, Michigan, USA.

Composite materials, sheet moulding compound in particular may be used for automotive body panels in place of sheet steel. However, the stiffness required from the part frequently necessitates the use of strengthening members which are fixed onto the inside of the panel. Ideally such stiffening members are also composite parts and the fixing method is adhesive bonding. The difficulties arise because of the need to use adhesives that cure well above room temperature. Local heating of the adhesive using radiant heating is generally too slow and may result in non-uniform bonds, while excessive heat generated by the curing adhesive may cause burning. The technique developed by Budd (European Patent Application EP 0 339 493 A2 and 0 339 494 A2) was dielectric heating. A high frequency signal is applied to electrodes mounted either side of an assembly to be bonded. This causes the adhesive to heat up and cure evenly. The method does not lead to overheating and consequent surface imperfections in the parts. The reasons for this are not totally understood but the change in dielectric properties of the adhesive on curing is cited as a possible explanation. Presumably the dielectric properties change such that the adhesive is no longer heated to the same extent by the applied signal.

Budd found that the method works best with a two part epoxy adhesive, a high frequency signal between 10 and 110 MHz and a voltage between electrodes of 300–8000 V RMS. The signal should not be applied for more than two minutes.

Budd says that it regards this technique as especially applicable to the production of composite hoods (bonnets), but doors and bumper assemblies may also be produced in the same way.

6.10 COATINGS

6.10.1 OXIDATION PROTECTION FOR CARBON–CARBON COMPOSITES

The need to protect carbon-carbon composites against oxidation during high temperature service is prompting considerable attention worldwide.

The accepted solution is to provide a ceramic coating on the surface of the carbon–carbon material but this is not as technologically straightforward as it sounds. Mismatches between the thermal coefficients of expansion of the ceramic coatings and the carbon–carbon composite give rise to delamination or crack formation at the interface between the two material types, leading to breakdown and potential failure.

Silicon carbide appears to be the favoured ceramic material for coating carbon–carbon. A recent paper by Stevan Dimitrijevic of the Boris Kedric Institut, Belgrade, presented at the Second Yugoslavinan Conference on Carbon Materials and published in *Hemijska Industrija* (Vol. 43, No. 10, October 1989), describes how chemical vapour deposition of silicon carbide onto a carbon–carbon composite resulted in a low weight loss in the materials after high temperature exposure (5.7% weight loss after three hours exposure to air flow at 1000°C). These properties were attributed to better crystal ordering and the reduction in microcracking as a result of increased chemical vapour deposition temperatures.

An alternative approach to improving the effectiveness of ceramic coating on carbon–carbon is the subject of a European Patent Application (EP 0 336 648 A2) from the Japanese National Aerospace Laboratory and Nippon Oil of Tokyo, Japan.

The Japanese process involves a combination of chemical vapour infiltration to fill the voids in a porous carbon–carbon composite, followed by chemical vapour deposition of a ceramic on the surface. The key feature is that the conditions are changed progressively after filling of the voids such that a gradient in the materials is effectively introduced.

An example of the process is given whereby a carbon–carbon composite is produced by relatively standard means, with a volume fraction of fibres of 50%, 30% of carbonaceous material and 20% voids. The composite was placed in a furnace and subjected to thermal chemical vapour infiltration with an initial gas composition of C_3H_8 which was progressively changed to C_3H_8, $SiCl_4$ and H_2 at 1150°C and 5 torr pressure to fill the voids. Subsequently, the temperature was increased to 1500°C and, the pressure to 300 torr to coat the surface. Studies of the filled and coated material revealed a gradual change in the material filling the voids from carbon to carbon/silicon carbide, while the surface was silicon carbide. A continuous change in texture was observed between the chemical vapour infiltration and chemical vapour desposition deposits, with no delaminations present.

The ceramics that may be used for coating the carbon–carbon composite by this route include silicon carbide, zirconium carbide, titanium carbide, hafnium carbide, boron carbide, niobium carbide, titanium diboride, or silicon nitride. The chemical vapour infiltration stage is best performed at temperatures from 1000 to 1500°C and at low pressures of 0.1 to 50 torr while the chemical vapour deposition stage is best performed between 1000 and 2000°C at pressures in the region 50 to 760 torr.

7. NON-DESTRUCTIVE EVALUATION

7.1 TECHNIQUES

7.1.1 MICROSCOPIC TECHNIQUES

The difficulties inherent in studying the fibre matrix interface in polymer composites may be eased by the advent of Confocal Scanning Optical Microscopy (CSOM).

This technique, which uses a laser to scan a specimen, only collects information from within the depth of focus, which is of the order of 0.8 µm. By scanning the beam across a specimen and storing the data, it is possible to construct an optical image of the material at a specific depth. Refocusing at different depths allows the material to be progressively examined without a series of intricate sectioning, polishing and etching stages, which would run the risk of destroying the structure, as well as being time consuming and difficult.

Preliminary results using this technique by Thomason and Knoester of Shell Research are presented in a paper, 'Application of confocal scanning optical microscopy to the study of fibre reinforced polymer composites, published in *Journal of Materials Science Letters* (Vol. 9, March 1990).

The authors describe their results from the study of Twaron Aramid fibre reinforced polypropylene, where transcrystallinity is highlighted, and glass fibre reinforced epoxy resins. They conclude that the technique holds great promise for interfacial studies, although more study is required to allow a greater interpretation of the images obtained.

7.1.2 X-RAY TECHNIQUE MEASURES CRACK GROWTH IN CERAMIC COMPOSITES

A study conducted as part of the European Community's BRITE programme by workers in France has concluded that X-radiography is a valid technique for measuring the real length of cracks developing in continuous fibre ceramic matrix composites (CMCs).

The study of fracture and fracture toughness in ceramic matrix composites has led to many theories regarding the nature of the toughening process which, in the case of continuous fibre composites, is greatly influenced by shielding effects resulting from fibre bridging of the crack tip.

Crack bridging has the effect of introducing the concept of an effective crack length and this may be measured relatively easily in practice by compliance measurements. The actual crack length, and therefore its relationship to the effective length, is not easy to directly measure.

The technique suggested in 'Oberservation of crack path in an SiC-SiC fibre composite by X-radiography and SEM', G. Navarre, J.C. Rouais and D. Rouby, *Journal of Materials Science Letters*, June 1990, involves the use of X-radiography coupled with an opacifant of zinc iodide. The zinc iodide diffuses thorough the material to the crack surfaces which are then revealed in the radiograph. According to the authors, a subsequent scanning electron microscopy study of fracture surfaces confirmed that the measurements of crack length taken from the X-radiograph corresponded to the true crack length which in most cases was about twice that of the measured effective crack length.

7.1.3 RADIOMETRIC ANALYSIS FOR QUALITY ASSURANCE

The demand for increasingly efficient plastics has led to the development of a great number of reinforced compounds and processes to produce them. In order to meet the specifications for the various product types, it is essential to control and monitor the concentration of the reinforcement with a high degree of accuracy. A paper entitled 'Radiometric analysis assures the quality of fibre-reinforced plastics' by D Mettlen, in *Kunststoffe*, Vol. 80, 1990, No. 2, pp. 190-192, reports that radiometric measuring methods provide the required data with high accuracy and speed. Depending on the measuring equipment used, additives that may also be incorporated can be analysed at the same time.

7.2 EQUIPMENT

7.2.1 FLAW IMAGER

A UK manufacturer of ultrasound diagnostic and non-destructive testing equipment, Diagnostic Sonar, has developed a product, the Flaw Imager, for use with composite structures and components. *Plastics and Rubber Today* says the resolution of the instrument, which can function with carbon and glass fibre composites, is 1 mm. Flaws can be detected to a depth of 25 mm in carbon fibre laminates. The apparatus displays a real time B-scan with a simultaneous A-scan projection.

7.2.2 IN-SERVICE OPTICAL INSPECTION METHOD

Detection of impact damage in-service components is as important as understanding the nature of impact damage itself, and a paper published in issue of *Composites* (Vol. 21, March 1990) presents some welcome results.

A team from the National Aeronautical Establishment in Canada has examined the potential of the optical surface imaging systems developed by Diffracto Ltd for detecting barely visible impact damage.

The Diffracto D-sight apparatus was developed for studying the surface quality of parts in real time and can detect depressions or distortions as small as 10 microns. Using computer based image processing, the Canadian team has been able to compare current images to previously stored images of a given area before and after impact. D-sight results have also been compared to C-scan images of areas with known impact damage. There is strong evidence that the D-sight equipment can detect damaged areas considered to be significant (e.g. according to US Air Force damage tolerance design requirements).

7.2.3 TRANSPORTABLE VERSION OF LORIA SURFACE ANALYSER

A transportable version of the Loria surface analyser has been announced by Ashland Chemical of Columbus, Ohio, USA.

The Loria system uses a scanning laser beam directed at a low angle to analyse surface characteristics such as long term waviness, bond-line read through, orange peel and distinctness of image. Ashland reports that the system is proving popular with the manufacturers of composites aimed at the automotive industry where surface quality is a critical factor. It claims that the Loria system can provide a quantitative assessment of surface properties (the Ashland Index number), allowing a 'Class A' finish to be obtained through systematic product development.

The transportable version of Loria has been developed at the request of customers who were looking for a unit that could be used for both production and laboratory environments and moved between the two, Ashland says. The analytical capabilities of the transportable system are identical to that of the original unit designed for laboratory use.

7.2.4 ANALYSER CAN MEASURE FILLER CONTENT

A range of process control and instrumentation technology has been launched in the UK by Schwing of Warley.

According to the company, its Compuglass Analyser is of particular interest. It is designed to provide a fast, accurate, non-destructive method for determining the glass or filler content in a composite. The unit, which may be used as a separate facility or as part of an on-line process control system, provides results in a few seconds with an accuracy of 0.2%, Schwing claims. The computer based system may be integrated with existing statistical process control or quality systems.

7.2.5 LASER MODELLING SYSTEM

Quadrax Laser Technologies Inc, a subsidiary of Quadrax Corp of Portsmouth, Rhode Island, USA, is producing laser modelling systems that can fabricate highly accurate models directly from computer aided designs in a matter of hours.

According to Quadrax, the Mark 1000 Laser Modelling System uses a low viscosity light – cured resin and a high powered 5 W argon ion laser to fabricate prototype parts at five times the rate of systems based on UV laser light technology, and at tolerances of 0.13 mm.

The system builds up a model of a part layer-by-layer, by scanning the surface of a light curable resin. The scan pattern of the laser light is controlled by the computer aided design of the parts. The use of the high powered laser ensures good curing and, therefore, a high green strength of the prototype part produced via this process. The diameter of the laser beam can be controlled from 0.089–3.18 mm. The maximum size of model that may be constructed in one piece is 30.5 x 30.5 x 30.5 cm. Larger models could however be built up from sub-units and bonded together.

The system is now commercially available.

8. TESTING AND STANDARDS

8.1 STANDARDS AND RELATED ACTIVITIES

8.1.1 ASTM STUDY GROUP TO HARMONIZE INTERNATIONAL STANDARDS

A new subcommittee of ASTM's D-30 High Modulus Fibers and their Composites Committee has been formed to pursue the international harmonisation of standards in the area. The new group, D30.02, will begin by reviewing current standards to detemine where non-compatabilities exist between different standards. The non-compatabilities may occur in relatively minor aspects such as terminology, specimen dimensions and conditioning, or in major respects such as the types of test used to measure a property, the data reported and the materials used.

Attempts will be made to accomodate minor differences by adjustments to ASTM standards where possible but major differences will require dialogue between key personnel on an international basis. The committee is therefore looking for volunteer members, particulary ASTM members overseas with contacts and affiliations with standards bodies in their own country. The Committee will, in addition, seek to encourage ASTM round-robin activities to become more international in their scope.

8.1.2 CODES OF PRACTICE FOR CONSTRUCTION INDUSTRY

A consortium of mainly UK companies has assembled to bid for funding from the European Community's EUREKA project in order to establish codes of practice governing the use of composites for the construction industry. Emphasis will be placed on pultruded glass fibre composites.

The consortium believes the potential market for structural pultruded sections is enormous, but is hampered by the absence of suitable design codes.

The existing members — Halcrow Polymerics, Fiberforce, Dow Chemical, Vosper Thorneycroft and Tech Textiles — have defined the project but will continue to look for further partners from both the UK and Europe.

8.1.3 DESIGN AND INSTALLATION OF COMPOSITE TANKS AND TUBULARS

A data manual which contains a section on 'European Recommended Practice' for the design and installation of composite fuel tanks and tubulars has been produced by the European Promotion Association for Composite Fuel Tanks and Tubulars (EPACT), based in Harderwijk, The Netherlands.

The manual lists all relevant International, European and ASTM standards, together with information on its member companies, EPACT says.

EPACT is trying to increase the size and influence of its organization which currently has 23 member companies and 5 associated member companies. Over half of those involved are fabricators, with certified installers and raw material suppliers making up the balance. One of the aims of EPACT is to establish a quality standard within the industry and create a common European Community standard.

In addition to the data manual, the organization holds an annual meeting (the last in the series was held in Maastricht, The Netherlands, on the subject of environmentally safe service stations) and training courses to introduce composites to potential users.

The data manual costs Dfl200 and the proceedings of previous EPACT conferences cost Dfl300.

8.1.4 BRITISH COMPOSITES SOCIETY INITIATIVES ON TESTING

The British Composites Society has identified the area of testing and standards for a number of special initiatives. These include a European meeting on the subject to be held in September 1992, probably in Amsterdam, and a series of workshop meetings addressing individual testing methods.

Both activities are aimed at providing, in different ways, easily accessible forums for the presentation of data and general debate on the subject of test methods for composites. The Society recognises that there is a need in Europe for activities similar to those promoted by ASTM in the USA. The standards bodies in Europe are generally not pro-active in stimulating research into testing in the same way as ASTM, and in particular ASTM's D-30 committee on high modulus fibres and their composites.

The major conference is to be run by the BCS together with the European Association for Composites Materials and other bodies such as the National Physical Laboratory in the UK are also to be associated with the meeting. The meeting will take place in 1992. The workshops are envisaged as a series of single day meetings, probably three a year, which will take place as part of the Society's general workshop programme. The testing workshops will also involve direct participation from the National Physical Laboratory where most meetings will be held. Dr Graham Sims of the National Physical Laboratory is to coordinate the testing workshop programme.

It is intended that the workshops will provide the opportunity for a UK National Forum on Testing to evolve - a tentative title for this activity being 'Forum for Advanced Composites Evaluation, Testing and Standards (FACETS)'.

The first meeting in this series, which attracted over 50 delegates, was held on October 18th 1990 and concentrated on Compression Testing. The second meeting, in February 1991, featured shear testing, while meetings on fracture toughness testing, metal matrix/ceramic matrix composites testing and high temperature testing are scheduled to take place.

8.1.5 TEST PROCEDURES ESTABLISHED BY AUTOMOTIVE COMPOSITES CONSORTIUM

The Automotive Composites Consortium in the USA has established a test procedure manual to standardize data collection for the research of automotive structural composite materials. The manual gives specific directions to the supplier community by defining the data needed for initial design of composite materials. The intent is that the manual serves as a standard for all automotive structural polymer composites testing, according to Bernie Swanson, who is the current chairman of the Automotive Composites Consortium (which is a pre-competitive research initiative of Ford, General Motors and Chrysler).

A set of standard tests is specified for the characterization of flat sheets of composites to be evaluated for use in structural applications

The manual assumes liquid composite moulding, is the manufacturing route for panels and, accordingly, some items are specific to liquid composite moulding materials. However, the intention is that the test procedures will become adopted for all composites, such as structural sheet moulding compound (SMC) and stampable thermoplastics, with additional items being added to allow for, for example, flow induced orientation in sheet moulding compounds and crystallinity in thermoplastics.

The test sample plates must be 610 mm square and 3.2 mm thick. A detailed cutting plan for individual test coupons is specified. The range of procedures includes tensile (ASTM D638), compression (ASTM D3410), and shear (Iosipescu) tests, with additional methods specified for assessing volume fraction (ASTM D2584), specific gravity and density (ASTM D792), linear coefficient of thermal expansion (ASTM D696), and dynamic mechanical properties (ASTM D4065). A total of six plates must be tested and a specific reporting procedure is included.

It is significant that ACC has adopted the Iosipescu double opposing V-notch test for measuring shear, with the specimen dimensions and details following the recommendations of the University of Wyoming. This test is not a standard test as yet, but is likely to be adopted by ASTM in the near future.

The most surprising omission is the lack of an impact test in the overall plan, given the automotive industry's interest in this property. However, ACC is addressing this issue along with other down-stream properties, such as durability, creep and fatigue. Recommendations will appear in future editions.

8.1.6 TEST METHODS FOR CARBON FIBRES

SACMA (The US-based Suppliers of Advanced Composite Materials Association) has issued a new set of five SACMA Recommended Methods (SRM) for testing carbon fibres. These include:

- SRM 13-90 Mass per Unit Length of Carbon Fibres;

- SRM 14-90 Sizing Content of Carbon Fibres;

- SRM 15-90 Density of Carbon Fibres;

- SRM 16-90 Tow Tensile Strength of Carbon Fibres;

- SRM 17-90 Twist in Carbon Fibres.

- SACMA is not a standards issuing body as such but, as it represents the composites raw materials industry, its recomended methods are tantamount to *de-facto* standards. SACMA is now collaborating with ASTM and SAE in the USA to assist in further evaluation of the methods.

8.1.7 STANDARD FOR MOISTURE EQUILIBRIUM OF COMPOSITES TO BE DEVELOPED

A task group to develop a uniform equilibrium standard for materials affected by moisture, with emphasis on structural polymeric composites, has been set up by the ASTM.

Task group chairman Richard Fields of Martin Marietta says, according to *Standardization News,* that the group is preparing for a round robin programme for a proposed test method which would include procedures for:

The determination of moisture absorption/desorption properties; and, the conditioning of test specimens prior to use in other materials test methods.

8.1.8 ASTM COMMITTEE FOR DATA FORMATS

The ASTM committee E-49 on computerisation of material property data has a new sub-committee, E-49.08, which is concerned with Data Recording formats for Non-metals.

The activities of this new subcommittee will be coordinated with the relevant materials committees including D-30, High Modulus Fibres and their Composites and D-20, Plastics.

8.1.9 REVISED UK STANDARD FOR UNSATURATED POLYESTER RESINS

The British Standards Institute of Milton Keynes, UK, has published a revised standard 'BS 3532: 1990: Method of specifying unsaturated polyester resin systems'.

This revised standard replaces the earlier 'BS 3532: 1962' and identifies four mandatory properties, but allows optional additional requirements to specify the materials according to their intended applications.

8.1.10 UK STANDARD FOR REINFORCED POLYTETRAFLUOROETHYLENE

The BSi (British Standards Institute) has published a new standard 'BS 6564 Polytetrafluoroethylene (PTFE) materials and products. Part 3:1990 specification for E-glass fibre filled polytetrafluoroethylene'. This standard specifies the requirements for glass-filled polytetrafluoroethylene with three levels of filler and two levels of performance. No standard is superseded.

8.1.11 SPECIFICATIONS AND STANDARDS FOR PLASTICS AND COMPOSITES

What promises to be a useful reference volume has been published by ASM International.

'Specifications and Standards for Plastics and Composites' is a 224 page book written by Frank T. Traceski, from the Organisation of the Secretary of Defense in the USA.

The book lists more than 2000 standards from organizations worldwide.

The standards and specifications are listed by number and title for quick reference. The scope and terms of reference of each standards developing body is included along with addresses, telephone numbers and fax numbers for obtaining hard copies of standards or further enquiries.

The book costs US$85 (ASM Members US$68).

8.2 TEST DEVELOPMENT

8.2.1 ROUND ROBIN RESULTS ON DELAMINATION TESTING

The first results from a round robin exercise conducted by the European Group on Fracture, aimed at developing standard methods for the determination of delamination resistance of composites, are published in *Composites Science and Technology* ('Glass/nylon 6.6 composites, delamination resistance testing', P. Davies and D.R. Moore, Vol. 38, No.3, 1990).

This particular set of results concerns testing on glass/nylon systems where some difficulties where found in applying a fracture mechanics approach. The study, which involved ten laboratories, had to contend with conditioning problems with the nylon, the semicrystalline nature of the material and different interpretations of the data. Despite this, reasonable agreement was found on measured values of GIC and GIIC. The experience gained in the exercise, and detailed in this paper, has been integrated into a draft protocol for delamination resistance testing which is currently being evaluated.

8.2.2 MIXED MODE DELAMINATION TEST

A test method has been proposed which its developer claims will allow the determination of pure mode I and pure mode II delamination toughness to be obtained from the same specimens, together with data on specified combinations of the two loading modes.

The method, which was devised at NASA's Langley Research Center in the USA, combines elements of the edge notch flexure test for mode II (sliding shear deformation) and double cantilever beam tests used for measuring the mode I (tensile normal stresses). A single lever is used to apply loads to a specimen, with the ratio of the various loading modes controlled by the positioning of the lever. NASA claims that during delamination crack growth in the specimens the ratio of mode I and mode II stresses does not change by more than 5%. Although the simple beam theory equations used had to be modified to take account of elastic interaction between the two arms of the specimen and the shear deformation, the resulting equations agreed closely with finite element results and provide a basis for both selecting GI/GII test ratios and computing the mode I and mode II components of measured delamination toughness.

8.2.3 MEASUREMENT OF INTERLAMINAR TOUGHNESS

The study of the interlaminar toughness of composites has always been a widespread activity in universities, but recently raw material suppliers are regularly quoting GIc and GIIc data as part of their standard sheets. The industrial interest is in part prompted by attempts to link essentially 'materials-based' property values such as toughness (particularly GIIc), to the more 'structure-based' properties, such as compression after impact strength, required for selection and qualification purposes.

Attempting to link toughness data to compression after impact strength is different; not least because the compression after impact test is an arbitrary method which may or may not reveal genuine material differences. These difficulties aside, it is recognized that reported values for GIc and GIIc differ widely for given composites, with much of the discrepancies resulting from both minor variations in specimen preparation, test procedure, and data reduction.

The European Group on Fracture has been mounting a coordinated study on testing for interlaminar toughness with a view to the eventual development of standardized method that would give meaningful data. A paper has emerged from this study in *Composites Science and Technology* (1990), 'Measurement of GIc and GIIc in carbon/epoxy composites', by P. Davies, C. Moulin, H.H. Dausch (Lausanne University) and M. Fisher (Ciba Geigy). The paper presents round robin results obtained from ten European laboratories on GIc and GIIc tests performed using double cantilever beam tests (for GIIc). The focus of this particular paper is the effect of starter crack on the measured properties of initiation and propagation toughness. The starter notch may be varied by changing the thickness of an insert (e.g. polytetrafluoroethylene film) and its length, or by performing, for example, an initial GIc test to grow a crack prior to a GIIc test.

The conclusions of the paper are significant. Mode-1, GIc propagation values were found to be reproducible but dependant on starter film thickness. Initiation values were dependant on starter film thickness for both GIc and GIIc. Mode-1 values are not affected by decreasing the width of specimens below the usual 20 mm, but mode-II values are width dependant. The need to standardize on starter crack configuration as well as overall test geometry is clearly demonstrated.

A second paper concerned with GIc measurements appeared in *Composite Science and Technology*; 'The use of thin DCB specimens for measuring mode-I interlaminar fracture toughness of composite materials' is by G. Caprino, of the University of Naples, Italy. This paper is more concerned with the methods of data reduction and Caprino suggests that errors may arise in common methods of calculating Glc from experimental data. However, the paper states that the use of simple equations, where the input data is directly measured values for critical load and critical deflection, permits an accurate evaluation of fracture toughness, even when large deformations occur. This allows thinner DCB specimens to be used than is normal practice without loading the simplicity of the data reduction scheme.

8.2.4 TABBING EFFECTS ON COMPRESSIVE FAILURES

Compression testing of unidirectional composites remains a controversial area, with the quest for a reliable measurement of the intrinsic compressive strength (if such a property exists) continually thwarted by artifacts introduced by the various test methods.

While there is continuing debate as to the relative merits of the shear loaded tests such as ASTM D3410 (IITRI and Celanese tests), edge loaded tests ASTM D695, and various combined edge and shear loaded variants (such as the Imperial College fixture) there has been less attention paid to the effects of specimen preparation and details such as tabbing.

Odom and Adams of the University of Wyoming, in a paper in Composites (Vol.21, July 1990, pp.289–296) consider the various failure modes identified for a large group of specimens produced with a variety of tab configurations. By careful consideration of the failure loads, mode of failure and tab configuration they suggest a number of possible failure processes.

The tab configurations studied include: steel tabs (with and without tapers); glass fabric epoxy tabs (again with and without tapers); and a number of debonded or partially gripped variants. Failure modes reported include transverse failures, split transverse failures, and branched transverse failures, all of which are considered to have similar initiation events - emanating from the free edge of the material and being representative of tab configurations that provided adequate support.

Where inadequate support was provided by the tabs either a shear failure preceded by noticeable buckling or a brooming failure was observed. The brooming failure was specifically linked to poor gripping of the tabs at one end of the specimen resulting in an unbalanced diffusion of load.

Steel tabs provided the best support for the specimens, but untapered glass fabric epoxy tabs, which represent current standard practice in the industry, were more or less equivalent. However, the tapered glass fabric tabs provided the worst results in all of the shear failures recorded.

The average strength results measured with all tab configurations and produced by all failure modes did not differ radically, with the best average for steel tabs being 1759 MPa and the worst for the tapered glass fabric tabs being 1400 MPa for specimens cut from the same test panel. However, the convincing link established by the authors between failure mode and tab configuration must be taken into consideration when attempts are made to develop models or compressive failure and cross correlate data.

8.2.5 SHEAR TEST METHODS

V-notched beams evaluated by ASTM

The results of an inter-laboratory round robin test programme to evaluated the Iosipescu V-notched beam test for measuring the shear properties of composites are published in a paper in ASTM's *Journal of Composites Technology and Research*.

Amongst the many shear tests available for measuring in-plane shear of composites, the Iosipescu and other related V-notched beam tests have become increasingly popular in recent years. The test method itself has been largely pioneered by workers at Wyoming University in the USA (Adams and Walrath) where the specimen and test fixtures have undergone considerable refinement for over ten years.

The interest in this methods and its widespread use has prompted ASTM to draft a new standard test method to cover the use of such V-notched specimens. As part of the process for issuing a new standard, ASTM routinely organises a round robin test programme in order to assess both the repeatability (within a given laboratory) and reproducability (between various laboratories) of the method.

The round robin in question here involved a total of 9 organisations in the USA, seven of which were industrial concerns and two (Wyoming and Delaware) universities. Tests were performed on carbon and Kevlar reinforced epoxies (in 0° '90° and $0/90^\circ$ configurations) and SMC.

The results indicated that acceptable repeatability and reproducability was obtained for both shear strength and shear modulus measurements made with Iosipescu tests specimens where the fibre had a 0° orientation. The coefficient of variation of results for strength ranged from about 4 to 6% within and between laboratories for these orientations while modulus results showed coefficients of variation between 2.6 to 30%. The tests on specimens orientated at 90° were unsatisfactory (failures were premature and prompted by spurious transverse tensile stresses) and this orientation is not recommended. Specimens with a $0/90$o lay-up were similar to 0° specimens, with a much reduced variation. Sheet moulding compound failures were not of the permissible type for the test method and as such could not be considered. A limited number of tests were also performed on the related asymmetric four-point bend test where acceptable failures were obtained for tabbed sheet moulding compound. Insufficient tests were performed to assess the correspondence between asymmetric four point–bend and Iosipescu tests.

The method should therefore be adopted by ASTM as a standard very soon. It is interesting to note that the method as proscribed in the draft ASTM standard has already been adopted by the Automotive Composites Consortium (Ford, GM and Chrysler) in the USA as part of its recommended test methods package for companies wishing to supply composite materials to the auto industry. Both Ford and GM were involved in the round robin programme.

Three in-plane shear methods evaluated.

A research team in Canada (Lee, Munro and Scott, NRC/University of Ottowa) have evaluated three in-plane shear tests with a view to assessing their value for calculating the modulus of complex lay-up laminates in tension.

The test methods were the Iosipescu test, as proscribed by the University of Wyoming (and, therefore, according to the draft ASTM standard, (see above), the 10° off axis and 45° test, according to ASTM D3518-76. The shear modulus for three materials (all carbon–epoxies) were obtained by taking either secant or tangents to the load deflection curve obtained from the tests. The authors reported little difference in measured shear modulus from all three tests using either method.

The resulting shear modulus data was used in conjunction with previously obtained experimental results for longitudinal and transverse modulus and poissons ratio to calculate, using a laminate analysis code, the tensile modulus of a 48 ply laminate, lay up $[+45^\circ/0^\circ/-45^\circ/90^\circ]6s$. The predicted results agreed very well with measured data for such a laminate, irrespective of the value of shear modulus used. Representative results included a predicted value of 53.8, 55 and 55.4 GPa for an IM6/1806 laminate for the three test methods, against a measured value of 54.3 GPa.

The conclusions here are that the real laminate construction was reasonably insensitive to shear modulus and as such it is of little consequence what shear test is used to generate test data for laminate analysis. Predictions relating to laminates where a greater proportion of the fibre are aligned at 45° to the loading axis are likely to exhibit a greater dependence on the measured value of G12a.

8.3 DATA GENERATION

8.3.1 DESIGN FRAMEWORK FOR BRAIDED COMPOSITES

Owens Corning Fiberglas Corp, Toledo, Ohio, USA, and Drexel University of Philadelphia, Pennsylvania, USA, have collaborated to establish design guidelines and an engineering database of braided fibreglass structural composites.

In order to facilitate the adaption of braided composites for structural applications, Owens Corning funded Drexel to produce an 'Integrated Design Framework' for braided tubular composites. This framework has involved the establishment of experimentally verified design curves relating braiding parameters to braid geometry and the modelling of braided composite properties using a fabric geometry model.

Preliminary test results obtained by Drexel seem to be in good agreement with its model.

8.3.2 COMPUTER AIDED DESIGN OF COMPOSITE STRUCTURES

A computer aided engineering system for designing composite structures is being developed by the National Aerospace Laboratory, Amsterdam, The Netherlands, in cooperation with the Fokker Aircraft Company.

The evaluation of alternative designs in the preliminary design phase of aircraft structures has become increasingly complicated with the greater use of composites. This, in turn, has significantly raised the number of design variables. The researchers say that the system being developed, COCOMAT, tackles this problem, making it easier for the designer to evaluate alternative designs.

COCOMAT supports the design of composite panel structures. These may be flat or cylindrical, stiffened with hat or blade stiffeners, and made of various laminates and composite materials.

The team says that the system has been planned to contain sufficient built-in expertise to help a non-specialist user create and modify composite structures and control special-purpose analysis programs. This should allow the designer to concentrate on designing rather than be faced with data handling and analysis problems. Moreover, the designer should also be relieved of many routine tasks as there is a high degree of automation when changing laminates, structure or analysis parameters, it is stated.

8.3.3 PREDICTING TENSION FATIGUE LIFE OF COMPOSITES

A method for predicting the life of composites under tension — tension fatigue life - has been developed at NASA's Langley Research Center in the USA. The method incorporates both a general fracture mechanics characterization of delamination and an assessment of damage on laminate fatigue life.

The researchers carried out tension fatigue tests on quasi-isotropic and orthotropic glass epoxy, carbon epoxy and glass/carbon epoxy hybrid laminates, and data on the onset of edge delaminations were used to generate plots of energy release rate as a function of cycles to the onset of delamination. These plots were then used, together with strain energy release rate analyses of delaminations starting at matrix cracks, to predict the onset of local delamination. Stiffness loss was measured experimentally to account for the accumulation of matrix cracks and delamination.

Total fatigue failure was subsequently predicted by comparing the increase in global strain resulting from stiffness loss to the decrease in laminate failure strain resulting from delaminations forming at matrix cracks through the laminate thickness.

NASA claims good agreement between measured and predicted lives indicating that the through thickness damage accumulation model can accurately describe fatigue failure for laminates where the delamination onset behaviour in fatigue is well characterized and stiffness loss can be monitored in real time to account for damage growth.

8.3.4 TESTING FOR SINGLE FIBRE PROPERTIES

Some reports have recently become available from US military establishments containing information and data on testing for single fibre properties.

A master's thesis from the US Air Force Institute of Technology at the Wright Patterson Air Force Base, Ohio, looks at compression testing of various fibres (T-50 and P-75S carbon fibres, Kevlars, PBO and polyphenylene benzobisthiazole using the elastica loop test (where a loop of fibre is pulled between microscope slides until failure) and the bending beam test (where a single fibre is attached to the compressive face of a cantilever beam subject to bending). These single fibre tests have to date involved assumptions of linear fibre elasticity to derive strength and strain data and accordingly have been subject to question. The thesis contains details of a Fortran program written for the numerical analysis of non-linear elastic problems such as bending of single fibres with large displacements. This may allow these test methods to become acceptable standard tests in the future.

In addition to obtaining strength data, the compressive failure modes of fibres, which occur as the formation of a kink band in polymeric fibres and as a fracture in carbon fibres, were investigated. It was concluded that for polymeric fibres the critical kink band formation represents the buckling of separated microfibrils due to plastic instability.

The tensile testing of single fibres was examined by the US Army Materials Technology Laboratory who measured the dispersion in the failure loads and diameters of various types of single filament graphite fibre.

A test procedure was developed that allowed the diameter and tensile properties of a given fibre to be measured without handling between the two tests. This enabled workers to determine that no correlation exists between the fibre failure load and the diameter of the fibre. Considerable variability was observed between both the fibre diameter and the failure load, with greater variability being observed in fibre failure load than fibre diameter; this is the subject of further study.

8.3.5 PRIMERS AND ADHESIVE BONDING OF CARBON REINFORCED COMPOSITES

Priming adhesive bonded joints between carbon fibre reinforced composites results in no effect on bond strength both after initial joint formation and after long term exposure to hot humid conditions, according to researchers at the UK's Royal Aircraft Establishment (RAE).

The RAE study involved comparing joint strengths, bondline void contents and modes of failure for primed and unprimed composites. Six epoxy film adhesives with their recommended primers were tested. The effects of using epoxy primers on carbon fibre reinforced epoxy composite bonded joints was studied for adhesives curing between 120˚C and 175˚C. Some of the recommended primers used were protective coatings dried at 70˚C while others were corrosion inhibiting primers cured at 120˚C. Primed and unprimed joint properties were then compared.

The scientists had believed that primers might act to reduce bondline void content by reducing the roughness of the pretreated composite surfaces. However, this was found not to be the case. It was considered more likely that void reduction, where it occurred, was a result of moisture removal during the primer drying cycle, with most voiding associated with moisture entrapment in the interstices of the adhesive carrier cloths. Although the primers did not influence strength, the mode of failure was altered somewhat with the extent of interfacial failure being reduced and fracture within the composite substrate itself being increased.

8.3.6 PYROLYSIS OF COMPOSITES AT HIGH HEATING RATES

The effect of bulk heating rates ranging from 0.16 K/s to 30 000 K/s on the pyrolysis of a carbon/phenolic composite has been studied at the US Naval Research Laboratory.

Hitherto, the effects of high heating rates (10 000 K/s) have been unreliably inferred by extrapolating low rate data. The behaviour of composite materials at these heating rates has recently become of great interest because of the need to predict high performance material behaviour in harsh thermal environments. There is also a need to model accurately the interaction of directed energy with polymeric and composite materials, especially with respect to material blow-off resulting from the generation of internal gas pressures.

High heating rates and temperatures up to 1200°C were achieved with a special resistance heating system designed at the Naval Research Laboratory. The researchers modelled the pyrolysis by assuming two mechanisms are operating: a process with a low energy of activation, which dominates at low heating rates; and a process with a high energy of activation which becomes important at high heating rates.

Kinetic parameters describing the pyrolysis were obtained using a modified Friedman approach. Kinetic and modelling parameters have been determined for the processes occurring at temperatures above 450˚C and have been used to produce accurate reconstructions of sample thermographs over a wide range of heating rates.

8.4 BACKGROUND SCIENCE

8.4.1 IMPACT

A number of papers have been published in the general area of impact behaviour of composites.

'On the inelastic impact behaviour of composite laminated plate and shell structures', *Composite Structures*, 14, No. 2, pp. 89-111, 1990 by H.L. Lin and Y.J. Lee, from the Institute of Naval Architecture, National Taiwan University, reports on finite element modelling of the impact of plates and curved shells. The work considers an inelastic impact by assuming elastic behaviour in the composite but the loading to be inelastic. The authors point to the importance of including shear deformation and rotary inertial methods in their analysis. Such techniques are not covered in classical plate and shell theory.

The theoretical predictions of their model are supported by experimental studies and show that curvature in a composite sheet via a plate stiffening effect increases the impact force, reduces contact duration and promotes multiple impacts.

One of the complicating factors associated with studies of impact of shell or plate structures is that damage processes occur as a result of both bending stresses and contact stresses. In an attempt to separate out these effects, S.R. Swanson and H.G. Rezaee from the University of Utah in the USA have performed impact testing where bending deformation is prohibited. In their paper

'Strength loss in composites from lateral contact loads', *Composites Science and Technology*, Vol. 38, 1990, pp. 43-54, they report on filament wound flat plates of carbon fibre epoxy, impacted whilst completely supported by a hardened steel block — thereby eliminating bending. This impact situation was modelled using finite elements and the progress of damage through the material, as defined by fibre fracture, was predicted. The residual strength of these laminates was assessed by considering the damaged zones to be equivalent to through thickness holes.

Experimental results showed good agreement to the predictions. The influence of contact stresses becomes more significant at high velocities of impact and for thick sectioned material.

W.J. Cantwell and J. Morton present a model for the impact failure of carbon fibre-epoxy type composites which ignores effects due to contact stresses, as well as stress waves and vibrational contributions to impact response. Nevertheless, the results presented in 'Impact perforation of carbon fibre reinforced plastic', *Composite Science and Technology*, Vol. 38, No. 2, 1990, show good agreement with experimental results for laminate thickness up to 4 mm. The model predicts the energy required to perforate a laminate, makes allowances for geometric parameters, and covers high velocity impact where the perforation energy is largely independent of areal geometry of the late and low energy impact where areal dimensions are important.

8.4.2 SPALLING IN CARBON COMPOSITES AFTER IMPACT

A paper by D.G. Dixon of British Aerospace (Bristol, UK), entitled 'Spall failure in carbon fibre reinforced thermoplastic composite', appears in *Journal of Materials Science Letters* (Vol. 9, 1990, p. 609). It looks at the development of damage in APC-2, carbon fibre polyetheretherketone composites subjected to impacts in the 36–520 ms^{-1} range.

The paper is interesting in that it considers high velocity impact or shock loading as opposed to the more commonly studied lower velocity impacts which occur during drop weight testing.

Dixon shows that a spall fracture occurs as a result of the intersection of shock waves generated during the impact and reflected back from the back surfaces of the specimen. In common with homogeneous continua, the spall failure are the result of the coalescence of a multitude of micro-fractures rather than the propagation of a single crack. However, in the case of the composite the spall is confined to well defined interlaminar regions.

The spalled area exhibits brittle fracture surfaces with no evidence of the plastic deformation found on APC-2 fracture surfaces after lower energy impacts. The spall zone itself was confined to an area corresponding to the diameter of the (flat) circular impacter, but the higher energy impacts studied also resulted in associated delaminations outside of this region.

8.4.3 DURABILITY OF COMPOSITES

Increasingly durability is being highlighted as a major area of composites performance where our knowledge is deficient. A paper from J.L. Sullivan of the Ford Motor Co 'Creep and Physical Aging of Composites' in Composites Science and Technology) is therefore most welcome.

The author has studied the creep of aligned, continuous glass fibre -vinyl ester composites in both transverse tension and shear ($45°$ tension). An important observation is that physical aging of the matrix plays a significant role in determining the long term response of the composite under conditions where linear viscoelastic creep operates. Short term behaviour or momentary creep curves are amenable to manipulation using time-temperature superposition, but this is not possible for long term results. Aging for times of up to 1000 hours was performed on various specimens but the maximum duration of 'long term' static loaded creep tests was only about 12 hours for any given specimen. This is unfortunate (given that data obtained at long time periods is rare) but it does not detract from the importance of the authors finding that without a long term aging effect many glassy polymer matrix composites would probably be rendered useless as structural materials for the automotive industry.

8.4.4 POISSON'S RATIO AS A DAMAGE PARAMETER

A simple process for monitoring the progressive damage in a laminate during some form of deformation is desirable. The nature of damage in most forms of fibre reinforced polymer is microcracking and attempts have been made to link stiffness changes to the development of that damage. In various laminate constructions, stiffness is not found to be very sensitive as a damage indicator. Particularly in 0/90 constructions with carbon fibres, where the stiffness of the 0/90° plies would dominate the response of the laminate and mask changes in the stiffness of the 90° plies.

Smith and Wood from the University of Surrey in the UK have examined the use of Poisson's ratio as an alternative parameter in such situations (P.A. Smith and J.R. Wood, *Composites Science and Technology,*, Vol. 38, 1990). In their paper they theoretically model the effect of transverse ply cracking on Poisson's ratio (v12) which accurately predicts the observed experimental changes. They show that Poisson's ratio reductions of 40% are obtained in both glass and carbon reinforced laminates whereas the corresponding changes in longitudinal stiffness are 20% in glass fibre composites and only 5% in carbon fibre systems.

Poisson's ratio as a damage parameter is also cited in a forthcoming paper from Lam and Piggot from the University of Toronto, Canada. The paper, 'The durability of controlled matrix shrinkage composites: Part 3, measurements of damage during fatigue', P.W.K. Lam and M.R. Piggot, *Journal of Materials Science*, Vol. 25, February 1990, looks at aligned fibre pultrusions in fatigue loading. In this case, the observed effect is a relatively linear increase in Poisson's ratio, v12, with number of cycles. The most interesting results from this paper refer to the effects of matrix modifying systems consisting of expanding monomers (dinorbornene spiro orthocarbonate or tetramethyl spiro ortho carbonate). These act to reduce microstresses in the composite that would otherwise have been introduced during cure, and, improve the fatigue performance. Lam and Piggott link the onset of fatigue failure to the possible presence of misaligned fibres which introduce off-axis stresses into the matrix.

8.4.5 SIMPLE EXPRESSIONS FOR $\upsilon23$, G23

A new route for calculating $\upsilon23$, G23 for unidirectional laminate is described in an article by T.W. Clyne in the March 1990 issue of *Journal of Materials Science Letters* (T.W. Clyne, 'A compressibility-based derivation of simple expressions for the transverse Poisson's ratio and shear modulus of an aligned long fibre composite', Vol. 9, March 1990).

It is convenient to be able to obtain simple expressions for the various elastic constants that define a composite lamina. Most moduli or Poisson's ratios of a unidirectional lamina may be calculated using simple relationships such as the rule of mixtures (for E1, $\upsilon11$) or the Halpin-Tsai expressions for E2, u21. But these simple expressions do not work for the $\upsilon23$, G23 terms which must be calculated using complicated numerical techniques such as the Eshelby equivalent homogeneous inclusion analysis.

Clyne's analysis invokes consideration of the overall compressibility of the lamina. Ultimately, he derives an expression for v_{23},

$$v_{23} = 1 - v_{21} - E_2/3K$$

and subsequently an expression for G23,

$$G_{23} = E_2/2(1 + v_{23})$$

The predictions of these expressions are compared to those derived from Eshelby's numerical technique for two extreme variants on the composite theme, glass fibres in epoxy and silicon carbide in titanium. The results agree very well and show that v_{23} has a relatively high value for a wide range of volume fractions which the author indicates may well have important implications for laminate stress analysis.

8.4.6 THROUGH THICKNESS STRENGTH OF CARBON FIBRE COMPOSITES

It is generally assumed that the through thickness strength of a unidirectional composite will be similar to the transverse strength in-plane. A letter in the May 1990 issue of *Journal of Materials Science Letters* (Vol. 9), 'Transverse anisotropy in the strength and fracture surface of unidirectional carbon-fibre epoxy composites' by T. Honzo, Y. Sawada and Y. Nakanishi (from the Government Industrial Research Institute, Osaka, Japan), shows that this is not always the case.

The Japanese results showed a 32% difference in the transverse in-plane strength compared to the transverse out-of-plane strength. The researchers attribute this difference to both fibre bridging effects and the generation of faults in the x-y plane during processing of the laminates.

These observations may not be too significant at present, as most composites are thin–walled structures; however, there is an increasing use of composites for thick section applications where through thickness stresses, and therefore strengths, are significant. Any introduction of out-of-plane weakness due to manufacturing effects will need to be examined further.

8.4.7 MODELLING TRANSVERSE THERMAL CONDUCTIVITY

A model for the transverse thermal conductivity of unidirectional fibre composites has been proposed by S. Grove of Polytechnic Southwest in the UK that allows for fibre dispersion. 'A model of transverse thermal conductivity in unidirectional fibre reinforced composites', *Composite Science and Technology*, Vol. 38, 1990, No. 3, pp. 199-209.

The use of spatial statistical techniques in combination with finite element analysis has been demonstrated by Guild, Davy and Hogg *(Composites Science and Technology,* Vol. 36, 1989, pp. 7–26) and is shown to allow accurate determination of transverse moduli for the case of a random dispersion of fibres. Grove has adapted the same approach to model the transverse thermal conductivity. His results indicate an 8% difference between calculated results for similar volume fraction (50%) composites; in one case the fibres, are assumed to exhibit square packing and in the other they are randomly dispersed. This change is nearly independent of the thermal conductivity ratio of the matrix and fibre.

The author hopes to develop this work for multidirectional laminates and particulate reinforcement and to include an allowance for voids.

8.4.8 TRIGGER MECHANISMS FOR ENERGY ABSORPTION

Long fibre polymer matrix composites are potentially efficient energy absorbing systems that could be used for automotive applications to provide crashworthiness. The major problem with composite component failure under impact conditions may take the form of a sudden catastrophic shear failure which, although requiring a high load, does not absorb much energy.

The progressive crushing process responsible for significant energy absorption needs to be triggered in some way and a number of geometric triggering devices have been explored. Two papers appear in *Composites Science and Technology* that specifically consider aspects of the triggering mechanisms for composite tubes.

The first, 'Comparison of bevel and tulip triggered pultruded tubes for energy absorption' by M.J. Czaplicki and R.E. Robertson, of the University of Michigan, and P.H. Thornton, from the Ford Motor Co, Dearborn, USA, appears in Vol. 40, No.1, pp. 31–46.

This work seeks to obtain a better insight into the relative performance of tubes with either a bevelled edge trigger or a tulip trigger.

The paper concludes that the energy absorbing capacity of tulip triggered tubes is consistently superior to bevelled tubes. The researchers link this to the development of different cracking patterns which are formed in the tubes during initial failure and maintained throughout the crushing process.

The second paper is 'Trigger mechanisms in energy absorbing glass cloth/epoxy tubes' by I. Sigalas, M. Kumosa and D. Hull, from the University of Cambridge in the UK., published in Vol.40, No.3.

The paper focuses on the initiation of the crushing process when the trigger is a chamfered edge and examines the various effects on crushing mechanisms of differing chamfer angles, both experimentally and by finite element modelling.

The results show that trigger mechanisms are dependent on chamfer angle. A combination of internal and external cracks within the chamfered part of a glass cloth/epoxy tube gives rise to the formation of an equilateral triangular cross-section and further loading generates a wedge of crushed material from the apex.

The internal stress generated by this intruding wedge of damaged material into the wall of the tube is relieved by nucleation of lateral cracks which propagate from the wedge tip to the nearest tube wall. This process repeats itself, with combinations of wedges and small rings of material being pushed aside towards the internal and external walls respectively and providing the damage process for energy absorption.

8.4.9 CRUSHING OF FIBRE REINFORCED TUBES

While the importance of triggering mechanisms on the damage processes in fibre reinforced tubes is well known, the mechanisms of energy absorption, and the factors controlling damage development in composite tubes is still unclear. The proliferation of composite constructions used in such studies has contributed to this situation. A paper entitled 'A Unified Approach to Progressive Crushing of Fibre Reinforced Composite Tubes', *Composites Science and Technology*, 1991, Vol. 40, pp. 377-421, attempts to apply unifying principles to the crushing behaviour of a wide range of composite tube constructions.

The author is Professor Derek Hull of Cambridge University in the UK who has pioneered much of the work in this field. He concentrates in his paper on the interactions of five variables namely: i) microfracture processes at the crush zone, ii) forces acting at the crush zone, iii) microstructural variables associated with the composite materials in question, iv) shape and dimensions of the component, and v) testing variables such as temperature and crush speed.

The many different crushing patterns exhibited by different constructions (angle-ply filament-wound tubes, glass-cloth fabric tubes, 0/90 symmetric composite tubes, sheet moulded compounds etc. are rationalized as belonging to two general categories — either splaying failures or fragmentation failures. The forces acting in each case and the microfracture events responsible for each mode of failure are discussed. Transitions between each failure mode may be induced in a given material by varying factors such as the ratio of axial to hoop reinforcement, together with the arrangement and symmetry of the reinforcement. It is not clear for any given construction which failure mode provides the greatest energy absorption as this depends on the specific microscopic strength properties of the material. Transitions between failure modes and/or considerable variations in specific energy absorption can be introduced by variations in testing rate and temperature, but the trends are not always consistent from one material to another.

The field is far from completely understood but this paper indicates that considerable progress is now being made in understanding a very complex but practically valuable composite structural response.

9. HEALTH, SAFETY AND THE ENVIRONMENT

9.1 HEALTH AND SAFETY

9.1.1 DOW TASK FORCE ON STYRENE EMISSIONS

A task force to help customers to comply with new US Occupational Health and Safety Administration (OHSA) regulations on styrene emissions has been set up by Dow Chemical Co of Midland, Michigan, USA. It is chaired by Susan Hearn, an environmental specialist in the Chemicals and Metals Group.

The regulations took effect in September 1989 and include a revised air contaminants rule which restricts the permissible time weighted average styrene exposure to 50 ppm — down from the previous figure of 100 ppm.

The first step for the task force has been to produce an information folder containing material safety data sheets, updated regulatory information, facts about safe handling of the company's styrene-containing products, such as Derakane vinyl-ester resins, and information on protective equipment and clothing.

9.1.2 THE HEALTH RISK FROM SILICON CARBIDE WHISKERS ASSESSED

Concerns regarding the health risks associated with whisker reinforcements have been widespread over the last few years, to the extent that ASTM in the USA has established a sub-committee of their E34 committee on Occupational Health and Safety to examine the issue.

The Chairperson of that committee, Sam Weaver gave his assessment of the risks associated with silicon carbide whiskers in an article in ASTM's *Standardisation News*.

Whiskers in general, and silicon carbide whiskers in particular, are regarded as potentially dangerous due to similarities with asbestos. Information received by ASTM E-34.70 suggests that whiskers of silicon carbide of similar dimensions to asbestos will probably provide similar health problems and *in-vitro* studies have confirmed a similarity in the effects of the two materials. However, Weaver indicates that the situation is not as bad as it initially seems and that there are a number of positive factors.

Firstly, in three separate studies involving hundreds of animals only one cancer was reported after silicon carbide exposure — which is not statistically significant.

Secondly, the greatest health risk from fibres is the development of mesothelioma which is caused by very small fibres of less than 0.2 μm diameter. The diameter of silicon carbide whiskers can be controlled during manufacturing to ensure that this size range is avoided.

Thirdly, the nature of silicon carbide manufacturing operations lend themselves to controls to eliminate airborne fibres (unlike the mining of asbestos).

Fourthly, the composites manufactured from silicon carbide whiskers bind the whiskers into the structure and even extreme treatment such as machining with cutting tools does not lead to individual whiskers being released into the environment.

The message from Weaver seems to be clear. Providing sensible precautions are taken during manufacturing, the risks from using silicon carbide whiskers are minimal.

9.1.3 FIRE AND SMOKE TOXICITY PERFORMANCE STUDY

The fire, smoke toxicity and corrosivity of polymeric composites are being studied by two US partners.

ICI Advanced Materials of Wilmington, Delaware, has joined with the Southwest Research Institute of San Antonio, Texas, to study composite materials and systems used in aircraft interiors, naval vessels, mass transit vehicles and building materials.

Southwest Research Institute is conducting a test programme in which the first phase will try to establish performance data on a range of thermosetting and thermoplastic composites. Test methods will include cone calorimetry (heat release measurements), radiant combustion and smoke exposure, as well as the latest US Federal Aviation Authority and US Navy test procedures, such as MIL STD 108(SH) for submarines. The composites being evaluated were chosen from input supplied by potential users.

In the second phase, the performance of barrier materials will be evaluated.

The initial work is concentrating on small scale samples, but, as optimum material coating configurations are identified, full scale fire tests will be performed for specific applications. ICI and Southwest Research Institute say they will work with a number of potential users in this area.

The project started in March 1990 and should take 18–24 months.

9.2 RECYCLING

9.2.1 RECYCLING OF SHEET MOULDING COMPOUNDS

Reports in *Modern Plastics* point to growing concern within the sheet moulding compound industry regarding the recycling of sheet moulding compound components.

General Motors in the USA, acting with the sheet moulding compound Automotive Alliance (a group of sheet moulding compound moulders and material suppliers), has released preliminary results from a US$100 000 research programme aimed at assessing the viability of pyrolysis as a method of recovering sheet moulding compound.

Apparently, in trials conducted to date, sheet moulding compound parts have been shredded and fed into a chamber which is heated to about 750°C. Under these temperatures the organic constituents are vaporized and the gas emitted is used to fuel the continued heating of the material which results in an ash of filler fibre and carbon. Possible applications for the ash include a filler for shingles or cement. No doubt GM and the Automotive Alliance would like to make use of the reinforcement in the ash and tests have been performed to determine whether the ash could be used to produce new sheet moulding compound.

The impetus for this work is the coincidence of a likely increase in the volume of sheet moulding compound consumed by automotive applications, coupled with increasing environmental concerns regarding recycling of plastics. The General Motors APV minivan is the first high volume sheet moulding compound skinned vehicle in the USA. GM is also to proceed with a Camaro/Firebird replacement in 1993-94 which will involve sheet moulding compound exteriors on a model with an anticipated production volume of 400 000/year. Meanwhile, the competitive thermoplastic resin producers are exploiting the theoretical recyclability of their materials as a valuable marketing tool.

9.2.2 RECYCLING/RECLAIMING 'APC-2'

A programme at the Welding Institute in the UK has been the investigation of the potential for reclaiming or recycling advanced thermoplastic prepregs. The study has examined both the mechanical properties of recycled material and the economic viability of the process.

The programme is of some significance given the off-stated remarks by suppliers of thermoplastic composites that their materials have an environmental edge over thermosets.

The studies at the Welding Institute focussed on reconstituting scrap APC-2 into sheet form and attempting to further process this sheet material into shapes of varying complexity. The scrap was taken through a process whereby it was cut up into well defined pieces produced by a commercial (Monomunching), fragmentation device.

In general terms, the results indicate that the larger the particle used to produce reconstituted sheet material the better the mechanical properties and the more closely they resembled those of quasi-isotropic APC-2 laminates. The monomunched materials provided the poorest mechanical properties but conversely exhibited far better formability than the other grades of material. By way of an example, sheets made from 6 mm particles of scrap are quoted as having a tensile modulus of 37.2 GPa compared to 48.96 for a quasi isotropic sheet of APC-2. Tensile strength was 236 MPa compared to 704 MPa while compressive strengths of 325 MPa were measured, compared to 361 MPa for the laminate. The corresponding properties of sheet produced from 19 mm particles of scrap are 41.6 GPa modulus, and strengths of 270 MPa in tension and 380 MPa in compression.

The economic assessment considered capital outlay, corporation tax, a project lifetime of 5 years (over which all capital equipment is amortised), a realistic output rate for scrap conversion, fixed wages costs, a zero raw material cost (assumed to be scrap available in house rather than recycled material), and running costs based on average electricity prices. Breakeven prices for the material were calculated on the basis of a variety of required rate of return, and load factors. At a load factor of 50% and a rate of return of 20%, a break even selling price of £32.46/kg was determined.

Much of this data is published in a paper by McGrath of the Welding Institute, Clegg of Sheffield Polytechnic and Morris of ICI, all in the UK, entitled 'Properties and economics of reclaimed long fibre thermoplastic composites', *Composites Manufacturing*, Vol. 1, No. 2, 1990 pp 85-89. They conclude in the paper that reclaimed APC-2 obtained from consolidating particles is a promising route that provides better properties than injection moulding ground scrap.

9.2.3 GERMAN SUPPLIERS LINK IN PLASTICS WASTE RESEARCH

Three major German plastics suppliers, BASF, Bayer and Hoescht, are to join forces in tackling the problem of recycling and disposal of plastics based waste materials. A new company is to be jointly established, capitalised at DM12 million, called Entwicklungsgesellschaft für die Wiederwerwertung von Kunststoffen (EWK) and based in Wiesbaden. About 8-10 people are expected to be employed by the new venture which will work closely with the relevant departments in the parent organisations. Initially, it is to act as a focal point for plastics waste information and technology in Germany. The types of industrial sector and therefore the sources of scrap and waste of concern to this body are industrial waste in general, automotive, domestic appliance, electrical appliances, building waste, and chemical containers — but not packaging.

10. RESEARCH INITIATIVES

10.1 GENERAL

10.1.1 EUROPEAN PROGRAMME FOR RESEARCH AND DEVELOPMENT

Details have emerged of the next R&D Framework Programme of the European Community. The Framework programmes are the mechanism through which Community runs its various specific research initiatives such as BRITE/EURAM. The new Framework Programme will be divided into two parts, covering the years 1990–92 and 1993–94. The total budget has been set by the Council of Ministers at 5.7 billion ECUs, split as 2.5 billion ECUs for 1990–92 and 3.2 billion ECUs for 1993–94.

The sums of money allocated to industrial and materials technologies totals 888 million ECUs over the whole period.

10.1.2 ROUND TWO OF BRITE/EURAM

The second round of the European Community's BRITE/EURAM programme was announced on 1 March 1990. The research programme is valued at 500 million ECUs (approximately US$550 million) over a four-year period. It aims to stimulate innovation in industrial technologies and advanced materials through international collaborative research.

The priority themes for round two have been revised to reflect changing industrial needs. They fall within the following technical areas:

*   advanced materials technologies

*   design and assurance for products and processes

*   application of manufacturing technologies

*   technologies for manufacturing processes.

Industrial companies, universities and higher education institutions and research organisations within the EC are eligible to take part. There are three classes of project that may be supported:

Industrial applied research projects where the total project costs need to fall within the range 1–3 million ECUs, covering more than ten man-years of effort. At least two legally independent companies from different member states must be involved and the EC funding may reach 50% of the total costs, with any university input funded at 100%.

Focused industrial research will also be funded (up to 10% of the total budget). This area does not require industrial participation directly but endorsement by industry is essential. The project costs should be in the range 0.4–1 million ECUs with at least ten man-years of effort involved.

A further proportion of the total budget will be used for coordinated activities.

10.1.3 EUROPEAN COMPOSITE FORUM LINKS WITH EUROPEAN COMMISSION

The European Composite Forum is steadily establishing closer links with the European Commission and is set to assume an important role in determining future priority areas for European research funding on composites.

European Composite Forum is a body created five years ago with the expressed aim of bringing together the European research community and industry. At that time it was considered that the research and industrial communities had a common and immediate interest in developing ties within Europe irrespective of the slower political moves towards cooperation. The Forum, which is funded entirely by industrial subscriptions, set up a network linking laboratories across Europe. The network supplies up to date information on research activities and interests to the Forum which distributes the information to industrial subscribers in the form of a European Composites Encyclopedia. In this way companies have a lead on the latest research coming out of the laboratories across the whole range of composite activities. In addition, the Forum has created a European computer network enabling members to interact and have access to information. Not least in importance is the annual meeting which is organised to identify future research trends and industrial requirements.

The Commission funds many research initiatives which have a relevance to composites and the overall budget is substantial. The budget for programmes such as BRITE/EURAM is set to increase, but the staffing levels of the Commission will remain fixed. This will inevitably mean the Commission turning to outside bodies such as European Composite Forum for help in areas such as determining likely future research trends, as a reservoir of experts and research groups for helping to formulate research programmes and training, for advice on standardisation, market trends and other issues.

In response to the approach of the Commission, the industrial companies supporting European Composite Forum are to meet to formulate a positive response to the offer for increased collaboration offered by Mr Würm of the Commission.

10.1.4 GERMAN INSTITUTE OF COMPOSITE MATERIALS

The Institute for Composite Materials GmbH has been established at the University of Kaiserslautern, Germany. The institute has been established as a limited company with the sole stakeholder being the state of RheinlandPfalz.

The executive director of the Institute is Professor Manfred Neitzel who formerly worked for BASF AG at Ludwigshafen. There will be three technical-scientific directors at the institute, responsible for specific aspects of the research programme.

Professor Neitzel will act as technical-scientific director of manufacturing research, while Professor Klaus Friedrich, formerly of the Technical University Hamburg-Harburg, has been appointed technical-scientific director for structure-property relationships. A third technical-scientific director for mechanics and design aspects has yet to be appointed.

The Institute will have a custom–built building of some 6000 m^2 of office and laboratory space located on the university campus which is due to open officially in September 1991.

The Institute intends to interact closely with industry and sees a role for itself in assisting small and medium sized enterprises in exploiting composites more effectively while also developing larger scale research programmes with the major industrial concerns. It will also interact closely with the University and students studying at the Institute for Composite Materials for their diploma or doctorate are likely to have a good insight into the practical needs of industry when they leave.

10.1.5 SOL-GEL PROCESSING OF CERAMIC MATRIX COMPOSITE SHAPES

A siginifcant research grant (worth nearly £300 000) has been awarded to Professor B. Harris and Dr R.G. Cooke of the School of Materials Science at Bath University, UK. The grant is for three years' work on the development of sol-gel processing methods for manufacturing components for chemical plant and other engineering systems from ceramic matrix composites.

The grant is under the advanced manufacturing technology remit of the UK Science and Engineering Research Council's ACME Directorate, and will be carried out in close collaboration with Ceramic Developments Midlands Ltd at Corby, whose Managing Director, Dr R.W. Jones, is himself a specialist in ceramics technology. Ceramic Developments Midlands Ltd says it is strongly committed to the development of sol-gel processing, and is already using the technique for the commercial fabrication of a variety of ceramic artefacts.

The programme, which will involve parallel manufacturing and evaluation work at both Bath and Corby by a team of some eight scientists end engineers, is aimed at the adaptation of resin-based composites manufacturing methods for the economic production of dense, net-shape components possessing appropriate levels of corrosion, wear and abrasion resistance, thermal shock resistance, and toughness, for a range of applications in plant and machines operating at ordinary temperatures.

10.1.6 MULTI-CLIENT PROGRAMME ON INVESTMENT CASTING OF METAL MATRIX COMPOSITES

A multi-client three year project aimed at optimizing cheaper manufacturing techniques for metal matrix composites is being launched by BNF Metals Technology Centre of Wantage, UK.

The project will concentrate on investment casting, a technique, which produces near-net-shape castings with high dimensional accuracy and is used for the production of complex shapes and thin-walled components.

The main material under consideration in the programme is particulate silicon carbide reinforced aluminium for improved stiffness, vibration damping, high temperature strength, increased wear resistance and reduced thermal expansion.

The investment casting of metal matrix composites will require significant modifications to accepted foundry practice. Optimization of melt preparation, mould filling techniques and the design of runners and filling systems for particulate reinforced materials have already been identified as major areas for study. Some work on the use of fibre preforms is also planned.

10.1.7 COPPER AND NICKEL-BASED METAL MATRIX COMPOSITES

BNF Metals Technology Centre, Wantage, UK, is now making progress in a study of metal matrix composites based on copper and nickel alloy matrices.

BNF has run programmes on metal matrix composites for some time but these were expanded during 1989 to cover copper and nickel alloy matrix systems. Initial results are now being generated on copper alloy/ceramic particle and copper alloy continuous ceramic fibre systems. The manufacturing method uses liquid metal infiltration of preforms and the technology is to be extended for use with nickel matrix composites in the near future.

BNF has plans to launch a multi-client project in order to attract the necessary funding to continue the work in collaboration with an appropriate industrial concern.

10.1.8 PERA LEADS CONSORTIUM ON MACHINING LIGHT METAL MATRIX COMPOSITES

PERA, Melton Mowbray, UK, has formed a consortium of European partners to collaborate on a project studying the machining of metal matrix composites based on light metal alloys.

PERA's partners in the programme include Aluminium Ranshofen and Leoben University from Austria, National Forge and Leuven University from Belgium, HTM and ABB from Switzerland and VTT from Finland. The total project value is of the order of £1.5 million and is part of the COST 506 initiative (COST Collaboration on Science and Technology). PERA's contribution to the three year programme is costed at £660K and will be partly funded by the Department of Trade and Industry in the UK.

Project objectives include obtaining data on machinability, effects of machining on substrates and effects on cutting machinery.

10.1.9 PROJECT STUDIES CAST METAL MATRIX COMPOSITES

The technology of casting particulate reinforced aluminium alloys will be developed by a four nation team to be funded by the European Commission (EC).

The three year project is valued at £1.5 million and will be funded under the Basic Research in Industrial Technologies for Europe/European Research in Advanced Materials (BRITE/EURAM) programme. It will be managed by Dr Andy Feest at the Harwell Laboratory, in the UK.

Apart from Harwell, the other participants in the project are Foseco International Ltd, based in Birmingham, UK, Hydo-Aluminium from Norway, Renault from France, and the German aluminium component manufacturer, Vereinigte Aluminium Werke AG.

European sourced ingots will be converted into near finished precision castings using six different techniques. The automotive industry is seen as the major end-user for the technology developed.

10.1.10 OFFSHORE COMPOSITES PROGRAMME

A £1 million university research programme in the UK studying the 'Cost effective use of fibre reinforced composites offshore' has led to a second, two year phase, which will began in October 1990.

The research programme has taken place under the control of a steering body, Marinetech North West, which was established by a group of five universities in the North West of England. Funding for the programme came initially from the UK's Science and Engineering Research Council, the Ministry of Defence, the Department of Energy, and a consortium of industrial organizations, including multinational oil companies and chemical firms.

The next phase of the study will continue to place emphasis on the areas of fire resistance, degradation due to water ingress, blast, and hot gas flame and impact resistance.

10.1.11 FIRST CONE CALORIMETER FOR UK UNIVERSITY

In a continuing move to establish a strong fire research group working on composite materials, the Department of Materials at Queen Mary and Westfield College, part of the University of London, has taken delivery of a cone calorimeter.

The cone calorimeter is a piece of fire research equipment originally designed by Dr V Babraskus of the National Institute for Standards and Testing in Washington, USA, with the prime objective of measuring heat release. This is the critical datum now required for composites in aircraft interior applications (and, by extension, other mass transit application areas). Current certification requires materials to pass heat release requirements measured on the Ohio State University apparatus which is extraordinarily expensive and equipment sensitive. Result from a given test machine are reproducible but unlikely to equate to results generated on the same material in a different laboratory. The cone calorimeter is considered to be more reproducible generally, and much more economical to operate. In the longer term it is anticipated that it will become the standard method for heat release testing adopted by FAA, ASTM and others.

The QMW apparatus is the first such facility to be installed in a British University. The composite fire research programme at QMW is guided by Senior Research Fellow, Steve Grayson, formally director of the Wolfson Fire and Materials Centre. Existing projects have included studies on composites for marine applications with BP, but the research group will now be looking to extend their activities into the aerospace and transport industries

10.1.12 DESIGN DATA INITIATIVE

The Polymer Engineering Group of the British Plastics Federation has completed the first phase of its Design Data Initiative with the publication of a report entitled 'Project Needs Identified by Working Groups'.

The Initiative was prompted by the need to establish a better framework for the provision and use of design data and to encourage better design procedures for polymeric materials, including polymer composites.

A number of working groups were established, including one concerned specifically with composite materials, while others covered general areas such as the effects of processing on materials properties, durability, and design of joints — all of which are relevant to the composites industry. The working groups were charged with identifying the precise needs of industry for design methods and data, current levels of capability, ongoing work in the area and future research needs.

All of these tasks have been completed and a strategy for progress in the field has been developed which includes the identification of a number of key research programmes. The report provides a concensus industry view of its needs and also the basis for a co-ordinated programme funded jointly by Government and industry which will form the second phase of the Design Data Initiative.

10.1.13 NOTTINGHAM UNIVERSITY LAUNCHES COMPOSITES CLUB

A composites club to bring industry and academia together to promote knowledge of modern composite materials was launched by Nottingham University, UK, in May 1990.

The university says that it has also formed a Composites Institute which will bring together the expertise of academics in several departments to promote research, short course provision, and postgraduate study. The Composites Club, according to Director, Professor Mike Owen, will be the interface between the institute and industry.

In addition to the organizing liaison between the university, industry, professional bodies and other educational establishments, it is envisaged that the club will: provide a forum for direct contact; publish a regular quarterly newsletter; organize seminars and discussions; identify needs for short courses and workshops; identify research opportunities; and promote collaborative research.

10.1.14 ADHESIVES AND CLEAN COMPOSITE SURFACES STUDIED

Two research initiatives related to polymer matrix composites have been announced by PERA, the Melton Mowbray UK-based technology centre.

The first relates to the surface quality of composites parts. The condition of the surface of a structural part often determines the success of a component, particularly if coatings or adhesives have been applied. Cleaning and conditioning surfaces effectively has therefore been identified by PERA as a vital area for study. A multi-client programme is to be established that will investigate the processes and mechanisms for cleaning and conditioning the surfaces of composites. The minimizing of substrate damage is a key objective.

This programme is a logical development for PERA which has been working with the Cryoblast cryogenic blasting system for over a year. This technique, which involves blasting surfaces with frozen CO_2 pellets, has been found to be very sensitive and will remove contamination without damagng the substrate.

The programme is to be run by the advanced metals and surface engineering department at PERA and has been costed at a total of £620 000 over three years. The contact at PERA for this programme is Laurence Archibald.

The second programme concerns the development and transfer of adhesive technology.

PERA has identified three areas where progress is needed before further exploitation of adhesives will be considered by industry. These include: design, where there is an absence of user friendly design theory procedures and software; manufacturing, where reliable, quality validated, automated systems to handle high volume production are needed, and; reliability, where there is an absence of on-line real time non-destructive testing techniques for joints.

Four individual projects comprise PERA's effort to redress the situation. The technology resulting from these projects (on design software, adhesive computer aided design, automated bonding concepts, and on-line non-destructive testing) will be made available to the 7000 small and medium sized enterprises linked to PERA's international network throughout Europe. Contact at PERA for this area is Gerry Boyce.

11. INDUSTRY NEWS

11.1 NEW COMPANIES

11.1.1 TECHNOLOGY TRANSFER COMPANY ESTABLISHED

A new company, Optimat, has been set up by five organizations in the UK to transfer materials technology from technology producers into profits for its clients.

The company will be based in East Kilbride, Scotland, and has been established as a result of a review of Scottish engineering by the Scottish Development Agency, which identified an increasing need for materials technology.

The founder companies are all heavily involved in developing materials technology and include RAPRA Technology, The National Engineering Laboratory, The Welding Institute, Coopers and Lybrand and the Scottish Development Agency itself.

11.2 EXPANSIONS AND INVESTMENTS

11.2.1 ACQUISITION TARGETS IN ADVANCED MATERIALS IDENTIFIED

While many within the composites industry are concerned about short term prospects, the long term outlook for composites and advanced materials in general is extremely promising. A report from Strategic Analysis will therefore be of interest to the many business concerns considering entering the advanced materials field as well as those companies already involved that wish to increase their range of products and/or establish a US operation. 'Investment opportunities in advanced materials' singles out 124 potential acquisition targets in the USA that are involved in advanced materials (from ceramics to adhesives, and of course composites).

The list was selected from a total of 500 companies that were surveyed, with the companies selected according to the criteria of size, primary business independence and novelty and future potential.

The report costs US$15 000.

11.2.2 BEKAERT ACQUIRES STAKE IN SPANISH COMPOSITES BUSINESS

Bekaert, the Belgian steel wire and cord producer, who moved into the composite pultrusion business in 1984, has acquired 40% of the shares of Bremen SA in Spain.

Bremen is a polymeric composites manufacturer, based at Minguia, Vizcaya in Spain, and concentrates on pultrusion, moulding and filament winding. The two companies became linked recently, with Bremen acting as distributor for Bekaert products in the Iberian market, while Bekaert distributed Bremen products in the rest of West Europe.

11.2.3 DARCHEM TO EXPAND

Darchem Composite Structures Ltd has opened a design and research and development facility at its Huntingdon works in the UK. The company is noted for its work in aerospace, motor racing, simulation, training aids and the production of full size aircraft facsimiles.

The facility includes Auto CAD computer aided design packages, and Nisa finite element analysis, together with computer controlled test equipment. Darchem says that these additions to its autoclave, machine shop and dedicated composite laboratory will enable it to provide customers with a conception-to-production service.

11.2.4 PULTRUSION FACILITY

John Shaw Ltd, Worksop, UK, has introduced a £0.5 million pultrusion facility to produce a range of glass fibre reinforced products.

John Shaw is at present the second largest producer of steel wire rope in the UK. The new product range is clearly aimed at similar markets to existing products (i.e. wire/rope/cables), with continuous pultrusions up to 15 km long being planned.

11.2.5 ASHLAND EXPANDS ITS FRP SUPPLY OPERATIONS

Ashland Chemical has further expanded its FRP Supply Division, by the acquisition of the assets of E.F.Laughlin and Co,of Bellvue, Washington, USA.

Laughlin was itself a distributor of reinforced plastics materials to the fabrication industry and in particular distributed the entire range of Reichhold thermosetting polyester resins such as Dion, Atlac and Polylite resins. Through this acquisition, FRP Supply will gain access to customers in the Pacific Northwest and Mountain states regions of the USA and increase its interaction with Reichhold.

Ashland seems to be supporting a vigourous expansion programme at FRP Supply which already claims to be the only national distributor to the reinforced plastics industry in the USA.

11.2.6 BENTLEY HARRIS EXPANDS COMPOSITES MANUFACTURING CAPACITY

Bentley Harris of Chester Springs, Pennsylvania, USA, is to expand the composites manufacturing capacity at its Lionville, Pennsylvania, plant. The expansion has been made possible by the transfer of the non-composites business (automotive and oven door seal production lines) to a new plant at Gordonsville, Tennessee.

New manufacturing equipment is expected to be installed by the company which specializes in braiding preforms and resin injection processes. The company says that its expertise in braided preforms has led to its selection by the Automotive Composites Consortium (comprising GM, Ford and Chrysler) to develop braided preforms for advanced automotive structures where resin transfer moulding is the expected production process.

Bentley Harris has also recently been involved in slitting and braiding thermoplastic prepregs.

11.2.7 THERMOPLASTIC COMPOSITES PLANT PLANNED

A plant to manufacture 'Taffen' structural thermoplastic composites will be constructed in Lynchburg, Virginia, USA, by the Exxon Chemical Polymers Group.

The plant is designed for expansion in line with market demand and has an initial capacity of 7000 t a year (15 million pounds a year).

Taffen structural thermoplastic composite is a glass fibre-polypropylene random in-plane composite produced in sheet form using a paper making process developed originally by Ajomari, from whom Exxon purchased the technology. To date, Exxon has produced the material at a pilot plant in Belgium.

The material is designed for stamping operations and is aimed at the automotive market. Potential applications include load floors, seat components, bumper beams, instrument panel retainers, under engine noise shields, complete front ends and battery trays.

The product is similar to materials produced using the 'Radlite' paper making route, originally developed by Wiggins Teape in the UK. Licences for that particular process have been taken out by GE in The Netherlands, Ahlstrom and recently by K-Plasheet a consortium of Japanese companies, which includes Kawasaki Steel, Sumitomo Chemical, Takiron and C. Itoh and Co.

11.2.8 INCREASED PRODUCTION OF POLYETHYLENE FIBRES

Production capacity of its high performance polyethylene fibre will double by the first quarter of 1991, Allied Signal Inc of New York, USA, has announced.

The 'Spectra' fibre is produced at the company's facilities in Petersburg, Virginia, in two varieties. 'Spectra 900' is a 1200 denier fibre primarily aimed at ropes, cordage and impact resistant composites, while 'Spectra 1000' is 650 denier and used for racing sails, plastic reinforcements, ballistic applications and protective clothing.

11.2.9 TEXTRON DOUBLES CAPACITY OF 'AVCARB' CARBON FIBRES

Textron Specialty Materials, Lowell, Massachusetts, USA, has announced the completion of a batch graphitizer furnace which will double the production capacity of the company's high purity, staple Avcarb carbon fibre to 34 000 kg a year.

The furnace pyrolyses the carbon fibres at temperatures of up to 1800°C for 12–24 hours, which Textron says reduces the levels of impurities such as alkaline metals to less than 50 ppm.

Avcarb carbon fibres, which have been produced by the company for some ten years, are used in carbon-carbon composite brakes and rocket nozzles. Qualification has been obtained for use in the composite brake systems of the McDonnell Douglas F-15 and F/A-18, Lockheed C-5A/B and Boeing 757 aircraft.

11.2.10 ZOLTEK CARBON FIBRE PLANT EXPANSION

Zoltek Corp, of St Louis, Missouri, USA, has announced plans to double its current polyacrylonitrile based carbon fibre manufacturing capacity by building a facility in the University Research Park, situated 25 miles west of St Louis.

Zoltek's existing carbon fibre plant is at Lowell, Massachusetts, and was acquired from the Stackpole Corp in 1987. The combined capabilities of the two plants will allow the company to produce more than 450 000 kg of Pyron oxidized polyacrylonitrile fibres and around 110 000 kg of Panex polyacrylonitrile based carbon fibres a year.

Facilities for carbon fibre manufacturing, batch carbonizing and carbon textile fabrication will be included at the new plant.

11.2.11 SAINT GOBAIN TO ESTABLISH GLASS FIBRE PRODUCTION IN BRAZIL

Saint Gobain, the French industrial concern that is the parent of glass fibre producers Vetrotex, is to extend its interests in Brazil with the construction of a glass fibre plant.

The company president has stated that it is the local market for composites in Brazil, particualry in the marine industry, that has prompted the move.

The new plant will represent an investment of US$80 million which will add to Saint Gobain's extensive investments in Brazil which already total US$1.2 billion.

11.2.12 DEMONSTRATION PLANT FOR HIGH STRENGTH POLYBENZOBISOXAZOLE FIBRES

The first demonstration production plant for the polybenzobisoxazole family of high performance fibres has gone into limited production.

CommTech International of Menlo Park, California, USA, owner of patents covering the polybenzobisoxazole technology, says that published test reports based on fibre produced at the plant show tensile strengths and stiffnesses comparable to carbon fibres.

Polybenzobisoxazole is a family of extended chain, liquid crystalline polymers which combine high strength and stiffness with resistance to high temperatures, oxidation, moisture, and ultra-violet radiation. The company believes that polybenzobisoxazole fibres embedded in resins will provide a new generation of high strength, lightweight composites for use in air and rail transportation, mechanical and electrical equipment, and many other applications. CommTech claims that polybenzobisoxazole can be produced at competitive prices for a worldwide market estimated to be as large as 11 million kg a year.

An option on the exclusive manufacturing rights for polybenzobisoxazole, in the Western Hemisphere, was granted by CommTech to Dow Chemical in 1989. Dow recently exercised this option and is operating a pilot plant for the PBO (paraphenylene polybenzobisoxazole) form of polybenzobisoxazole fibre.

11.2.13 CIBA GEIGY TO EXPAND PREPREG CAPACITY IN EUROPE

Ciba Geigy has installed two identical lines capable of producing 1500 mm wide unidirectional prepregs in its European plants. One has been installed at the company's bonded structures unit at Duxford near Cambridge in the UK while the other has been installed in the Brochier SA factory at Dagneux-Montluel in France. According to Ciba Geigy, these machines are identical to plant already operated by the company at its facility in Anaheim, California, USA.

The Duxford machine will produce solvent-based prepregs, with the matrix films introduced to the carrier on a new reverse roll coater. In contrast, the plant at Dagneux will use hot melt coating and automated continuous manufacturing techniques developed at Anaheim. The Duxford and Dagneux plants will be capable of exchanging matrix films to maximize flexibility.

11.2.14 EASTMAN CHEMICAL TO BUILD NEOPENTYL GLYCOL PLANT IN EUROPE

Plans to build a neopentyl glycol plant in Western Europe have been announced by Eastman Chemical Co, of Kingsport, Tennessee, USA. The plant will have an annual capacity of more than 20 000 t and is scheduled for completion towards the end of 1992. As yet no final decision has been made on the location.

Neopentyl glycol has many applications including the production of neopentyl glycol-polyester resins which are used extensively for gel-coats and for certain corrosion resistant applications of glass reinforced plastics.

11.2.15 SHEET MOULDING COMPOUND FACILITY FOR BUDD

A major manufacturing plant for sheet moulding compound components in Kendallville, Indiana, USA, has been opened by Budd Co of Troy, Michigan, USA. The plant will produce seamless body panels for a range of General Motors multi-purpose minivans. Each van contains about 170 kg of sheet moulding compound and a total production of all models of 225 000/year is expected.

According to *Modern Plastics International*, facilities at the Kendallville plant include computerized fast-acting presses in the 9000–48 000 kN range, with bed sizes up to 4.2 x 2.7 m. All presses are also equipped with mould vacuum-degassing features.

Whatever the long term impact of processes such as resin transfer moulding on the automotive industry, sheet moulding compound is definitely the important high volume composite material at this moment.

11.2.16 AIRTECH TO OPEN MANUFACTURING PLANT IN EUROPE

The construction of a European manufacturing facility has been announced by Airtech International Inc of Carson, California, USA. Airtech produces a range of tooling materials including vacuum bagging film, tooling prepregs, composite back-up tubes for support structures and various other ancillaries.

The company says that the extra manufacturing capacity to be provided by the new plant in Differdange, Luxembourg, is needed to satisfy the growing demands for its products in Europe.

Vacuum bagging materials (Airtech produces a range suitable for cure temperatures up to 422°C) and tooling prepregs will be produced at the plant.

11.2.17 IM-TECH ACQUIRES COMPOSITES MOULDING FACILITIES

German sheet and dough moulding compound moulder, Kolbus Kunststoffwerk of Bassum, was acquired by Im-Tech, Eschborn, a subsidiary of Deutsche Shell AG, on 1 April 1990.

Kolbus was founded in 1972 and produces high performance moulded parts from sheet and dough moulding compounds, injection moulding resins, and fibre reinforced thermoplastic sheet. Im-Tech says that it intends to actively expand production of specialized parts for the automotive and electronics industries. The fibre reinforced thermoplastics sector of the business is also to be strengthened and the reaction injection moulding of structural composite parts will be undertaken.

In 1989 Im-Tech acquired filament winding capability which it uses to supply high performance components to the aerospace and machine building industries. The company says that the purchase of Kolbus will enable it to expand its ability to offer customers a complete problem solving package for high performance parts.

11.2.18 EXXON INSTALLS LARGE DEVELOPMENT PRESS FOR AUTOMOTIVE COMPOSITES

Exxon Chemical has installed an 800 tonne vertical hydraulic press at its Machelen Technology Centre in Belgium. The new press will be primarily used for developments projects in the automotive field using Exxon's 'Taffen' structural thermoplastic composite.

Taffen is the name now given to the thermoplastic product originally developed by Ajomari and acquired by Exxon. It consists of random glass fibres and polypropylene in sheet form, with the unconsolidated sheet produced using a paper making process.

Exxon claim that Taffen has exceptional flow moulding characteristics, making it ideal for large complex structures such as automotive front ends, dashboards and the like. The new press will allow fundamental studies to be performed aimed at optimising the moulding process as well as a prototyping capability.

11.2.19 DOWTY BUYS PRESS FOR COMPOSITE PROPELLER BLADES

A 650 t press for the forming of composite propeller fan blades has been installed at Dowty Rotol. The glass or carbon fibre reinforced fan blades are for use in the unducted-fan jet engine under development at GE Aircraft Engines in the USA.

The press, supplied by Mackey Bowley International, Gravesend, UK, is a refurbished 650 t downstroke hydraulic press with five rams, a stroke of 610 mm and a daylight of 760 mm. An extensive automatic operating cycle allows maintenance of a constant pressure throughout a cure period which may last up to six hours.

11.2.20 AUTOCLAVE INSTALLED AT WESTLAND AEROSPACE

Westland Aerospace of the Isle of Wight, UK, has just commissioned a £2 million autoclave — is the eighth autoclave to be installed at the company's advanced composite facility.

The autoclave is manufactured by Sholtz GmbH of Germany and weighs 53 t with internal dimensions of 3 x 9 m. It requires a 23 t air receiver. The unit operates at 200˚C and is designed for large thermosetting resin components.

11.2.21 MONTEDISON TO ESTABLISH COMPOSITES CENTRE

The Tencara shipyard at Porto Marghera near Venice, Italy, has been chosen by Montedison as the location of its centre for composite materials.

The centre will develop new materials and applications and has been sited at the shipyard because of the company's involvement there with the development of a composite hull for the Moro di Venezia, a competitor in the Americas Cup.

The hull project has been rated by Montedison as technologically comparable to current achievements in the aerospace and Formula 1 motor racing sectors. It has involved a range of Montedison companies including Structural Polymer Systems, Mofrini, Montefibre, Texindustria, SIR, and Himont.

According to Italo Trapasso, vice-chairperson of Montedison, the company is now spending US$250 million a year on materials research.

11.2.22 HIMONT OPENS RESEARCH CENTRE IN HONG KONG

Himont Inc, Wilmington, Delaware, USA, has opened a research and manufacturing technology centre in Hong Kong, to develop advanced materials including composites.

The facility, which cost US$ 7.7 million, will provide technical back-up in the development of polymer based composites and will work to meet environmental standards in manufacturing. The focus is to be directed towards the use of plastics and not their manufacture.

According to an article in the *Journal of Commerce*, the Hong Kong government views the choice of Hong Kong for this centre as "..indicative of the strategic location advantages.." the colony has to offer. A US$3 million contribution to the centre by the government was probably equally influential in the choice , particularly now that many Hong Kong companies, faced with high wages and land costs, are looking to alternative locations such as Thailand for similar research ventures.

11.3 JOINT VENTURES AND AGREEMENTS

11.3.1 HERCULES AND RHONE-POULENC COLLABORATE ON CERAMIC FIBRES

A joint agreement covering the development, marketing and sales of ceramic fibres has been announced by Hercules Advanced Materials and Systems of Wilmington, Delaware, USA, and Rhone-Poulenc SA, France.

The companies say that a memorandum of understanding has been signed which will allow Hercules to market and sell Rhone-Poulenc's silicon carbonitride fibre 'Fiberamic' in North America.

At present, Fiberamic fibre is produced in France at a pilot plant. The agreement allows for production in the USA by Hercules if the demand is sufficient or if other factors lead to an increased requirement for a domestic source.

The fibre itself comes as a multifilament tow of 250 or 500 fibres, and as short fibres 0.8 mm long, each with a nominal 15 microns diameter and a density of 2.4 g/cm^3. The composition by weight is Si (55%), N (22%), C (15%) and O (8%). The fibre remains amorphous up to temperatures of the order of 1400°C, and the company claims that this makes it suitable for high temperature applications in metal and ceramic matrix composites. Room temperature properties quoted by Rhone-Poulenc include a tensile modulus of 220 GPa, a tensile strength of 1800 MPa and a coefficient of thermal expansion of 3.1 x 10^{-6}/°C.

11.3.2 DU PONT AND LANXIDE JOINT VENTURE

Lanxide Corp of Newark, Delaware, USA, and Du Pont Co of Wilmington, also in Delaware, have announced a joint venture which will commercialize high performance microelectronic components using Lanxide's patented Primex reinforced metals technology.

The company will be called Lanxide Electronic Components LP, and will offer hybrid circuit electronic packages, carrier plates, substrate chassis and support structures. Products will be tailored to meet specific requirements, e.g. thermal management, material compatibility and structural demands, with initial products based on ceramic particle reinforced aluminium. Future products under development will use fibre reinforced metals. A 37 000 square foot production facility has been constructed. This is the third such at Newark to be formed between Lanxide and Du Pont. The previous companies formed were Lanxide Armor Products Inc, and Du Pont Lanxide Composites Inc, both of which were established in 1987.

11.3.3 QUADRAX AND HEXCEL LINK ON COMPOSITE PANELS

Quadrax Corp of Portsmouth, Rhode Island, USA, has signed a definitive agreement with Hexcel Corp, Dublin, California, to develop, manufacture and market cored structural panels for aerospace, naval and rail applications.

According to a report in the *Journal of Commerce*, the panels are to be made by combining a core of Hexcel's metallic or non-metallic honeycomb with a facing of Quadrax's thermoplastic prepreg fabric. The products are to be manufactured and marketed by Hexcel when commercialized. The two companies will work together to develop the necessary bonding technology.

11.3.4 DOW–UNITED TECHNOLOGIES JOINT VENTURE

A joint venture company was established by US companies Dow Chemical of Midland, Michigan, and United Technologies, in December 1989, to exploit composite technology.

The company is called Dow–United Technologies Composite Products Inc and currently employs 500 people — 350 in production and 150 at an engineering development centre. The business is essentially a specialist composite manufacturing company, merging the activities of Dow Chemicals and Sikorsky Aircraft in this area. It offers design, testing, qualification and production services. The parent companies say that Dow–United Technologies Composite Products is to be able to work with its customers from the outset of a project and as such concentrates on selling composite parts rather than composite materials.

The formation of the company allows Dow and UTC to concentrate on their core businesses (materials development in the case of Dow and helicopter production in the case of UTC) while still retaining a financial interest in the development of the composite manufacturing business. Freeing the composite business from the parent activities is also expected to allow a greater concentration of resources and the development of targeted expertise in a way that was previously impossible.

In anticipation of future growth, Composite Products has announced a US$20 million investment that will be used to double the production capacity of its current Tallassee plant in Alabama and create a new headquarters and research and development facility at Wallingford, Connecticut.

11.3.5 GKN ENTERS JAPANESE MARKET

The UK-based engineering group GKN recently announced a contract to supply its lightweight composite leaf springs to the Japanese truck market. The leaf springs will be manufactured at GKN Composites', Telford, UK, plant and then assembled into steel leaves in Japan, forming hybrid springs for the Mitsubishi Motors Corp.

The contract for the springs, which are to be used in the front and rear suspension units of Mitsubishi's 8 t 'Fighter' truck, was won by Translite KK, a joint venture company set up by GKN and Mitsubishi Steel Manufacturing Co in 1985, specifically to develop the Japanese market. Mitsubishi Steel will assemble the spring units in Japan.

A weight saving of some 40 kg per truck is claimed by GKN through the use of the hybrid springs which are interchangeable with existing multi-leaf steel springs. Additional benefits are reported including improved ride and handling, together with reduced interior noise.

GKN Composites has been a limited company since January 1990 and operates within the GKN Automotive Group. It has already supplied composite leaf springs to Leyland Daf, Iveco, Mercedes Benz and London Taxi International.

11.3.6 AUSTRIAN –JAPANESE JOINT VENTURE FOR ALUMINA FIBRES

The Austrian company, Rath Co of Vienna and Tokyo-based Denki Kagaku Kogyo KK have formed a joint venture with the intention of producing and selling up to 200 t/year of alumina short fibres, primarily in the USA.

A plant has been built in Austria by the two companies for the manufacture of alumina fibres and Rath has created Rath Performance Fiber Co, based in Delaware, USA. The latter has already received its licence to sell alumina fibres in the USA. The final part of the package concerns Denki Kagaku Kogyo which is about to start a market survey for alumina short fibres in the USA.

11.3.7 WIGGINS TEAPE LICENSES 'RADLITE' PROCESS TO JAPAN

Wiggins Teape has licensed a consortium of Japanese companies to use its Radlite process for producing stampable reinforced thermoplastic sheet, according to a report in *Metals and Materials*.

The Radlite process is an adaption of paper making technology and combines chopped glass fibres and thermoplastic polymer powder into a sheet form. When consolidated this reinforced stampable sheet material has improved properties compared with materials produced by alternative routes, such as melt impregnation of fibre mats. Fibre dispersion and sheet consistency are particularly notable properties, it is stated.

The Japanese consortium has established a joint venture called K-Plasheet and the partners include Kawasaki Steel, Sumitomo Chemical, Takiron, and C. Itoh and Co. A plant costing US$10.3 million is being built at Kawasaki Steel's Chiba works near Tokyo. The polypropylene matrix powder will be supplied by Sumitomo. Applications for the material include automotive interior panels, bumpers, beams, engine undercovers and seat backs.

Current licensees of Wiggins Teape include GE Netherlands, who produce an Azdel product using this process, and Ahlstrom of Finland.

11.3.8 NATIONAL PLASTICS TAKES BALLISTICS LICENCE FROM OWENS CORNING

National Plastics, a subsidiary of Courtaulds Advanced Materials, has negotiated the first European licence to produce ballistic laminates using technology developed by Owens Corning, Toledo, Ohio, in the USA.

The technology covers both patented advanced materials systems and process technology, with the key element being the use of Owens Corning's S-2 high performance glass fibres.

National Plastics intends to use the new technology to gain access into vehicle protection markets. This move will complement its existing range of personal protection systems which includes helmets and fragmentation vests.

11.3.9 AMERICAN CYANAMID OBTAINS LICENCE FOR 'PMR-15' POLYIMIDE RESIN

Rohr Industries, Chula Vista, California, USA, has licensed American Cyanamid to use a Rohr-developed synthesizing procedure to produce an improved PMR-15 polyimide resin.

PMR-15 is a high temperature polyimide resin that is capable of operating at service temperatures in the range of 320°C (600°F). The exclusive licencing agreement will enable Cyanamid to exploit the resin for use in adhesives and composite prepregs where one of the major customers is likely to be Rohr Industries itself for advanced nacelle structures.

The key element in the improved technology (the ownership if which will be retained by Rohr) is a reduction in the amount of methylene dianiline involved in the resins production to below 0.1 %, without altering the final composition, shelf life or processing characteristics. This is of considerable value given the great health concerns surrounding methylene dianiline.

Cyanamid reportedly plans to carry out scale-up programmes at its Stanford, Connecticut, chemical research facility. Ultimately, the production of the prepregs is planned for the company's Saugus plant in California.

11.3.10 GORHAM TO LICENSE CERAMIC TECHNOLOGY

A proprietary technology for fabricating fully dense ceramic matrix composites and advanced ceramics is to be licensed worldwide by its developer Gorham Advanced Materials Institute of Gorham, Maine, USA. The technology involves sinter/HIPing, high pressure reactive sintering, and sintering processes in general and has developed out of a three year, US$2 million research effort into processing of aluminium oxide, silicon nitride, aluminium nitride, partially stabilized zirconia, and silicon carbide.

11.3.11 GLACIER TO SUPPLY 'HY-LOAD' BEARINGS TO EUROPE

Glacier Industrial Bearings is to act as exclusive European distributors for the Hy-Load composite bearings manufactured in the USA by Engineering Plastics Inc.

The Hy-Load bearings a consist of carbon and glass fibre filled polytetrafluoroethylene backed with perforated stainless steel sheet. This combination of materials provides a creep resistant system capable of operating without lubrication and in most caustic or acid environments up to 280˚C.

The Hy-Load bearings come as wrapped and lagged bushes, thrust washers and sliding plates, and complement Glacier's existing range of bearing materials based on polytetrafluoroethylene lined with materials such as lead, polyetheretherketone and acetal copolymer.

11.3.12 PPG SELECTS FRP SUPPLY TO DISTRIBUTE PRODUCTS

PPG Industries of Pittsburgh, Pennsylvania, USA, has selected FRP Supply to act as distributor for its complete line of glass reinforcements in the state of Wisconsin, according to the latter's parent company, Ashland Chemical Inc of Columbus, Ohio.

FRP Supply distributes PPG's products in other parts of the USA and this recent development is part of a general expansion into the Midwest, following a successful entry to the West Coast market last year. Ashland Chemical claims FRP is the largest distributor of resins, catalysts, reinforcements, and other materials to the fibre reinforced plastics industry in the USA.

11.3.13 PULTREX AND EUROCARBON APPOINT US AGENTS

The US agencies for UK-based company Pultrex, manufacturer of advanced pultrusion, pullforming and filament winding machines, and Eurocarbon BV of Tilburg in The Netherlands, producer of flat and tubular braids and speciality fabrics, have recently been awarded to C3 International of El Dorado Hills, California.

Pultrex machines are currently in service in 22 countries worldwide. The largest pultrusion machine currently offered by the company, the P1200M, is said to boast a pulling force of 12 000 kg and a profile envelope of 1000 x 250 mm.

Eurocarbon's products, which use a variety of high performance fibres including carbon, aramid, glass and ceramics, found application in medical items such as prostheses and in sporting goods.

In addition to operating as company agents, C3 International acts as a consultancy company specialising in marketing and sales support to the composites industry.

11.3.14 RESEARCH INC RANGE AVAILABLE IN UK

Astro Technology of Fareham has been appointed the sole distributor for Research Inc of the USA. The range includes short wave infra-red heaters which are successfully used in composite manufacturing processes such as filament winding and in preheating, welding and materials testing. The model 5070 Multi-Zone Load Test Heater is designed for use in conjunction with mechanical testing equipment and should fit between the jaws of most standard testing machines. The unit is capable of heating a 1.25 mm thick specimen to 1500°C in less than 90s, according to Research Inc.

11.3.15 COLLABORATIVE AGREEMENT BETWEEN CANADA AND JAPAN

The Reinforced Plastics/Composites division of the Society of the Plastics Industry of Canada has signed a collaborative agreement with the Japan Reinforced Plastics Society.

This move follows a visit by representative of the Canadian reinforced plastics/composites industry to Japan in 1989 and aims to improve industrial cooperation between the two countries.

Specific actions that will result include an exchange of membership lists between the two organizations and exchange of publications, reports, conference papers and any literature of mutual interest.

11.3.16 ASHLAND LOOKS TOWARDS EUROPE

Ashland Chemical, of Columbus, Ohio, USA, is looking to increase its business in Europe following its success in applying Arotran resin transfer moulding resins to the car maker Lotus for use in its new Elan model. According to *Plastics and Rubber Weekly*, this may result in a major investment over the next few years although the form of this investment is not clear at present — a green field site or a joint venture are possibilities.

Ashland has a major position in the US automotive industry, particularly with respect to supplying General Motors (nb: Lotus is now a partly owned subsidiary of GM).

11.4 TAKEOVERS AND MERGERS

11.4.1 MITSUBISHI RAYON ACQUIRES NEWPORT

Mitsubishi Rayon America Inc, the US subsidiary of Mitsubishi Rayon Co Ltd, of Japan, has purchased Newport Composites Inc, and Newport Adhesives Inc of Santa Anna, California, USA, for a figure believed to be in the region of US$330–350 million, according to *Composite Market Reports*.

Newport Composites is a material supplier of prepregs, mainly to the sports and non-aerospace industries. The link with Mitsubishi will probably lead to an increased effort aimed at the aerospace sector and, accordingly, some of Mitsubishi's products, including resins and fibres, may be added to the Newport product list in time. The two formally separate units will now operate under the combined name of Newport Adhesives and Composites Inc.

11.4.2 ACQUISITIONS BY CIBA GEIGY

Ciba Geigy, Basle, Switzerland, has made a number of acquisitions to consolidate its composites business in Europe.

The Austrian company Danutec Werkstoff GmbH, Neumarkt and Linz, has been purchased in a joint venture with the Austrian Petrochemical Industry (Petrochemie Danubia) which is a 49% stakeholder.

Danutec specialises in prepregs for sporting goods and industrial applications which it sells under the tradename 'Strafil', and also produces laminates ('Polyspeed') and foaming systems for panels, ('Modipur'). Sales of Strafil prepregs amounted to US$28 million in 1989.

Danutec will now become the industrial prepreg unit within the Ciba Geigy composite business, allowing the other businesses at Duxford in the UK and Dagneux in France to concentrate on aerospace applications.

The second company to be acquired is the Salver company which is based in Brindisi, Italy.

Salver, which employs about 100 people, is a composite parts producer whose clients include Aeritalia, Augusta, Fiat, Aviazione, Boeing, McDonnell Douglas, Piaggio, Aermacchi, Dassault and Sonaca. The purchase of Salver in Europe is part of the same overall strategy that saw another supplier of composite to the aerospace industry, Heath Tecna, of Kent, Washington, USA, being bought by Ciba Geigy in 1988. The existing Salver management will continue to be responsible for Salver which will operate within Ciba's worldwide Composite Products organization.

11.4.3 ADVANCED REFRACTORY TECHNOLOGIES ACQUIRES AMERICAN MATRIX INC

Advanced Refractory Technologies Inc of Buffalo, New York, USA, has acquired the assets and technology of American Matrix Inc, Knoxville, Tennessee.

Advanced Refractory Technologies is continuing to develop the various product lines of American Matrix which includes whiskers and platelets of materials such as titanium nitride and silicon carbide whiskers.

Advanced Refractory Technologies itself produces refractory oxide powders and other ceramics, such as aluminium nitride, titanium and zirconium diboride, boron carbide and beta silicon carbide.

The expended organization now claims an integrated business with capabilities in advanced materials, production of powders, whiskers and platelets, and component fabrication. The latter involves production of parts from both monolithic ceramics and ceramic matrix composites, metal matrix composites and cermets.

11.4.4 CRAY ADVANCED MATERIALS SOLD

Cray Electronics (Holdings) plc has sold its entire shareholding in its subsidiary company Cray Advanced Materials Ltd, Yeovil, UK, to private British investors. The change in ownership is reflected by a change of name to Advanced Materials Systems but all staff and directors have been retained.

The particular expertise developed by Advanced Materials Systems is in the liquid pressure forming route to net shape metal matrix composite parts. The technology is protected by patents held by the UK's Ministry of Defence and licensed to Advanced Materials Systems and essentially consists of the infiltration of a fibre preform with molten metal.

11.4.5 TENMAT ACQUIRES LAMINATED PLASTICS PRODUCTS

Tenmat Ltd of Manchester, UK, has acquired Laminated Plastics Products, formerly part of the Glaxo Group.

Laminated Plastic Products specializes in the manufacture and assembly of corrosion resistant composite items, such as agitators and baffles for the chemical industry, under the tradenames of 'Lamanil' and 'Fluoranil'.

Laminated Plastics' manufacturing facilities and personnel have been transferred to Tenmat's factory at Trafford Park where Tenmat also specializes in high performance engineering composites for the chemical industry.

11.4.6 DSM EXPANDS IN THE UK

The Dutch multinational DSM is continuing its expansion in the European composite moulding field by acquiring ERF Plastics, one of the largest UK producers of sheet moulding compound parts, in a £4.5 million deal. ERF Plastics was originally established by ERF Ltd, an independent UK truck producer, to supply the composite moulded parts for ERF truck cabs. The agreement between the two companies guarantees supplies of cab components for the future.

The sale has been agreed with DSM Resins UK Ltd, formerly Freeman Chemicals Ltd, a supplier of resins and moulding compounds to the UK market.

In a further development, the company has purchased 46 acres of land at Ellesmere Port (with an option on a further 10 acres) in order to establish a common site for all UK production, development and office facilities.

11.4.7 PILKINGTON SELLS GLASS FIBRE PROCESS TO SAINT GOBAIN

The Pilkington Group of St Helens, UK, is selling its alkali-resistant glass fibre process, 'CemFil', to Compagnie de Saint Gobain of France. The move is expected to make the French company the world leader in the field of glass reinforcements for cement.

11.4.8 UK RESEARCH INSTITUTES COMBINE

Fulmer Research Institute and BNF Metals Technology Centre are to combine to form a single R&D organization, working primarily in metallic and composite materials development and associated enabling technologies.

The new grouping has arisen from the purchase of Fulmer Research by BNF Metals Technology Centre, from the former owners, the Insitute of Physics. This organization retains control of remaining parts of Fulmer Ltd, i.e. Fulmer Yarsely and Yarsley Quality Assurance. The new organization will trade under the name of Fulmer Materials Technology.

11.4.9 UK PROFESSIONAL INSTITUTES FAIL TO MERGE

The three leading UK based materials institutes, the Institute of Metals, the Plastics and Rubber Institute and the Institute of Ceramics, have failed in their attempts to merge to form a single Institute of Materials.

This leaves the three bodies in some disarray, at least in the short term. A Federation of Materials institutes has already been formed by the three bodies which has paved the way for merger talks and begun the process of cooperation between them. Given the support within the membership for merger it is likely that the Federation will be strengthened and another attempt at full merger may be made in a few years time. In the meantime, bodies such as the British Composites Society that had looked towards an Institute of Materials as a possible future home will have to bide their time and continue as separate independent bodies, for the time being at least.

The move to amalgamate professional bodies in the UK to form ever more powerful organisations is however continuing, with a recent announcement that the much larger Institution of Mechanical Engineers and Institute of Electrical Engineers have entered into merger negotiations.

11.4.10 DSM SUBSIDIARIES SEPARATE

Two former DSM subsidiaries based at Tilburg in the Netherlands have parted company. Eurocarbon Tilburg BV and Eurocord BV were both part of DSM's High Performance Fibers BV group, but now Eurocarbon has been integrated into the Compounds and New Developments business unit of DSM while Eurocord has undergone a management buyout and is under the independent control of managing director C Croon.

Eurocarbon will continue to concentrate on processing of technical high performance fibres such as carbon, glass, aramid, and DSM's Dyneema polyethylene, and thermoplastic compound yarns. Eurocarbon specialises in twining and weaving techniques and their application in composites (particularly overbraiding, preforming, 3D and triaxial weaving).

Eurocord is primarily involved with the processing of textile yarns via knitting braiding and ribbon weaving techniques.

11.5 RESTRUCTURING

11.5.1 DU PONT CEASES PRODUCTION OF PITCH-BASED CARBON FIBRES

Du Pont de Nemours and Co of Wilmington, Delaware, USA, has decided to cease producing its range of pitch-based carbon fibres.

Sources at the company say the decision was prompted by the large investment deemed necessary to sustain and develop activity in this area. The long term returns from the market for high and ultra-high modulus carbon fibres were not considered sufficient to justify further outlays — despite the considerable sums of money already invested.

Du Pont's pitch-based fibres have tensile moduli of 724 GPa ('E-105') and 894 GPa ('E-130'). E-105 and E-130 both maintain a strain-to-failure of 0.55% and have tensile strengths of 3.3 and 3.9 GPa, respectively.

Existing stocks of the fibres will be used to honour current commitments to customers and research laboratories but production has now ceased, according to the company.

11.5.2 RATIONALIZATION FOR CIBA GEIGY OPERATIONS IN EUROPE

Ciba Geigy is to rationalize some of its operations in Europe. The Duxford plant near Cambridge in the UK has been identified as the centre of excellence for the production of formulated epoxy products, while the plants in Switzerland and Spain will become the sole production centres for basic liquid and solid epoxy resins for the coating industry. The phasing out of the production of the basic epoxies from Duxford will be accompanied by a £9 million investment in a plant which it is claimed will be the most modern example of its type in Europe. Duxford will produce products for the UK adhesives, electrical and tooling industries, composites matrices and structural casting resins and aerospace adhesives for international markets.

11.5.3 ADVANCED MATERIALS DEVELOPMENTS AT BP

The development of the UK-based group BP's advanced materials business is continuing, with a series of acquisitions, restructuring and new ventures.

The metal matrix composite business has been strengthened by the purchase of DWA Composites, a specialist US organization which will complement activity centred in the UK on both particulate and silicon carbide monofilament reinforced metals. The 'Sigma' process for producing silicon carbide monofilaments was purchased in 1987 and commissioned at the BP Corporate Research Laboratories in Sunbury, UK. The Sigma fibre is uncoated and therefore susceptible to attack by molten metals (although BP claims, excellent final properties despite this).

Currently, most Sigma fibre produced at Sunbury is consumed by research programmes (with various consortia in the USA and Europe) aiming to develop coated or modified fibres. BP hopes to be able to supply coated monofilaments and their metal matrix composites sometime during 1990. An agreement has recently been signed with the Atlantic Research Corp of Alexandria, Virgina, USA, to supply coated Sigma monofilaments for evaluation in the US National Aerospace Plane project (NASP).

Particulate metal matrix composites are also produced in the UK, routinely in 9 kg billets, with billets of 34 kg being demonstrated and even larger billets being planned. Most effort has been put into aluminium matrix (2124 Al/Cu and 8090 Al/Li alloys) with silicon carbide particulate reinforcement. As part of the scaling up of these interests, a facility is now being established at the Royal Aircraft Establishment (RAE), Farnborough, UK. Collaboration with RAE will, it is hoped, help to accelerate product development. Initial quantities of metal matrix composites are being released to customers in the form of sheet, plate and extrusions.

BP Metal Matrix Composites Ltd is still under the control of the BP Ventures Division, while BP Advanced Materials has moved under the wing of BP Chemicals. The Advanced Materials group includes Hitco and US Polymeric in the USA, together with the Carborundum Co.

Total sales in the advanced structural materials area by BP companies are now reported to approach US$0.5 billion, according to Kevin Gordon, manager of metal matrix composite development, who sees BP's goal as covering the whole of the advanced materials spectrum.

11.5.4 SHELL TO DIVEST IN THE USA

Shell is to divest itself of a number of its composites interests in the USA. The companies involved are Compositek and Xerkon Inc, both of Brea in Southern California (these are largely integrated operations), and Winding Technologies Corp of Springville, Utah. Compositek and Xerkon are both wholly-owned Shell operations, while Shell only has a 30% stake in Winding Technologies. All of these interests were under the umbrella of Shell's SPACE outfit (Shell Polymers And Catalysts Enterprises Inc) which itself is a subsidiary of Shell Oil Company. The businesses are not being sold as a result of poor performance and all report busy workloads and optimism for the future. It would appear that Shell, is viewing the divestitures as a way of investing extra resources in other parts of the SPACE business. Companies in the composites field remaining inside SPACE include Morrison Moulded Fiber Glass and the AFC division which are more heavily involved in non-aerospace work. The automotive sector has been a long term target for Shell, with a considerable number of successful experimental applications evolving from collaboration between the Westhollow Corporate Research Laboratories in Houston, Texas, and various large and small automotive companies.

11.6 NAME CHANGES AND RELOCATIONS

11.6.1 AMOCO PERFORMANCE PRODUCTS RELOCATES

Amoco Performance Products of Ridgefield, Connecticut, USA, says it is relocating its corporate offices, together with its research and development facilities, to Atlanta, Georgia.

The 29 800 m^2 research centre is located at 4500 McGinnis Ferry Road, in suburban Alpharetta. The corporate offices will move from Ridgefield, while the research and development was formerly located in various sites, including Bound Brook, New Jersey, Parma, Ohio, Naperville, Illinois, and Augusta, Georgia.

11.6.2 ICI TO OPEN EUROPEAN APPLICATIONS CENTRE

ICI Advanced Materials is to open a European Application Centre in Oestringen, Germany.

The centre, to be led by Dr Eberhard Doering, will consist of a computer aided design office, a laboratory, and a machine hall. It will provide services on part and mould design, materials selection, part development, performance assessment, and pre-production runs, with an emphasis towards the automotive industry, ICI says.

The Oestringen site currently includes a plant for ICI Fibres and ICI Fiberite.

11.7 CONTRACTS AND COMPANY PERFORMANCE

11.7.1 REVENUES UP 10% FOR DU PONT ENGINEERING FIBRES

Revenues for Du Pont Engineering Fibres of Le Grand-Saconnex rose 10% in 1989 to a total of DM597 million, the Swiss-based company reports.

Above average gains were recorded in the automotive field, where 'Kevlar' is continuing to make inroads in the asbestos replacement market for brakes, clutches and gaskets, as well as its increasing use in the high performance tyre market. The growing aircraft market has also boosted demand for both Kevlar and 'Nomex' honeycomb, it is stated.

Following the recent commissioning of the Kevlar plant in Maydown, UK, work started on a plant for production of Nomex in Asturias, Spain, with start-up scheduled for the end of 1992. Total European investment in Engineering Fibres Systems' production facilities over the period 1985–90 has now passed DM800 million, according to the company.

11.7.2 BASF PREDICTS GOOD GROWTH FOR COMPOSITES

A German chemical company is predicting good prospects for its composites operations.

In its annual report, BASF of Ludwigshafen announces overall pre-tax profits for 1989 of DM4384 million, an increase of 17.5% compared to 1988. An example of the growth recorded in 1989 was a 50% increase in sales of structural prepregs in the UK, mainly servicing the boom in the civil aircraft market.

The report informs shareholders that a considerable research and development effort is being maintained in composites, with some emphasis on high temperature bismaleimides. An expansion of the US manufacturing capacity is underway to meet increasing demand, and capacity should be doubled by the end of 1990.

The company is also concentrating its effort on its polyester-based composites ('Palatal' and 'Palapreg') for the automotive market. A pilot plant for the production of composite leaf springs for trucks has been built at the company's Worms site.

11.7.3 LANXIDE REPORTS GROWTH

Lanxide Corp of Newark, Delaware, USA, has reported results showing growth in sales to US$905 000 during the third quarter of 1990.

This is an increase of 470% on the same quarter last year. Total sales for the first nine months of this fiscal year amount to US$1.6 million, an increase of 437% on 1989.

Sales are primarily to customers for test and evaluation purposes, holding out the prospect for further substantial growth in the future, the company says.

11.7.4 RECORD EARNINGS FOR PPG

Second quarter earnings of US$141.0 million on sales of US$1.57 billion have been reported by PPG Industries, the US glass and coatings concern. Corresponding figures for 1989 were US$127.3 million earnings on sales of US$1.49 billion.

The company says that this stronger performance is despite lower sales of its glass fibres during this period. The slight downturn in glass fibre sales has not dissuaded PPG from continuing its acquisition policy.

11.7.5 LEYLAND-DAF CONTRACT FOR BTR PERMALI RP

BTR Permali RP, Gloucester, UK, has won an initial £600 000 contract for hot press moulded glass reinforced plastic (GRP) components for a Leyland-DAF –4 t military truck.

An order of 5350 of these trucks has been placed by the UK's Ministry of Defence and the truck goes into production in 1991. BTR Permali RP will be supplying a variety of interior components for the truck cab.

11.7.6 METAL MATRIX CONTRACT FOR ADVANCED COMPOSITE MATERIALS

An important contract from the US Air Force for the production of discontinuously reinforced aluminium metal matrix composites has been awarded to Advanced Composite Materials Corp (ACMC) of Greer, South Carolina, USA.

The contract, which is valued at US$23.9 million, is of a type which guarantees a company a minimum market for an emerging technology rather than for the production of a quantity of specific parts. This comes under the Title III provisions of the US Defense Procurement Act, designed to provide an incentive for companies to expand their industrial facilities by means of a purchase commitment.

The ACMC discontinuously reinforced aluminium composites are produced from aluminium powder blended with silicon carbide whiskers or particles. The Title III contract is for a three phase programme spanning five years. The first phase, of 18 months, will refine production techniques and design new plant for a scale-up in processing capabilities. The second phase, lasting 27 months, will allow installation of a commercial scale manufacturing plant producing billets of more than 270 kg. In phase three, ACMC will manufacture and sell 33 400 kg of discontinuously reinforced aluminium billet to the Air Force.

ACMC is a subsidiary of Tateho Chemical Industries Co Ltd. Tateho will fund the design and construction of the plant during phase two of the programme.

11.7.7 GORHAM AWARDED FOLLOW-ON DARPA CONTRACT

Gorham Advanced Materials Institute, Gorham, Maine, USA, has been awarded a twenty four month Phase II Small Business Innovation Research contract by the US Defense Advanced Research Projects Agency (DARPA). The contract is to develop silicon carbide whisker-reinforced silicon nitride matrix composites.

The contract is a follow-on to a Phase I programme which demonstrated full densification of silicon nitride with up to 4 wt% silica, reinforced with 15 vol% silicon carbide whiskers using containerless Sinter/HIP processing. A chemical vapour deposition-applied coating on the whiskers improved the densification and the resulting composites were found to retain more than 90% of their room temperature four point bend strength at test temperatures up to 1200°C. The room temperature fracture toughness was reported to be of the order of 6 MPa.m$^{0.5}$

11.7.8 ROLLS ROYCE MATEVAL WINS ORDER FROM FIBERITE

Rolls Royce Mateval, of Warrington, UK, has won a US$400 000 order for a multi-axis composite material scanning system from ICI Fiberite (Composite Structures) based in Tempe, Arizona, USA.

The order, which the company says was won in the face of considerable competition, is for a system based on the Micropulse 2 programmable inspection controller which uses ultrasonic water jets to detect defective components.

The key feature of this particular system, according to the manufacturer, is its ability to accommodate panels of varying geometry.

11.7.9 SCHAPPE LOOKS FOR MARKETS FOR SPUN YARNS

S.A. Schappe, the French textile company, is embarking on a world wide marketing initiative, seeking composite applications for its stretch broken yarn spinning process.

Schappe's process involves taking continuous filaments and stretch breaking them into long fibres. This process effectively removes weak spots along the fibres and the company claims that this will improve tenacity and processability. The long fibres are then spun into a variety of yarn products. Material options include carbon, glass aramid and thermoplastic fibres.

Erskine-Johns Company is supporting Schappe's existing US business and is also helping to identify composite applications for such products.

11.7.10 ADVANCED COMPOSITES MEETS NATO STANDARDS

The quality system operated by the Advanced Composites Group of Heanor, Derby, UK, has been re-assessed and approved to the requirements of NATO standard AQAP-1 by the Ministry of Defence in the UK. The Advanced Composites Group is a supplier of mouldings, prepregs and tooling materials to the composite industry.

11.7.11 QUALITY AWARD FOR UK COMPANY

The British Standards Institution (BSI) of Milton Keynes, UK, has certified to BS 5750 quality standards the UK manufacturer Insulation Equipments Ltd of Oswestry.

Insulation Equipments specializes in producing fire-safe composites for applications in the mass transit, marine and off-shore industries, using phenolic resin systems in particular.

A recent innovative application of phenolic glass reinforced plastic produced by Insulation

Equipments is the structural casing for a series of emergency telephone installations, to be introduced throughout the UK by the Royal Automobile Club (RAC). These units need to be fire resistant, durable and able to resist weathering, conditions which favour its Melaform phenolic, the company claims.

11.7.12 AWARD FOR COMPOSITE ARTIFICIAL LIMBS

The technical excellence of the Endolite artificial limbs, produced with extensive use of composite materials, has been recognised by a Queen's Award for Technical Achievement 1990. Endolite's manufacturer Chas A. Blatchford and Sons, UK, currently supplies 65% of the market for lower leg amputees in the UK and the second largest market for its products is now the USA.

The Endolite artificial limbs exploit composites for many novel features, including the Multiflex ankle joint, as well as carbon fibre composites for the main load bearing components.

Blatchford is currently training staff for an artificial limb factory which is to be established in China under the auspices of UNIDO.

CHAPTER 3

PLASTICS

1. **CONTENTS** 252

2. **EXECUTIVE SUMMARY** 263

3. **MARKETS**

4. **MATERIALS**

9. HEALTH, SAFETY AND THE ENVIRONMENT

2. EXECUTIVE SUMMARY

Within the plastics market, two crude categories of innovation became apparent during 1990, distinguished by their scale and degree of persistence. On the one hand, there are areas where change is managed and institutionalized and has gathered momentum over the years; the main example of this is the automobile industry. The other category is characterized by the change being made in response to events external to the industry; during 1990 this latter category was mainly driven by environmental considerations, such as the ones made necessary in foam technology as a result of the phasing-out of chlorofluorocarbons (CFCs). Further, interactions are possible between events in the two categories; an example of this is the extra momentum given to the progress in metal replacement in cars by the increased fuel efficiency obtained via weight saving.

In terms of market scale, the automotive sector dominates the scene. Here the combination of mass market, pressure on costs and, in many cases, relatively modest engineering requirements provide the conditions for continuing encroachment by large volumes of plastics. Similarly, opportunities continue to be exploited in the electronics industry. Another large market is in polyurethane foams, but here the dominant feature for the near future is not industry-led but dictated by outside forces, namely the elimination of chlorofluorocarbons on environmental grounds.

Outside the high tonnage areas, there are interesting medical, optical, scientific and sporting applications. The high-tech tennis racket is the most obvious example of materials innovation in a sporting area, and this now presents a new application of polyethylene fibre. Bioengineering continues to be a fruitful area where high cost polymers can be used. The polymer gel that shrinks to a tiny fraction of its original volume on exposure to light seems to be a classic case of a solution looking for a problem — a description in former times applied to the laser.

Some genuinely new materials have been developed, while the routine introduction of new grades of established polymers for specific applications continues. Notable are new liquid crystal polymers and the continued activities in blends. An exotic high performance polymer, polyaryletherketone with melting point higher than that of polyetheretherketone, has promise for yet more testing environments, with a current application in bioengineering. The influence of the electronics industry is again apparent in the form of new conductive polymers, useful for screening and other purposes.

One of the spurs to the introduction of plastics components is the saving in labour which can be achieved by the manufacture of large components in a single step. The development of ever larger injection moulding machines is a reflection of this. Larger components are, however, more difficult to make, and so better understanding and control of the moulding process is sought. Computing is essential to both the understanding — in the form of simulation software — and control — in the form of microcomputer systems. The latter are also now in evidence for control of equipment for testing.

Few industries are immune from the pressures which result from their environmental effects. An obvious example of this for the polymer industry is the elimination of chlorofluorocarbons. The degradability debate is perhaps more interesting, as it illustrates the complexity of environmental issues. The growing evidence that global warming is being caused by so-called greenhouse gases has led to a rethink about the environmental consequences of using degradable polymers, which emit such gases as they degrade. With degradable polymers going out of favour, attention has shifted back towards the recycling of polymers.

One report cites plastics, of all solid waste materials, as having the highest growth in their rate of recycling. Recycling is in some instances, particularly in the US, being helped by the use of new technology for separation in recycling plants, and there is a growing market for hardware such as granulators, shredders and separators. The feasibility of recycling plastic car bumpers has been demonstrated, and this could become a big new area if it were to be adopted by the automobile industry. The environmental problems associated with polyvinyl chloride (which makes up about 30% of the polymers used in the world) were discussed at the PVC '90 conference. These include the disposal problem, which is exacerbated by the corrosive gases and acid rain produced by the incineration of polyvinyl chloride. One of the conclusions was the need for increased recycling; polyvinyl chloride, because of its cheapness, is recycled in small quantities. This raises the question of the efficacy of market forces in this context — is this to be addressed by making the polluter pay?

On the research front, a significant development in the UK is the creation of an interdisciplinary Research Centre (IRC) in polymer science and technology. This is a consortium involving the universities of Leeds, Durham and Bradford in which expertise in polymer physics, chemistry and engineering has been combined.

News from industry tells of bustling activity in investments, takeovers, joint ventures and restructuring. Some of this is rationalisation and consolidation in preparation for the European single market set to start up in 1992. New companies to promote environmental services are reported. Some of the investments and acquisitions can be seen to fit in with the rational strategies of particular companies, but the overall picture is complex. The only clear impression is that of a busy market place.

3. MARKETS

Here we encounter important end-user industries whose dominance will be reflected throughout the plastics chapter of this Source Book. The automobile industry is a major end-user, one where continued replacement of metal by plastics has provided growth over and above that implied by inherent growth in car manufacture. Increased electronic and electrical applications follows the increased use of microelectronics. The chlorofluorocarbon question continues to provide a level of uncertainty for the size of future markets in polyurethane foams.

3.1 GENERAL

3.1.1 POLYVINYL CHLORIDE — PRESENT AND FUTURE

A study by Phillip Townsend Associates documents 93% of the reported consumption of polyvinyl chloride raw materials in seven European counties, pinpointing quantities, grades and suppliers of raw materials consumed, as well as processing methods used and end-use markets served.

'Plastic buyer profiles — Europe' documents the consumption of 1 012 000 t of polyvinyl chloride at 395 sites in Germany. France reports a slightly lower consumption of polyvinyl chloride, with 805 000 t processed at 432 sites. The United Kingdom ranks third in PVC consumption among the countries studied, counting 640 00 t processed at 349 sites. Also included in the study are The Netherlands, reporting 188 000 t (71 sites); Belgium, reporting 169 000 t (96 sites); Austria, reporting 82 000 t (68 sites); and Switzerland, reporting 62 000 t (47 sites).

The report lists the common methods of processing polyvinyl chloride in the countries studied. In order of popularity, as evidenced by the number of sites using them, these are: profile extrusion; pipe extrusion; blow moulding; wire and cable extrusion; coating; film/sheet extrusion; and calendering.

Building materials, such as door and window parts, and consumer products and furnishings head the list of end-user markets served by PVC processors. Other markets reported are bottles and rigid packaging, pipe and pipe fittings, industrial components, automotive parts, and wire and cable.

The UK future for polyvinyl chloride is the subject of a study by Corporate Development Consultants (CDC), who have started to investigate the prospects for thermoformed packaging in the UK up to 1994. This will evaluate what substitutional trends might develop in the UK, how far suppliers of alternative thermoforming materials such as amorphous polyethylene terephthalate (APET) and oriented polystyrene (OPS) can exploit this situation, and how demand for unplasticized polyvinyl choride (uPVC) film might be affected.

The period to 1994 will be an especially challenging time, given the creation of the Single European Market. To safeguard business development over this period, it will be important for thermoformers to strengthen their marketing effectiveness; this is an essential prerequisite for the growth of turnover and profits. As part of this investigation, therefore, CDC is also undertaking a detailed assessment of the image and reputation of leading thermoformers, which will be reported confidentially to participating subscribers. This will enable subscribers to measure how their image and reputation compare with those of their main competitors. The study will also provide a comprehensive analysis of trends in the demand for thermoformings by end-user sector.

3.1.2 UK PLASTICS INDUSTRY IN THE 1990s

Perhaps more optimistically, a recent report from market analysts Key Note predicts a 30% per annum growth in the market for British-made plastics during the 1990s. This forecast is based on the fact that in many categories such as packaging, building and construction, and general mouldings, sales have doubled over the last five years.

The UK plastics processing industry is preparing itself for 1992, when the European Economic Community will become a free-trading area. Already a host of mergers, acquisitions and rationalizations have taken place; these are intended to form the basis of a stronger, more Europe oriented marketing effort.

According to Key Note, the basic challenge facing plastics processors is to gain the ability to manage complex systems at a time when more intense pressures can be expected as a consequence of the competition between companies, markets and countries. The key markets for plastics processors will be packaging, with an imposed obligation for easier recycling of waste; and the automotive, telecommunications and business machines industries.

A primary requirement for success in ousting other materials will be the ability for processors to be able to produce large and small items accurately and precisely, with very fast machine times, hot or cold worked.

Key Note's 67-page report, 'Plastics processing', costs £155.

3.1.3 NYLON IN THE USA

A market research report, B124 'Nylon resins', by Ann Kowalski of the Freedonia Group, concludes that North American nylon resin consumption approached 300 000 t in 1988, thus establishing its fifth decade as the leading engineering resin;

nylon production and consumption will grow at a rate of 6% per annum between now and 1993, with electrical/electronics applications presenting the most significant growth opportunities;

transport will remain the largest nylon market, with niche opportunities available in the construction, filament, film and medical markets.

Approximately 90% of total nylon consumption is accounted for by nylon-6 and nylon-6/6, with the remaining 10% comprising speciality types (nylon-12, nylon-4/6, nylon-11 etc). Grades reinforced with glass and/or minerals account for one-third of consumption. The versatility of nylon, in respect of its compatibility with additives and other resins with which it can be blended, accounts for its ability to find new applications.

The nylon resin industry in North America is a mixed bag of large, multifaceted chemical and plastics companies and medium to small specialized operators. Competition is growing, but supply and demand remain balanced. Du Pont remains the world's leading supplier, although companies such as Allied-Signal, Monsanto and Hoechst Celanese are significant contenders.

3.1.4 POLYPROPYLENE IN EUROPE

A report by Corporate Development Consultants, 'The European market for polypropylene compounds' states that the combined consumption of mineral filled, glass fibre reinforced, elastomer modified and flame retardant polypropylene compounds in all end-use sectors was 470 000 t in 1989. This represented a market sales value of £505 million. As regards the future, highest growth is anticipated in mineral filled types, forecast at around 6% per annum until 1995, at which time they will account for some 71% of a total polypropylene compound consumption of 604 000 t.

However, a current growth rate of 8% a year is reported for oriental polypropylene in a four-volume report being published by IAL Consultants Ltd on the markets for the material in the UK, Germany, France and Italy.

IAL predicts that supply and demand will be in balance by 1994, and states that in 1989 there was almost 35% spare capacity in Western Europe following a surge of investment by all the major oriental polypropylene producers.

The report shows significant differences in the levels of penetration and patterns of consumption in each of the four countries studied, reflecting the cultural differences between them.

About 60% of all oriental polypropylene produced is used in food packaging. Italy is the largest European market for oriental polypropylene, with a 1989 consumption figure of 74 000 t out of a total of 260 000 t for the area considered. A major Italian use is in the backing for adhesive tape. The UK consumed about 54 000 t in 1989, and West Germany 38 000 t.

The use of oriental polypropylene, particularly in the UK, is regarded as an opportunity for enhancing the visual appeal of convenience food packaging. Surprisingly, this is not the case in France, where oriented polypropylene is used extensively for flower wrapping but is considered somewhat 'down market' for food products. In Europe generally, environmental concerns are restricting the growth of oriental polypropylene coated with either polyvinylidene chloride or metal foil.

IAL is intending to extend its coverage of the market to Spain and Benelux. The existing report is available either in full, price £4000, or in individual country volumes, price £1250 each.

3.1.5 EUROPEAN POLYURETHANES

IAL Consultants Ltd has estimated that the European market for polyurethane chemicals amounted to 1.414 million t in 1988. This corresponds to 33% of world demand. Growth is forecast to average 2.7% per year until 1993, to reach a total of 1.615 million t.

Growth rates for the major polyurethane product groups fall in the range 0–5% per year, but interesting niche markets will show a growth of 10–15% per year. Aliphatic isocyanates, for example, are showing dramatic growth rates of 10–20% per year. Future growth is expected to come primarily from coatings and paint applications.

Flexible foam remains the largest sector of the polyurethanes market, but it has declined from 55% of total consumption in 1979 to less than 41% in 1988. Rigid foams account for 26.5% of polyurethane consumption, and future growth is expected to average 3% per year. The large applications for rigid foam — building and appliance insulation — have reached a relative maturity. While certain building applications such as sandwich panels for industrial building are showing growth rates of 10% per year or more, polyurethane is not expected to significantly increase its share of the thermal insulation market beyond the present level of 5–6%.

The non-foam market is characterized by a wide range of products and applications, and accounts for 27% of the total European polyurethanes market.

The market for reaction injection moulding, and reinforced reaction injection moulding is forecast to grow at an average of some 4% per year. Bumpers and bumpers fascias account for 40–50% of the European automotive reaction injection moulding market, and there is considerable scope for further penetration. However, reaction injection moulding polyurethane faces increasing competition from engineering thermoplastics, and improvements in automation and mould design are needed to increase productivity and bring down costs.

Alternative estimates of European urethane chemicals can be found in a report by Frost and Sullivan, which forecasts the expansion of the European market from 1 546 000 t in 1988 to 2 082 000 t in 1994. The industry is expected to conquer the threats to its progress posed by concerns over toxic fumes, chlorofluorocarbons and toxic chemicals, although a great deal of time will be spent trying to find replacements for chlorofluorcarbons as secondary blowing agents.

The greatest growth is likely to come in microcellular applications, rising from 71 000 t in 1988 to 109 000 by 1994. Rigid foam will grow from 498 000 t to 760 000 t over the same period, whereas flexible foam, while remaining the largest single sector, will only grow from 813 000 to 998 000 t. Non-foam applications will rise from 164 000 to 215 000 t over this period.

Germany will continue to hold its position as the largest single market by 1994, with 558 000 t consumed, but the UK will be jostling Italy for second place, each with a market of 368 000 t. A substantial export market is expected to grow in the Eastern European countries — provided the finance can be found.

Report E1249/P, 'The European market for urethane chemicals', published 1990, is priced at US$5500.

3.1.6 POLYMER FOAMS IN THE USA

A Business Communications Co report (P-120, price US$2850), titled 'Polymeric foams', estimates that, in the USA, foams, accounted for about 2.45 million t of polymeric materials in 1989, and usage is expected to grow 4.1% annually to 1995. Polyurethane foam alone accounted for about 1.36 million t. However, the future depends heavily on the availability of environmentally acceptable substitutes for chlorofluorocarbons as blowing agents, and on measures to reduce the volume of solid waste produced, particularly for packaging.

3.1.7 VINYL ACETATES IN THE USA

A report by the Federation Group Inc, 'Vinyl acetate and derivatives' (ref. B166, price US$1200) concludes that the demand for vinyl acetate in the USA will expand at over 4% annually to reach 2.3 billion pounds (1 million tonnes) by 1993. This will result from good growth in vinyl acetate's principal derivatives, especially vinyl acetate emulsion polymers. Stimulated by continuing technological improvements, a multitude of available polymers and numerous market applications, vinyl acetate emulsion polymers for adhesives are expected to sustain strong growth, expanding over 5% annually to 740 million pounds (340 000 t).

3.1.8 THERMOPLASTIC ELASTOMERS IN THE USA AND EUROPE

A Frost & Sullivan study, 'The US market for thermoplastic elastomers' (ref. A2250), notes that these materials have several advantages over vulcanized rubbers and are displacing them in a growing number of traditional markets. They are melt-processible, which enables scrap to be recycled to make virgin-quality parts; they require little or no compounding or blending; fabrication is short and simple; tolerances are tighter and density usually lower. In addition, total production costs of parts made from thermoplastic elastomers are usually only one-half to two-thirds the cost of vulcanized rubber parts.

The report estimates a US market value of $835.9 million in 1989, and forecasts a growth to nearly $1.2 billion in 1994. Automotive applications will expand from a $173.5 million market in 1989 to one of $255.5 million in 1994.

Frost & Sullivan has followed up its survey of the US market for thermoplastic elastomers by examining the European outlook for these materials (report E1347, priced at $5750). As in the USA, thermoplastic elastomers are steadily displacing rubber in many applications, with Germany leading the way. Nearly one-third of all thermoplastic elastomers consumed in Western Europe are used by the automotive industry, with footwear coming second.

The biggest group of high performance thermoplastic elastomers are the copolyester-ether materials which have excellent high tear strength and fatigue life and outstanding fluid resistance. They are increasingly used in under-the-bonnet car components, wire and cable, and in hosing and tubing.

Higher grade olefinic thermoplastic elastomers will continue to strengthen their position in many of the traditional rubber markets, and the introduction of glass-reinforced thermoplastic polyurethanes is expected to open up new markets in competition against engineering thermoplastics.

The report predicts an overall growth in sales from $897 million in 1989 to $1.25 billion in 1994.

3.1.9 FUTURE OF SPECIALITY FILMS

Thermoplastic films will become a critically important commodity in the 1990s, according to a Business Communication Co market study (ref. P-063U, priced at $2450). The consumption of thermoplastic films is expected to rise from 3800 t in 1989 to about 4500 t in 1994. Market leaders will be biaxially oriented polypropylene, linear low density polyethylene and high density polyethylene, partly at the expense of low density polyethylene film. But the fastest growing sectors of the thermoplastic film market will be polycarbonates, polyimides and high temperature speciality films, the report predicts.

3.1.10 US OPPORTUNITIES IN CONDUCTIVE POLYMERS

According to a report from the Freedonia Group (Report B190, 91 pages, US$1200), the consumption of conductive polymer-based resins in the USA will increase at over 14% per annum, from 54 000 t in 1989 to 107 000 t in 1994 and to 218 000 t in the year 2000. The value of the products made from these polymers will rise from $430 million in 1989 to $995 million in 1994 and to $2475 million in 2000.

These figures reflect a growing demand for conductive polymers in the electronics industry. In 1989 they accounted for 5.1% of all plastics used in electronics in the USA; this figure is forecast to rise to 8.1% by 1994 and to 13.1% by 2000. By far the fastest growing category of material is intrinsically conductive polymers, which were available in negligible quantities in 1989 but are predicted to swell to about 450 t in 1994 and 11 500 t in 2000. The more conventional filled conductive polymers will expand in quantity from 54 000 t in 1989 to 106 000 t in 1994 and to 198 000 t in 2000, an annual growth rate of 14.3%.

More sophisticated circuitry demands better shielding

According to Freedonia, the proportion of conductive polymers in the total of resins consumed in electrical and electronic goods is expanding due to the increasing sensitivity and sophistication of integrated circuits and electrical assemblies, the rising level of corporate expenses associated with static-caused damages, and the strengthening of FCC regulations regarding EMI/RFI (electromagnetic and radio-frequency interference) shielding.

Acrylonitrile-butadiene-styrene has emerged as the most widely employed polymer for the fabrication of antistatic and EMI/RFI shielded products. Conductive acrylonitrile-butadiene-styrene demand will increase 13% from 13 500 t in 1989 to 25 000 t in 1994 and 45 000 t in 2000 as improved grades of acrylonitrile-butadiene-styrene are brought onto the market.

Polyvinyl chloride composites, blends and alloys comprise the second largest category of conductive polymers. Technical advances are creating new opportunities for polyvinyl chloride in high impact and high friction applications. Emerging applications, coupled with widespread end uses in business machine housings and worksurface protection accessories, will raise the consumption of conductive polyvinyl chloride by 15% per year from 9000 t in 1989 to 18 000 t in 1994 and 34 000 t in 2000.

Polycarbonate composites and blends are becoming increasingly popular as materials for electronic equipment components, including computer and copier housings, breaker boxes, gear shields and compact disc covers. The range of end uses for conductive polycarbonate will continue to broaden, and consumption will rise by 13.5% annually from about 8000 t in 1989 to 15 000 t in 1994 and 25 000 t in 2000.

Polyphenylene-based alloys, particularly polyphenylene sulphide, are emerging as important resins in the fabrication of moulded parts for explosion-proof containers and machines exposed to high impact and high temperatures. Demand for conductive polyphenylenes will grow by 10.2% per year from 7000 t in 1989 to 12 000 t in 1994 and 20 000 t in 2000.

Other widely used filled conductive polymers include polyethylene, nylon and polystyrene. Polyethylene demand will be stimulated by growing applications in antistatic packaging. Broadening opportunities in high impact packaging and rotating machine parts will increase conductive nylon consumption by 12% per year from 3600 t in 1989 to 6400 t in 1994 and 11 000 t in 2000. Increased demand for static dissipative storage and handling equipment will stimulate a high growth rate of 20% for conductive polystyrene, which is likely to reach 6800 t by 1994 and 16 000 t by 2000.

Opportunities in intrinsic conditions

Intrinsically conductive polymers such as polyacetylene, polypyrrole and polycarbaxole will provide growth opportunities for plastics manufacturers in the long term. Potential applications include superconductors, semiconductors, and magnets. However, the really significant developments expected over the next five years are the arrival of plastic rechargeable batteries and substitute materials for fibre optics.

Demand for device and equipment components will expand 23% per year to $620 million in 1994 and $1800 million in 2000, reflecting the advancing sophistication and broadening use of integrated circuits in industrial and consumer equipment. The market for antistatic packaging will escalate to $185 million in 1994 and $310 million in 2000, a 10% increase over current levels. Intensifying efforts by electronic component producers to minimize damage to products in transit will be the major impetus behind this growth.

The market for worksurface accessories will increase over 15% per year to $120 million in 1994 and $235 million in 2000 as manufacturers seek to strengthen static control systems within their production facilities.

Inventory losses attributable to electrostatic damage will stimulate a 15% annual growth in demand for materials storage and handling products, and a near 13% growth in other conductive polymer-based products, including static-dissipative apparel and coatings.

The structure of the conductive plastics industry

The conductive polymer segment of the plastics industry is relatively concentrated, with major resin producers such as General Electric Plastics, B.F. Goodrich and BASF dominating the market from a materials standpoint. The fabricated products sector is more fragmented and incorporates a few large companies and numerous small-to-medium sized processing firms. Research into intrinsically conductive polymers represents a separate segment of the field, encompassing a wide variety of participants.

Overall, the industry is in an early stage of growth, characterized by rapid technological change and a frequently shifting competitive structure.

3.1.11 FLAME RETARDANTS

Freedonia's 108 page Business Research Report B150, 'Flame retardants to 1993', priced at US$1000, predicts that the demand for flame retardants in the United States will expand at 8% per annum between now and 1993, reaching over $1 billion. Flame retardants for plastics will comprise over 90% of the market.

Coinciding with this study, BCC has published a business opportunity report on flame retardant chemicals. The report quantifies and qualifies market information on flame retardant and smoke suppressant chemicals for the US market.

Plastics account for 77% of the flame retarding and smoke suppressing chemicals used in the United States. Among these, alumina trihydrate is the dominant material, with phosphorus compounds coming second. Predicted annual growth rates for these two additives, between now and 1994, are 6.5% and 5.0%.

Interest in chlorine compounds as flame retardant additives remains high, even though they are toxic halogenated hydrocarbons. Bromine compounds tend to be more expensive but less toxic. Growth rates of 6.6% and 6.0% are predicted for chlorine and bromine compounds respectively.

The market for antimony oxide continues to recover after a downturn in the early 1980s. Producers are predicting an annual growth of 5%. Concern over its toxicity is being alleviated by the production of wetted grades and dustless concentrates.

Lesser used chemicals (boron, molybdenum, nitrogen etc.) continue to find a market as partial replacements for higher priced chemicals, and as synergists. Demand by major end-use markets will keep the annual growth of miscellaneous chemicals around 5% between now and 1994.

3.1.12 PLASTICS USAGE IN EUROPEAN AUTOMOBILES

A report from Frost & Sullivan predicts a shift away from steel and cast iron towards plastics and aluminium in the manufacture of West European cars. The report forecasts that the weight-percentage of these materials will rise from 9% in 1983 to 18.5% by 1993.

Western Europe is now the world's most important car producing region, with output in 1988 totalling a record 13.03 million units. All European car manufacturers are concerned with the need to produce cars which are more fuel-efficient, both to satisfy customer demands and to meet legislation on emissions levels. Consequently, they are increasing their use of light materials such as plastics and aluminium.

The most important applications area for plastics is car interiors, which in 1988 accounted for an estimated 56% of all the plastics used. However, the fastest growing area for the use of plastics is in engine components and electrical/electronic applications — each account for around 10% of use. Over the next few years, electrical and electronic applications for plastics are expected to virtually double.

3.1.13 HEALTH AND MEDICAL PLASTICS PRODUCTS IN THE USA AND THE UK

Sales figures for the US medical supply industry will rise from about US$24 billion in 1989 to some US$32 billion by 1994, according to a recent Business Communications Co (BCC) study. Report P-121, 'Plastics in non-packaging medical applications', predicts that implants, prostheses, catheters, intravenous devices and tubing will be the industry's fastest growing segments.

This growth is largely due to the replacement of traditional materials by plastics. Thermoplastic elastomers and engineering thermoplastics, particularly polycarbonate, will take the lead in the expansion. Polyurethane, copolyester and amide block copolymers will have exceptionally high growth rates.

'Plastics in health care', from NEDO books priced at £90, analyses the UK health care market for plastics processors, and concludes that £36 million of new business could be created by displacing imports of labware and invasive devices. There are good prospects in the supply of moulded components to original equipment manufacturers, and in the production of complete injection moulded, disposable, surgical and examination instruments which can then be supplied direct to distributors.

3.1.14 MEDICAL BIODEGRADABLES IN EUROPE

A recent Frost & Sullivan report on biodegradable polymers confines itself to their use in the medical and surgical fields. Report E1227, 'The European market for biodegradable polymer products', predicts that the market for these materials will grow from an estimated US$85.5 million in 1988 to US$253 million in 1994. Germany, France, Italy and the UK will account for about 70% of the West European market, Germany being the largest consumer.

Orthopaedic implants show the greatest potential for growth. These products will emerge from clinical testing in 1992 and 1993 and the market is expected to be then worth $52 million, occupying a 20.5% share of the market.

The wound care and dressing markets present good development opportunities, both in the hospital and the home care sectors, with sales rising to $73 million by 1993. A limiting factor is their higher cost in comparison with traditional dry dressing, but this is offset by faster wound cleaning and healing. The report forecasts that biodegradable sutures and orthopaedic devices which release an anti-bacterial agent as they degrade could help considerably in repairing traumatic injuries or infected wounds.

Vascular grafts with biodegradable components are now worth $9.5 million per year and this is set to increase to $35.4 million in 1993. Developments in vascular grafts will have an influence on demand. The report speculates that there will be competition between biodegradable and non-degradable materials, the issue depending on the results of clinical trials.

According to the report, the present market for medical and surgical products employing biodegradable polymers is highly fragmented. Development of new products is hampered by the exhaustive regulatory and legislative procedures required. In consequence, development is largely funded by the major international healthcare companies who have substantial R&D resources and experience in channelling products through the complex regulatory affairs maze.

In spite of this, all product sectors are expected to expand dramatically between now and 1993 and provide worthwhile returns for investors in new technology and product applications.

3.1.15 EUROPEAN ADDITIVES AND MASTERBATCHES

Three reports from Frost & Sullivan give guidance to the producer of additives or masterbatches for the European market. One of them pinpoints Germany as being the most logical choice for siting a new masterbatch plant in Europe.

Plastics masterbatches

Despite boasting two of Europe's biggest plastics companies, BASF and Hoechst, Germany still imports nearly one-third of its present consumption of masterbatches. The rationale for siting a masterbatch plant in Germany is strengthened by the recent reunification of that country. Scandinavia and Italy are also suggested by Frost & Sullivan as possible sites for new plants.

While the report forecasts steady growth from 1989 to 1994 in the market in Western Europe — with consumption rising from 355 000 to 407 000 t — a question mark still hangs over the export possibilities to Eastern Europe.

The report notes the dominance of the market held by the top ten producers, seven of which are multinationals. Operating from 30 separate sites in Western Europe, they account for 65% of total production.

Report E1367, 'The European market for plastics masterbatches', is priced at US$5750.

Fillers, extenders and opacifiers

Exploring the market for these materials throughout Europe, Frost & Sullivan forecasts a rise in sales from US$3.8 billion in 1988 to US$4.7 billion in 1994. Improved quality is one reason for the expansion of the market.

The report looks at natural minerals such as kaolin, calcium carbonate, talc and silica as well as chemical products like titanium dioxide, antimony oxide and lithopone. The paint and paper industries are the largest markets for these materials, but applications in plastics and rubber are also discussed. Titanium dioxide, for instance, is an important ingredient in decorative laminated paper, which has a polymer resin matrix, and the report pinpoints sizeable European markets, especially in France and Germany.

Talc is an important ingredient in polypropylene, which is one area of the plastics market set to enjoy good rates of growth in the early 1990s. However, in terms of tonnage, the filler used most within the European plastics industry is still calcium carbonate, which is used in polyvinyl chloride products. In 1988, demand stood at 1.2 billion t with an additional 70 000 t being used in polyolefins.

Despite possible health risks, the report says the market for antimony oxide, which is used as a fire retardant additive in polymers and copolymers, is strong. Fibrous and platey fillers, such as wollastonite, will play an increasingly significant role in both fibre-reinforced polymer composites and the engineering polymers of the future.

The total demand for mineral fillers in the rubber industry in 1988 is estimated at 506 000 t, worth US$221.9 million. This is set to rise to US$256.3 million by 1994. The rubber industry is a major user of silica and calcium carbonate as well as talc, barytes and titanium dioxide.

Germany has the largest share of the market for all these materials.

Report E1340, 'The European market for fillers, extenders and opacifiers', costs US$5750.

Heat and light stabilizers

The Frost & Sullivan report on stabilizers maintains that future markets are inextricably linked with the prospects of the plastics industry, which is under pressure from rising raw materials costs and the environmental lobby. These pressures could prompt a move away from the use of plastics in the packaging industry, which accounts for 30% of West European consumption and would be very sensitive to rising raw materials costs.

Thus, despite the rising demand for plastics, only marginal growth is expected for polyvinyl chloride heat stabilizers up to 1994, although the outlook for the light/UV stabilizer market is more encouraging.

The most significant of the heat stabilizers are lead compounds, which accounted for almost 69% of the total in 1989 — although by 1994 this will drop to 64%. In volume terms the barium/cadmium compounds are the next most important, followed by organotin compounds, but in value terms the organotins come second. Recent developments in heat stabilizers have made handling and dispersion easier; masterbatches are becoming increasingly widespread.

A major development in the light/UV stabilizers market was the introduction of the hindered amine light stabilizers in the mid 1970s. They provide superior protection against the harmful effects of ultraviolet radiation, and the latest versions appear to give long-term thermal stability to polyolefins. The report expects the growth in demand for hindered amine light stabilisers to remain strong till 1994, by which time they will account for over 40% of total consumption by volume.

By 1994 Western Europe will use 4300 t/yr of light/UV stabilizers. Germany will account for 22%, with France using 21%, the Benelux countries 19%, Italy 16.5% and the UK 12%.

Report E1349' 'The European market for heat and light stabilizers', is priced at US$5750.

4. MATERIALS

New grades of existing materials are routinely introduced to suit particular processing methods. This causes no excitement scientifically, but small advances of this nature can be commercially significant. On a more exotic level, we see in this section the introduction of new liquid crystal polymers, and may wonder whether polyarylether ketone (PAEK), with its higher melting point, will come to rival polyetheretherketone (PEEK). Developments in blends continue to be a sizeable area. The increased use of microelectronics, where screening materials are required, is one of the driving forces behind the interest in conductive polymer.

4.1 POLYETHYLENES

4.1.1 POLYETHYLENE GRAINS FOR INJECTION MOULDING

Hoechst has launched an ultra high molecular weight polyethylene in granular form, suitable for injection moulding. The company believes that the new grade, Hostalen GUR GX 579, will make the production of large quantities of complex moulded parts economically possible by injection moulding, because it has the characteristic of being free flowing as well as having molecular weight.

Hostalen GX 579 is claimed to exhibit outstanding features, including excellent wear resistance, good slip behaviour, high resistance to chemicals and high notched impact strength even at low temperatures. Table 1 lists the chief mechanical properties. Processing temperatures are between 200 and 250°C.

Application areas for the material are diverse, ranging from machinery and conveyor equipment for the chemical, textile and paper industries to sports goods and medical applications, particularly in the orthopaedics field.

TABLE 1: Properties of Hostalen GX579.

Property	Unit	Test method	Value
Density	kg/m^3	DIN 53 479	940
Viscosity number	l/kg	DIN 53 728	1800
Flow value F (150/10)	MPa	DIN 53 493	0.15
Bulk density	kg/m^3	DIN 53 468	≥400
Tensile stress at yield	MPa	DIN 53 455	>20
Tensile stress at break	MPa	DIN 53 455	>20
Elongation at break	%	DIN 53 455	>350
Notched impact strength (double notch)	kJ/m^2	DIN 53 455	>100

4.1.2 POLYMER FOR BLOW MOULDING

Blow moulding requires materials with high melt strength. A new material which is said to meet this requirement is 'Fortiflex K52-05-159' from Soltex Polymer Corp of Houston, Texas, USA. This high molecular weight polyethylene has a density of 9520 kg/m^3 and a high load melt index of 5. According to Soltex, this imparts excellent impact strength, tensile strength and elongation at break to a polymer which also meets the FDA (Food and Drug Administration) food contact requirements. Products include 30 and 50 gallon drums.

4.1.3 GIANT MOLECULES OF POLYETHYLENE

Molecular weights of 2–7 x 10^6 are claimed by DSM of Heerlen, The Netherlands, for its new grade of polyethylene. Designated Stamylan UH 610, this ultra high molecular weight polyethylene has a quoted flow value of 0.56 MPa and an intrinsic viscosity of 24 dl/g. Limited amounts of the material are currently being produced in a pilot plant at Geleen, where a 5000 t/yr plant is under construction and is scheduled to start production very soon.

4.2 POLYPROPYLENES

4.2.1 RESIN PELLETS

Improvements in processing are claimed to lie behind Himont's launching of new grades of Valtec polypropylene. A post-reactor stabilization method, known as the Spheripol process, has been used to impregnate the resin pellets with stabilizer and other additives without remelting; according to Himont, this improves the material's processing efficiency.

Five of the Valtec resins are designed for extrusion, and one — MG-412 — is also suitable for injection moulding. They are all claimed to meet the FDA regulations covering food contact, including cooking. However, this will probably prove irrelevant to their main applications; Valtec HL-406 and ML-409 are intended for sheet and profile extrusion, MP-401 has applications in the future under the car bonnet, HP-420 will find applications in netting, strapping and sheet extrusion, and MG-412 is intended for general purpose moulding and extrusion. Some properties of these resins are given in Table 2.

TABLE 2: Properties of Himont's resins.

Designation	Type	Flexural modulus (GPa)	Notched Izod strength (at 23°C)	Melt flow rate (g/10 min)
HL-406	homopolymer	1.6	134 N	0.8
HP-420	homopolymer	1.6	53 N	2
MG-412	copolymer	1.3	190 N	2
ML-409	copolymer	1.2	590 N	0.5
MP-401	copolymer	1.3	190 N	2

A sixth grade is designed primarily for non-wovens spinning. Valtec HH-442H is claimed to have an ultra-high melt flow rate and to offer the non-wovens industry — particularly the meltblown sector — processing and performance characteristics superior to those of comparable polymers.

Valtec HH-442H polypropylene has a melt flow rate of 400 dg/min and a very narrow molecular weight distribution. According to Himont, it is distinguished from conventional visbroken grades of the same viscosity by greater productivity, substantial energy savings, better process continuity, longer production runs, reduced die pressure and less build-up of char. The resulting fibre exhibits higher tensile strength, improved absorption, increased uniformity and greater softness, it is claimed.

HH-442H is being made in Himont's factory at Ferrara, Italy, and is already being used by a number of meltblown manufacturers. Himont says that the new grade opens up new opportunities for polypropylene to compete with other materials, particularly cellulosics, for medical and hygienic applications.

4.2.2 HIGH CRYSTALLINITY RESINS

Significant advances are claimed by Quantum Chemical Corp for a new range of polypropylene resins. These are among the products resulting from an interchange of technology between Quantum, BASF and ICI.

Products now being marketed by Quantum's USI Division include higher crystallinity polypropylenes (8802HO, 8820HO and 8815ZR) which balance impact strength against stiffness for use as bottle caps and closures. There are also two extrusion grade polypropylenes, 1406HF and 1406ZF, with very high impact strength; these are considered suitable for microwave containers and boil-in-bag applications which require FDA certification in the USA.

4.2.3 POLYMER FOR STRETCH BLOW MOULDING

Soltex Polymer Corp has produced a range of resins for stretch blow moulding. Fortilene 4507 is a high clarity random polypropylene copolymer with good impact properties.

4.3 POLYVINYL CHLORIDE

4.3.1 RIGID PELLET COMPOUND

Vista Chemical has launched a family of rigid pellet compound products for injection moulding, called Vistel rigid vinyl compounds. These products are commercially available to manufacturers of parts for pipe fittings, exterior weatherable products or general building components.

Applications for Vistel may include such products as furniture fittings, plumbing valves, couplings and fittings, electrical appliance control boxes, swimming pool filter fittings and housings, junction boxes for telephone systems and appliance parts.

4.4 NYLONS

4.4.1 NYLON 6 FOR ROTATIONAL MOULDING

Nylatek Ltd of Yeovil, UK, has brought out a grade of nylon (polyamide) 6 suitable for the rotational moulding process. The company has called the new grade Rotamid and claims that it is far easier to rotomould than nylon 12. It melts at about 200°C and can therefore be used to produce articles at temperatures of about 230˚C. This should enable the rotomoulding technique to produce engineering parts with good physical properties which may be enhanced by using fibre reinforcement.

4.4.2 INJECTION MOULDING GRADES

Two new Grilon nylon 6.6 injection moulding grades have been added to EMS's range of types 6, 12 and 6/12 nylons. These unreinforced grades are T300GMH, which offers increased resistance to heat ageing, and T300FC, a fast-cycling, easy-flow formulation.

4.4.3 MONOFILAMENT POLYAMIDE FIBRE

Du Pont has given the name Hyten to a monofilament polyamide fibre which, it is claimed, is 10% stronger than conventional nylon and 38% stronger than polyester. It is intended as sidewall reinforcement for rubber tyres, replacing conventional twisted cords. According to Du Pont, this simplifies tyre manufacture, saves 5–15% of material and reduces the tyre running temperature.

4.5 LIQUID CRYSTAL POLYMERS

4.5.1 AROMATIC POLYESTERS

Rhône-Poulenc has introduced a range of liquid crystal polymers. The liquid crystal polymers belong to the family of aromatic polyesters, and are sold under the brand name of Rhodester CL. According to the company, they have the essential characteristics of aromatic polyesters — heat stability, a low coefficient of linear expansion, high mechanical strength, chemical inertness and excellent flame retardant behaviour. They will be supplied in pellet form and processed by injection or extrusion.

Rhône-Poulenc expects these liquid crystal polymers to be used in applications where strength, stability and resistance to high temperatures are required, such as electronics, electrical engineering, aeronautics, automotive and chemical engineering.

4.5.2 PLATABLE LIQUID CRYSTAL POLYMER

Hoechst Polymers Division has introduced a new platable liquid crystal polymer designated Vectra C 810. It can be used to produce three-dimensional circuit boards using injection moulding, and its high temperature performance and chemical resistance reportedly allow it to be plated with copper, nickel and gold.

4.5.3 RIGID ROD POLYMERS

Researchers at Los Alamos National Laboratory are reported to have synthesized rigid rod polymers with unique structures that could lead to improved properties. Unlike most rigid rod polymers which have rigid backbones, these materials have flexible backbones based on *p*-phenylene terephthalamide and rigid side chains based on diphenylthiazolothiazole.

4.5.4 REINFORCING FIBRE

A lightweight reinforcing fibre called PBZ has been announced by CommTech International. PBZ is a family of liquid crystalline, extended chain, rigid rod polymers which are claimed to have high strength and stiffness and are resistant to high temperatures, oxidation, moisture and ultraviolet radiation. According to the manufacturers, stiffnesses of 400 GPa, tensile strengths of 5 GPa and compressive strengths of 0.8 GPa are available.

The fibres can be used alone or as reinforcements when embedded in resins. Other applications include the use of PBZ film or three-dimensional shapes in composite materials.

4.6 IMIDE POLYMERS

4.6.1 POLYCARBODIIMIDE

Polycarbodiimides are polymers containing the group -N=C=N- in the polymer chain. They are usually synthesized by self-addition polymerization of a diisocyanate, using organophosphorus compounds. Properties are similar to those of polyamide.

The Japanese firm Nisshinbo has begun to market a grade of polycarbodiimide with a molecular weight of about 15 000, far higher than that of previous grades. Named Carbodilite, this polymer is said to be transparent, with a slight brown tint, stable at room temperature, even in the presence of water, and suitable for either moulded components or films.

The physical properties of Carbodilite depend on the level of crosslinking. Highly crosslinked film has a tensile strength of 140–150 MPa, a tensile modulus of 3.0–3.5 GPa, and an elongation of up to 300%. Crosslinked moulded items have a flexural strength of 53 MPa, a tensile modulus of 3.6 GPa, and a deflection temperature of 261°C under 1.82 MPa load.

Electrical properties include a volume resistivity of 4.5×10^{16} ohm.cm, and a dielectric constant ranging from 3.42 at 60 Hz to 3.25 at 1 MHz.

4.6.2 IMIDE OLIGOMER

Sumitomo Chemical has developed what is described as an original amine terminal-type imide oligomer, having a high thermal resistance and a glass transition temperature of 200–230°C. It behaves as a thermosetting imide resin, and is used in conjunction with an epoxy resin.

The compound is recommended for use as a heat resistant laminated plate, as an encapsulating material for integrated circuits, and for selected type of moulding.

Sumitomo states that the material exhibits good adhesion at high temperatures, dimensional stability, and good insulation properties.

4.6.3 HIGH TEMPERATURE POLYIMIDES

A range of high temperature polymers based on polyimide compounds has been introduced by ICI Fiberite. Designated as PMR-15 (polymerized monomeric reactants), the range is claimed to withstand operating temperatures up to 325°C, with good thermo-oxidation properties, due to the 'addition' type curing mechanism. The materials are available as a free flowing powder, chopped carbon fibre bulk moulding compound, palletized chopped moulding compound for long-flow transfer and chopped E-glass fibre bulk moulding compound for electrical and thermal stability.

4.7 SULPHONES

4.7.1 POLYSULPHONE RESINS

Two polysulphone resins have been introduced by Amoco. (Radel R-5000 is designed for the medical market, particularly for sterilizable surgical trays which require a high degree of chemical and steam resistance to withstand the sterilization process. It also has excellent toughness and impact resistance, it is claimed.

R-500 is the transparent V-O general purpose grade. Opaque, black and fibre filled grades are also available.

The other Radel polyphenyl sulphone, R-7000, is formulated for the commercial aircraft interior market. It has inherent impact resistance and complies with the 1990 FAA (Federal Aviation Administration) regulation requiring low heat release, low smoke generation and low toxic gas emission. According to Amoco, it shows excellent resistance to aerospace fluids, especially under stress.

Amoco is also introducing two new Radel polaryl sulphone products formulated specifically for the automotive market. Radel A-200MR is a higher melt flow grade for very high current-handling automotive electrical fuse applications. A heat deflection temperature of 204°C is quoted for this material.

The other polaryl sulphone, Radel AG-360, is specially formulated for high heat applications in automotive lamp sockets. It was developed as a low cost alternative to polyether sulphone, polyether imide or polyphenylene sulphide. Amoco says that its processibility permits faster production cycles and ultimately lower part cost. With a heat deflection temperature of 201°C, Radel AG-360 has other potential automotive electrical applications such as battery insulators, alternator and starter motor components.

4.8 POLYSTYRENES

4.8.1 HIGH TEMPERATURE POLYSTYRENE

BASF's latest addition to the polystyrene family is a range of high temperature resistant materials with a Vicat softening temperature of above 100°C. BASF claims to be the only company to supply such products.

A special modification means that BASF can offer a range of polystyrene grades that can withstand higher temperatures than the normal maximum of 100°C. These grades are intended for applications in the automotive and electrical industries.

Six new grades are available. Polystyrol (HT) 2733 is claimed to be the best material, with a combination of high temperature distortion resistance, toughness and rigidity. With a Vicat softening temperature of 111°C and a silky finish, it is recommended for in-car components such as column claddings, rear shelf supports, door shelves and centre consoles.

Polystyrol 2735 has a marginally higher Vicat softening temperature (112°C) and is transparent. A suggested application is audio cassette windows.

Other potential applications for the range include loudspeaker housings and cassette boxes in cars, and housings for hot air blowers.

4.9 STYRENE ACRYLONITRILE

4.9.1 RANGE OF STYRENE ACRYLONITRILE PRODUCTS INCREASED

During 1989–90 there were extensions to Monsanto's range of styrene acrylonitrile copolymers. The new grade Lustran SAN 32 is said to have very good chemical resistance, rigidity and optical clarity. In addition, it claimed to offer processing ease at low cost together with resistance to heat deformation and scratches in the end product.

Key typical properties for Lustran SAN 32 include a tensile strength of 65 MPa; a Vicat A softening point of 108°C; an Izod impact strength of 1.5 kJ/m^2; and melt flow index of 19 g/10 min. According to Monsanto, the grade of common acids and alkalis, to edible oils and fats, fuels such as petroleum and kerosene, and to foodstuffs. It is available in a range of standard colours.

Another recent introduction is Lustran Sparkle which is claimed to feature improved optical clarity, low specific gravity (1.07), high chemical resistance and good processability. The new material is seen as a replacement for acrylic, polycarbonate and styrene/acrylic polymers for articles such as tumblers, dishes, medical and cosmetic packages and plumbing fixtures.

4.10 COMPOSITE MATRIX MATERIALS

4.10.1 DOW EXPANDS RANGE OF VINYL ESTERS

Three additions to its Derakane 411 range of vinyl ester resins have been made by the US chemical group Dow. The company says that the products have been developed in response to industry demands for increased toughness and demonstrate 30% greater tensile elongation properties than the existing Derakane 411 resins.

Derakane resins are recommended for fibre reinforced plastics storage tanks, vessels and ducts, and can be used with a wide range of fabrication techniques including hand lay-up, spray-up, filament winding and resin transfer moulding.

Derakane 411-350 is said to offer resistance to a broad range of acids, alkalis, solvents and bleaches.

A thixotropic agent is added to Derakane 411-700T to facilitate lamination in difficult locations by lessening the sag when the material is applied to vertical surfaces. This makes the resin suitable for applications such as tank interiors, marine vessel exteriors or swimming pool liners.

The 411-700 PAT grade of thixed resin contains a pre-measured promoter and accelerator which reduces variation in gel time.

4.10.2 EASY-TO-PULTRUDE VINYL ESTERS AND PHENOLICS

New resins recently announced by the Reactive Polymers Division of Reichold Chemicals Inc, Research Triangle Park, North Carolina, USA, include easy-to-pultrude vinyl esters and phenolics and a resin transfer moulding product which is said to cut cycle times. The easy-to-pultrude resins are a vinyl ester, Dion 31-034-01, which offers a smooth surface in tubing rods, and a phenolic pultrusion resin developed jointly with the company's Canadian subsidiary Reichold Inc. The resin transfer moulding resin, Polylite 31-512, is claimed to cut cycle time to less than 5 min.

4.10.3 PHENOLICS WITH HIGH HEAT RESISTANCE

Phenolic resins claimed to have high heat resistance and low smoke development and toxicity are now being sold in North America by Norold Composites of Mississauga, Ontario, Canada. These acid-cured phenolics can be used in composite applications such as sheet moulding compounds, filament winding, pultrusion and spray-up, and are marketed under the trade-name Norsophen.

4.11 BLENDS AND ALLOYS

4.11.1 NEW RANGES OF TRIAX ALLOYS

Monsanto has extended its range of polymer alloys marketed under the Triax label. In addition to the existing '1000' series of polyamide/acrylonitrile–butadiene–styrene blends, the range now includes alloys of polycarbonate and acrylonitrile-butadiene-styrene, polyvinyl chloride and acrylonitrile-butadiene-styrene, and others based on polyamide (nylon) and polyester.

The '2000' series of polycarbonate/acrylonitrile-butadiene-styrene alloys is reported to have high impact resistance, very good heat resistance, high rigidity and very high strength. The alloys are suitable for injection moulding, extrusion and blow moulding. Quoted applications include power tools, telecommunications equipment, domestic appliances and general industrial mouldings.

The '3000' series of polyamide/polyolefin blends, are used for automotive hoses and tubing, including fuel lines, air lines and hydraulic hoses. The alloys extrude easily, have good low temperature impact resistance, and are resistant to automotive fuels, organic chemicals and zinc chloride. They are intended to compete with nylons 11 and 12.

'Series 4000' designates a range of polyester-based alloys intended for garden equipment, domestic appliances and general industrial mouldings. Again, these materials are seen as a replacement for nylon in some applications.

A partnership between Monsanto and Vista Chemical Co has resulted in the production of an acrylonitrile-butadiene-styrene/polyvinyl chloride alloy intended for housings of business machines. Designated Triax CBE/1, the alloy is claimed to provide higher impact strength, stiffness and UV resistance than flame retardant acrylonitrile-butadiene-styrene, with which it competes. The alloy reportedly gains its processibility and heat resistance from the incorporation of alphamethyl styrene.

4.11.2 POLYCARBONATE/ACRYLONITRILE-BUTADIENE-STYRENE BLENDS

GE Plastics has added four grades, two of them flame retardant, to its Cycoloy product range of polymer blends, making a total of nine. These blends of polycarbonate and acrylonitrile–butadiene–styrene copolymer have been developed specifically for automotive interior parts and electronics equipment housings, although their properties make them suitable for a wide variety of demanding applications which include light fittings and electrical enclosures.

New grades HF1100 and HF1200 reportedly have better flow characteristics for enhanced productivity and appearance of the finished parts. Cycoloy C2950, incorporating a flame retardant, has been specially tailored to meet new heat resistance standards in the business machine industry and has a Vicat softening point of 113°C. C2100 is another flame retardant grade which has been developed to meet the more stringent requirements of mains current carrying applications; this material softens at 138°C.

4.11.3 DEVELOPMENTS IN ACRYLONITRILE-BUTADIENE-STYRENE

GE's acquisition of Borg-Warner appears to have been the spur to some recent developments in the acrylonitrile-butadiene-styrene resin field. The company has been upgrading its range of Cyclolac acrylonitrile-butadiene-styrene resins and is now marketing them as engineering materials tailored to particular applications.

What GE calls the G series includes four injection moulding grades of acrylonitrile-butadiene-styrene for general purpose, one transparent material, and three extrusion grades intended for sheeting.

The general purpose resins are targeted for uses in cars, garden equipment, cosmetics packaging and furniture.

4.11.4 MODIFIED POLYPHENYLENE ETHER/NYLON ALLOYS

Mitsubishi Gas Chemical Co is marketing two new grades of modified polyphenylene ether/nylon alloy. NX-7000, based on nylon 6, is intended for automotive components such as wheel caps which have to withstand the heat generated by braking as well as the wear and tear of everyday road use. The other, NX-9000, is based on nylon 66 and is recommended for other automotive parts and for thin connectors.

4.11.5 POLYBUTYLENE TEREPHTHALATE/ACRYLONITRILE-STYRENE-ACRYLATE

A blend of semi-crystalline polybutylene terephthalate and amorphous acrylonitrile-styrene-acrylate copolymer, reinforced with glass fibre, has been launched by BASF. The new material, known as Ultrablend, is claimed to have a high heat distortion temperature, good surface finish from mouldings, good flow characteristics and excellent stress cracking and weathering properties.

4.12 ENGINEERING POLYMERS

4.12.1 POLYARYLETHERKETONE — A NEW POLYMER

Polyaryletherketone resins are now available from BASF under the tradename Ultrapek. These are high performance materials specially designed for electronic and mechanical components subject to high temperatures, mechanical loading, aggressive chemicals or intense radiation.

According to BASF, Ultrapek is inherently flame retardant and has a UL94 classification of V-0 at 1.5 mm thickness, generating the lowest smoke density of any thermoplastic when heated. It melts at 381°C (polyetherether ketone melts at 334°C), and requires higher processing and mould temperatures than polyetherether ketone.

4.12.2 POLYETHER NITRILE

Idemitsu Kosan Co Ltd of Tokyo, has developed a polyether nitrile engineering plastic. The material, designated ID-300, has been produced on a trial basis at the company's central research laboratory in Chiba.

Figures quoted for ID-300 include a tensile strength of 140 MPa, a limiting oxygen index of 42, and glass transition, deformation and melting temperatures of 145, 165 and 350°C respectively.

4.12.3 FILLED POLYBUTYLENE TEREPHTHALATE

GE Plastics has given the name 'Heavy Valox' to a new range of polybutylene terephthalate engineering thermoplastics, characterized by very high loadings of mineral filler. Despite these high loadings, claims GE, Heavy Valox retains a flow rate during processing that is almost as good as that of unfilled Valox.

The key to this combination of properties (see Table 3) lies apparently in the shape and size of the mineral particles, together with a proprietary compounding technique and the presence of a special additive. In its hardened state, Heavy Valox behaves and feels almost like china, and is therefore targeted for tableware and microwave cooking containers.

TABLE 3: Properties of Heavy Valox from GE Plastics.

Property	Value
Melt temperature	266°C
Melt flow	80g/10 min
Tensile strength	45 MPa
Elongation	1.0%
Flexural strength	90 MPa
Flexural modulus	5.2 GPa
Izod impact strength	
notched	32 N
unnotched	160 N
Vicat temperature	218°C

The first commercially available member of the series, HV7065, contains 65% of mineral filler. Forthcoming grades are expected to contain up to 90% of filler. In addition to polybutylene terephthalate, GE is said to be developing 'heavy' grades of polycarbonate, polyether imide and polyphenylene sulphide.

4.13 FOAMS

4.13.1 HIGH TEMPERATURE FOAM

Rogers Corp has introduced a high temperature version of its Poron cellular urethane. Developed for shock and vibration control applications, this material — designated Poron 4723-23 — can be used in temperatures up to 125°C. It is recommended for automotive applications in the engine compartment, the firewall and the instrument cluster. It is available in thicknesses from 1.13 to 12.5 mm and in densities of 240, 320 and 480 kg/m^3.

4.13.2 HIGH STRESS FOAMS

Bayer has introduced two glass-mat reinforced integral-skin polyurethane foam systems. Baydur STR-C (compact) and Baydur STR-F (foamed) have been developed for the production of automotive parts which need to withstand heavy loads while being lightweight, dimensionally stable and stiff.

Both products are intended for reaction injection moulding processing. The process allows large, glass-mat reinforced parts, complete with anchoring elements and decorative facing material, to be produced in one operation.

Typical applications for the expanded Baydur SRT-F system are self-supporting parts for automotive interiors: headliners, post finishers, interior door panels, parcel shelves and instrument panel supports, for example. The solid Baydur STR-C system is used for the production of load bearing structural elements with a non-cellular polyurethane matrix, such as luggage compartment covers, spare wheel wells and engine shroud sections.

TABLE 4: Properties of Caril expandable beads.

Grade	Bead size (μm)	Temperature resistance (°C)	Recommended minimum density (kg/m³)	Suggested wall thickness (mm)
EX 307	300–500	105	60	1–2.5
EX 308	500–700	105	45	2.5–10
EX 402	500–1200	120	60	—
EX 403	500–1200	112	45	—
EX 404	500–1200	105	30	—

4.13.3 STIFF ENGINEERING FOAM

Packaging, building and construction, furniture and vehicle components are among the applications claimed for a new type of engineering foam on the market. Known as Caril, the material is being made by Shell Nederland Chemie BV and marketed by GE Plastics Europe.

Caril expandable beads are made from a blend of modified polyphenylene oxide and expanded polystyrene. They are characterized, it is claimed, by a balance of processibility, good thermal properties for an engineering foam material, and an excellent strength-to-weight ratio.

Caril beads are spherical and are expanded in manufacture by means of pentane. They are available in five grades; some of the properties are listed in Table 4.

Grades EX 307 and 308 are recommended for applications such as microwave resistant drinking cups and steam cleanable, reusable packaging. With its high temperature resistance, grade EX 402 is regarded as particularly suitable for automotive applications, which require the material to withstand the prolonged high temperature of the paint oven. For less demanding automotive applications EX 403 can be used, while grade EX 404 is suggested as being ideal for the insulation of hot water pipes.

When processed, Caril is claimed to be superior in both strength and durability to conventional foamed polystyrene. The final strength depends on the pressure and duration of steam treatment. Generally speaking, higher steam pressure means higher strength but also higher density.

Grades EX 307, 308 and 404 are said to be easy to process using conventional expanded polystyrene (EPS) foam bead moulding techniques, with the recommended precaution of good ventilation. However, for grades EX 402 and 403, varying degrees of modification are required depending on the type of machine used and the intended application. In addition to higher steam pressure, modifications may also include polytetrafluoroethylene-coating of moulds for easy release.

4.13.4 POLYPROPYLENE FOAM

Conventional polypropylene is generally known to have a low melt strength which is thought to be due to a low elongational viscosity. This has made it impossible to use PP for foaming applications, which require that the melt strength should be high enough for the cell walls to remain intact until the foamed polymer solidifies.

New developments at Himont, USA, have resulted in a grade of polypropylene, described as 100% homopolymer with a modified rheology. The melt has an elastic characteristic which allows a processing 'window' for foaming between 17 and 165°C.

4.13.5 SUPERPLASTIC FOAM

A Tokyo company has filed over 50 patents relating to a polyetherpolyol with a molecular weight of over 10 000, and to its associated production technology.

Asahi Glass Co Ltd says the material will be used to produce superplastic polyurethane foam. It can be injection moulded and is expected to lead to the development of a series of sealants.

4.14 CONDUCTIVE POLYMERS

4.14.1 ELECTRICALLY CONDUCTIVE POLYANILINE

An electrical conductivity level approaching that of copper is claimed for a polyaniline being developed by the Ministry of International Trade and Industry (MITI) in Japan. Using a special catalyst, the MITI team has raised the conductivity of polyaniline to about 400 000 S, roughly the same as that of gold.

Another conductive polyaniline has been produced in a fibril state by Mitsubishi Chemical Co Ltd of Tokyo, Japan, and will be used in combination with metallic lithium in the construction of a thin, flexible battery.

4.14.2 CONDUCTIVE ACETAL COPOLYMER

Hoechst has announced the introduction of a new grade of electrically conductive Hostaform acetal copolymer with a low surface resistance and low volume resistivity.

The new grade, Hostaform C9021 ELS, is an injection moulding polymer suitable for applications where electric charges must be neutralized or avoided, such as sensitive electronic components and software, or in areas like mines where explosions can be a hazard.

There is a reduction in some of the property values when compared with a standard grade of Hostaform C9021, but the new grade still offers good mechanical, thermal and chemical properties, it is claimed.

4.14.3 POLYSTYRENE FOR ELECTRONIC APPLICATIONS

LNP Engineering Plastics, a unit of ICI Advanced Materials of Exton, Pennsylvania, USA, has developed two partially conductive polystyrene compounds for electronic applications such as component bins and surface mountings.

Stat-Kon PDX-C 88320 and 89204 (the former being the more conductive) both contain carbon black and have a notched Izod impact strength of 160 N.

4.14.4 ELECTROMAGNETIC INTERFERENCE SHIELDING

Faradex is the name given to a range of thermoplastics for electromagnetic interference shielding and electrostatic discharge protection. These properties are increasingly in demand for casings of electrosensitive equipment such as office computer terminals.

The range is being offered by DSM Engineering Plastics North America Inc, a subsidiary of the Dutch group DSM. Faradex compounds contain about 1% by volume of stainless steel fibres, giving a volume resistivity of 0.1–1.0 ohm.cm, depending on the product formulation.

4.14.5 CONDUCTIVE POLYPYRROLE/POLYIMIDE COMPOSITE

Researchers at the Akzo laboratories at Obernburg, Germany, are reported to have produced a composite of polypyrrole, which conducts electricity, and polyimide. The latter component adds heat stability and flexibility to the polypyrrole. The composite consists of a layer of polyimide sandwiched between two polypyrrole layers, and is prepared by vapour deposition.

4.15 ELASTOMERS

4.15.1 POLYURETHANE ELASTOMERS

Macpherson Polymers is launching a range of engineering polyurethane elastomer systems — the Diprane 54 series — with Shore A hardnesses ranging from 55 to 95.

Diprane polyester-based systems were first developed and introduced by Macpherson over ten years ago. The Diprane 54 series represents the third generation of these products, and brings several additional advantages, such as lower viscosity and reduced demould time, without loss of abrasion or chemical resistance.

Macpherson claims that the new range has brought 4,4'-diphenyl-methane diisocyanate (MDI)-based systems closer than ever to tolylene diisocyanate (TDI)-based systems in processing characteristics, whilst retaining the benefits associated with MDI-based formulations of superior health and safety characteristics in both manufacture and processing.

Special research projects at Macpherson Polymers have led to formulations with fire retardant properties and others which resist fungal attack.

4.15.2 THERMOPLASTIC ELASTOMERS

Monsanto Chemical Co has launched two additions to its range of thermoplastic elastomers. Vyram is a mid-range performance material for general purposes, while Dytron XL is aimed specifically at wire and cable jacketing and insulation.

Vyram is claimed to be ideally suited for applications which traditionally use styrene-butadiene rubber or ethylene-propylene-diene monomer thermoset rubber compounds. It offers good cost/performance benefits in applications where the operating temperature does not exceed 110°C, it is stated. Vyram is said to combine good wear and compression-set characteristics with good hydrocarbon and excellent aqueous fluid resistance, along with ease of processability. Parts can be made from Vyram by extrusion, injection moulding and blow moulding.

Dytron XL is an alloy of a thermoplastic and a thermoset rubber produced by dynamic vulcanization technology. The resultant product can be processed as a thermoplastic but exhibits most of the performance characteristics of a fully cured thermoset rubber compound. It has the ability to be further crosslinked by the application of external radiation such as an electron beam.

Dytron XL is intended to replace a variety of thermoset rubbers and crosslinkable thermoplastics. The product line, says Monsanto, has been specifically developed to meet the requirements of the European wire and cable industry for products capable of maintaining performance and integrity following periodic exposure to extremes of temperature. Such extremes can occur under emergency power conditions or momentary high temperatures. The grades currently available are used for primary insulation or cable jacketing. Non-halogen flame retarded grades are under development.

Monsanto's production and marketing of these elastomers will be considerably helped by its recent acquisition of the Levaflex line of thermoplastic elastomers from Bayer AG of Leverkusen, Germany. Monsanto plans to incorporate the Levaflex products into its existing range.

4.15.3 ETHYLENE-PROPYLENE-DIENE MONOMER

Exxon Chemical has used what it describes as new technology to produce a new range of Vistalon ethylene–propylene–diene monomer rubbers. These are said to be characterized by low crystallinity, very high crosslink efficiency and tailored molecular weight distribution, leading to excellent collapse resistance, low viscosity, and improved performance in mixing and extrusion.

The first member of this range, Vistalon 7500, was introduced in 1989. A new grade, Vistalon 8600, has now appeared; this is specifically designed for sponge and solid profiles and is claimed to be ideal for complex automotive body seals. Both grades are being made by the Exxon subsidiary Socabu at Notre-Dame-de-Gravenchon, France.

4.15.4 FLUOROELASTOMER

3M announced a new fluoroelastomer, 3M Fluorel II. It is claimed to have a higher resistance to amines than its predecessor, 3M Fluorel, and is designed for use in seals for engine oil, gear lubricants and transmission fluids. 3M is also introducing a computer-based design kit to enable users to identify the materials that best meet their needs.

4.16 COATINGS

4.16.1 FLUOROPOLYMER COATING

Conventional fluoropolymers are difficult to use as coating materials because of their main virtue — they are generally insoluble. However, ICI researchers at Runcorn, UK, have now developed a series of novel resins based on alternating monomers of fluoro-olefins and vinyl ethers (including alkyl, hydroxyl and carboxyl). These amorphous copolymers are soluble in most organic solvents and are claimed to have potential for coating architectural materials, steel, aluminium and plastics.

4.16.2 ACRYLIC EMULSIONS

Two new grades of acrylic emulsion paint have been announced by Japan Catalytic Chemical Co.

Acryset is cured by ultraviolet irradiation and does not need a photopolymerization initiator. The paint is said to have a long shelf life and to be easy to apply, while the dried film is claimed to be stable to sunlight and heat.

Acryset emulsion paint is used as a coating which is crosslinked by spraying with water at room temperature. The resulting film is said to be strong and to have good water resistance properties.

4.17 FRICTION MATERIALS

4.17.1 FRICTION POLYMERS

BP Chemicals at Barry, South Wales, UK, has launched what is described as a new generation of polymers specifically designed for the friction industry. Launched under the trade name Cellobond, the polymers have been designed to give improved mechanical and physical properties in non-asbestos composites, especially with regard to friction and wear characteristics.

The Cellobond CFP500 series is said to feature a low level of free phenol.

4.17.2 POLYPHENYLENE SULPHIDE/POLYTETRAFLUOROETHYLENE

Lubricomp 189 is the name given to a composite of polyphenylene sulphide and polytetrafluoroethylene developed by LNP Engineering Plastics of Exton, Pennsylvania, USA.

The material is claimed to be injection mouldable, inherently flame resistant. stable at high temperatures, and wear resistant. In addition, it has a low coefficient of dynamic friction due to its polytetrafluoroethylene content.

5. APPLICATIONS

Not surprisingly the automotive sector was the dominant end-user during the year. The main driving force behind the replacement of metal by plastics is fuel efficiency via weight saving. One of the fastest growing areas of use of plastics in cars is electronic and electrical. This is a reflection of the general growth in electronic and electrical applications for plastics. Another important area is that of pipelines, where further opportunities are provided by systems for renovating and relining. Many of the items in this section are examples of ingenuity which are not explicable solely in terms of economic trends, but present interesting and potentially useful concepts.

5.1 AUTOMOTIVE

A survey of some of the recent uses of plastics in vehicle manufacture shows an increasing employment of materials specially developed for such applications. The environment within, say, a car bonnet (hood) calls for a degree of resistance to high temperature, vibration, abrasion and oil contamination which makes it unsuitable for commodity plastics. However, the enormous size of the potential market has spurred plastics manufacturers to develop varieties of materials that not only meet the car maker's performance specifications but are available as and when required, and at reasonable cost.

In some instances, car manufacturers have found that a high performance material already developed for another purpose satisfies their needs. Whether or not a new material is required, the success of any application usually depends on close cooperation between the two industries.

An example of such cooperation is evident in a developmental car engine developed by the Ford Motor Co. This makes extensive use of plastics and incorporates metals only for the moving mechanical parts, combustion chambers and cylinders — admittedly the most crucial components. The new engine is claimed to be lighter, more thermally efficient and quieter than its conventional counterparts.

5.1.1 NYLON (POLYAMIDE)

Nylon, in various grades, has been used for many years in under-bonnet applications such as water reservoirs. A recent, rather more critical use is as a brake vacuum assembly tank for the high performance Aston Martin Virage. Using nylon 6, Amber Plastics of Chesterfield, UK, is producing this triangular shaped tank to what are claimed to be the most stringent quality standards, including vacuum testing to 630 torr. The tank is 1kg lighter in weight in comparison with an existing metal cylinder.

Excellent heat and chemical resistance, high stiffness and low creep at high temperatures, toughness, good fatigue behaviour and fast cycle times are claimed to be the key characteristics of DSM's Stanyl nylon 4/6. Compared with 'conventional' nylon 6/6, this material contains a larger number of amide groups per length of chain, and its symmetrical chain structure leads to high crystallinity and a high melting point. Eight injection moulding grades are available; applications are being sought primarily in the US market.

Also in the USA, DSM's Nyrim reinforced nylon is being used by the Standard Car Truck Co of Park Ridge, Illinois, for wheel chocks. These are used to stabilize vehicles being transported by rail. In Europe, people whose lives are at risk from terrorists may feel safer in consequence of another use for Nyrim — as a safety device developed by the Defence and Security Division of the French Hutchinson group. The ASH (Anneau de Sécurité Hutchinson) has been designed to be fitted onto all kinds of vehicles whose tyres are likely to be hit by bullets or other projectiles. It permits high speed and offers road holding qualities that allow the vehicle to escape from the danger spot.

5.1.2 PHENOLICS AND POLYESTERS

Because of their better fire resistance, phenolics — usually reinforced with glass fibre — are replacing polyesters in many transport applications, particularly in aircraft and trains. However, glass-reinforced polyester is still in wide use; one item, described as a frontal parabola for Audi cars, is being made by a Chinese company on a machine supplied by PTS-Meico of Monza, near Milan, Italy.

Rogers Corp of Rogers, Connecticut, USA, is introducing a range of phenolic moulding compounds tailored for the automotive industry. Five grades are available.

BP Chemicals is conducting a joint R&D programme with Scandura Seals Co of Cleckheaton, West Yorkshire, UK, to obtain comparative data on the performance of phenolic glass reinforced plastics and other materials at high temperatures. This follows the launch of a heat shield on which the two companies worked closely together.

The heat shield is designed to protect sensitive electrical and engineering components from high temperature radiant heat sources. The problems have become more pronounced with the widespread introduction of turbochargers with their associated high running temperatures, and platinum catalytic convertors which tend to run extremely hot.

According to BP, the advantages of phenolic matrix glass reinforced plastic are that it can be moulded into complex shapes, and design changes can easily be made. Additional benefits are its excellent corrosion resistant and heat resistant properties, and its light weight. These performance criteria mean that it finds use as an alternative to asbestos and metal.

The 1989 Ford MN 12 Thunderbird/Cougar has a grille opening retainer made of Du Pont's Rynite 935 unreinforced polyester resin. Du Pont claims that this thermoplastic is sufficiently strong to make unnecessary the use of glass fibre reinforced thermosets for this application. Moreover, the use of thermoplastics means a weight saving of around 15% per grille; thermoplastics are also recyclable — "an increasingly important design consideration", says Du Pont.

The Ford Escort and Fiesta cars are likely to have inlet manifolds produced in polyester by the lost core technique. In this process, used by the German Frudenberg Group, a high performance polyester resin is formed around a metal core that is inserted into the mould. The core consists of a low melting point tin alloy which is removed by induction melting once the manifold is demoulded. The plastic manifold is reported to be 40% lighter than its metal counterpart and can withstand constant temperatures of 160°C.

In terms of quantity, polypropylene is the dominant plastics material in automotive use. Figures prepared by the Italian Ferruzzi Group show that an estimated 359 000 t of polypropylene will be used in Western Europe this year, followed by polyurethane at 270 000 and polyvinyl chloride at 170 000 t.

The 1990 Rover Metro has its front and rear bumpers made of a 'controlled rheology' grade of ICI's Propathene and Procom elastomer modified polypropylene. ICI claims that the bumpers provide a worthwhile meaure of 'zero damage' protection against low-speed impacts up to 2.5 km/h — the typical parking nudge — and the polypropylene materials retain their properties down to very low temperatures.

A grade of Procom polypropylene, reinforced with 40% glass fibre, is also used for a protection shield for the Volvo 440, 460 and 480 series. By enclosing the underside of the engine compartment, the underpanel improves the aerodynamics on these latest models, reduces drag and provides protection against water ingress. The product is said to be cheaper to produce than the resin-impregnated glass mat previously used, since the costly finishing operations needed for apertures have been eliminated by the use of injection moulding.

The Scottish company Royalite, now owned by British Vita, is moulding its PP R95 sheet into wheelarch liners for Lotus's latest front wheel drive version of the Elan sports car. According to the company, this wheelarch liner is a particularly demanding application and the Royalite R95 material is well suited to the task because of its extreme abrasion resistance, impact strength and light weight.

Another company interested in automotive PP compounds is Solvay, the Belgian plastics group, which has formed a partnership with US compounder Dexter Corp to develop, manufacture and market these materials. Solvay is keen to gain a commercial foothold in the American market, and will benefit from the use of Dexter's newly opened Automotive Applications Center near Detroit.

5.1.3 POLYURETHANE

The majority of polyurethane is used in the form of foam. Much research over the past year has gone into methods of producing this foam without using chlorofluorocarbons, which are believed to be causing the protective upper atmosphere ozone layer to be damaged and are therefore being phased out under the Montreal Protocol.

Bearing in mind that most polyurethane is manufactured *in situ* rather than in advance, manufacturers have striven to produce polyurethane systems that will keep pace with the demanding timescale of automotive mass production. Sanyo Chemical Industries Ltd of Kyoto, Japan, claims to have developed a polyurethane resin which cures in 20 s, in comparison with the current 45–60 s required for many systems. Designed for car bumpers, the resin is reportedly heat stable up to 120°C and does not create dimples or pinholes in the final product.

Whereas polyurethane foam is mainly used as a cushioning or insulating medium (for both sound and thermal insulation), solid (non-foam) polyurethane is itself a useful material. Bridgestone's Australian subsidiary has devised a way of combining the two types in one component — a car door panel — which can be manufactured on a pair of carousels to synchronize with the mass production schedule. The equipment is supplied by Elastogran Polyurethane GmbH of Strasslach, West Germany.

Dow Europe's Specflex range of polyurethanes (replacing the Urecor and Polyurax brandnames) covers flexible polyurethane systems, formulated components and performance polyols. These are used for seating, interior soft trim, sound and vibration damping, structural and semi-structural applications. Growth is forecast in all sectors, particularly in damping.

A novel application of polyurethane foam is in air filters. Specialist air filters from Ramair are finding rapidly increasing favour in racing circles, it is claimed. Reynard and Lola both fitted Ramair filters as original equipment on more than 70 Formula 3000 cars in 1990.

Conventional filters are generally regarded as unsatisfactory because they reduce engine efficiency — increasingly so as they gather contaminants from the air. The alternative design developed by Ramair uses a two layer foam to filter the air. The manufacturer says this is highly efficient, longer lasting, and does not clog in the same way as paper filters. The foam filter element is held in a polyurethane frame, which also holds the supporting mesh or cage and any fastenings needed to attach the filter.

Ramair has now developed a new heavy duty series of replacement filters, bringing foam filter technology to heavy commercial vehicles such as trucks, buses and off-road vehicles. The traditional paper filters used in these vehicles are prone to develop rapid soot plating, especially in areas with a high level of pollution. Hyperplast polyurethane elastomers are used for encapsulation of these filters.

5.1.4 POLYETHYLENE

The incorporation of Selar barrier resin from Du Pont, USA, reportedly enables high density polyethylene fuel tanks to meet Federal vapour emission requirements. The dry blended resin, added at 4–6% by weight, forms large, well separated platelets within the high density polyethylene matrix and so reduces the material's permeability to gasoline four-fold.

The incorporation of a compatibilizer enhances the adhesion between the resin and high density polyethylene, increasing the burst strength by 10%, so that a tank filled with glycol can withstand a 6 m drop test at −40°C without cracking.

The integral barrier cannot be worn off, it is claimed, and the improvements in barrier properties should last six years.

5.1.5 OTHER MATERIALS

In addition to rubber tyres, elastomers — both natural and synthetic — are used for such items as hoses, O rings and shock absorbers. Dowty O Rings International (DORI) recently collaborated with Ford's Climate Control Division in helping to solve a severe corrosion problem on the aluminium feed tubes used in vehicle air conditioners. DORI's solution was to develop a rubber compound for O rings which seals against refrigerants and is corrosion free. The company's technicians are now working on an internally lubricated fluorocarbon O ring material which has significantly smaller dimensional tolerances and should be much easier to fit.

Fuels with a high alcohol content attack components made from conventional elastomers. GE Silicones has launched two fluorosilicone heat cured elastomers for use with such fuels. Fluorosilicones are not only chemically inert but offer the highest levels of dielectric strength and arc track resistance of any elastomer.

Phillips Petroleum Chemicals claims successful under-bonnet applications for its Ryton PPS (polyphenylene sulphide), including fuel induction systems, engine and transmission components, lighting units, electrical and electronic components and ash trays. Phillips has just introduced a new grade of Ryton, R-4 XT, with higher toughness, impact strength and weld line strength than the standard grades.

What is claimed to be the first all-composite fuel pump has been made by Sofabex from Solvay's IXEF 1022 polyarlyamide. The choice of this material was decided largely by its very low creep when hot. According to Solvay, replacing the aluminium in the pump body produces an improvement in its thermal insulation. The internal temperature attained during use is notably lower, and this reduces the percolation and 'vapour-lock' effects largely responsible for the problems encountered when a hot engine is restarted.

Mitsubishi Gas Chemical Co Ltd has developed a heat resistant liquid bisthiazole resin for automotive applications, designed for temperatures up to 200°C. The resin is based on a polyimide series compound. It is available in a liquid state without the necessity for solvents, and, according to Mitsubishi Gas, is easy to process.

There is a filament winding grade (BT-3040) and a casting grade (BT-3003). BT-3040 is a two solution type; curing at 200°C, it yields a resin with a heat distortion temperature of 250°C. BT-3003 is a three-solution type which is pre-cured at room temperature, followed by a further cure at 200°C.

5.1.6 PLASTICS AND STEEL CAR

Vector II, GE Plastics' hybrid plastics and steel car, was first displayed at the K'89 exhibition at Düsseldorf, Germany, in November 1989. Since then it has undergone a comprehensive programme of tests, including dynamic crash testing of the thermoplastics bonnet, panel vibration assessment, door slam and body durability testing.

In addition to the bonnet, engineering thermoplastics have been used for front fenders, dashboard carrier and side doors, front-end support panel, tailgate, rear quarter panels and two underbonnet applications.

The issue of material recycling, now a major concern within the automotive industry, is claimed to have been an underlying consideration in the design of Vector II. GE Plastics believes that by giving full consideration to the problem of recycling at the draughting stage, it is possible to design parts to make them more easily recoverable from a vehicle when it reaches the end of its useful life.

5.2 AEROSPACE

5.2.1 POLYIMIDES

'Envex' polyimides from Rogers Corp have been chosen for a variety of applications in the aerospace field. The properties of interest are conformability, low coefficient of friction, high wear rate, high strength, good dimensional stability, satisfactory outgassing performance at high temperatures, and resistance to fuels and lubrication fluids.

One application is for poppet valves in the hydraulic actuator lines of the landing gear on the F-16 multimission aircraft. This component has to seal against an aluminium seat while holding a pressure of some 20 MPa. The material selected was Envex 1000, which reportedly can withstand the aggressive chemical environment, mechanical shock loading and the high fluid flow rates.

Another application involving Envex 1330 polyimide for the bearings in swivel joints was used in the deployment of the antenna arrays on the 1982 Space Shuttle. The material had to withstand temperatures from −130 to +90°C, exposure to hard vacuum and intense ultraviolet radiation, and also had to have slight electrical conductivity. A low graphite component in the material satisfied this latter requirement.

5.2.2 SEALS FOR TURBOFAN ENGINE

Dunlop Precision Rubber Division has developed a new fire-stopping seal for the latest version of the Rolls-Royce RB211-535E4 turbofan. This engine has been in service on the twin-engined Boeing

757 since 1984, and is designed to provide the reliability required for long over-water flights. Advanced technology features make it exceptionally quiet and economical in operation, it is claimed.

Dunlop is supplying a wide variety of seals throughout the RB211-535E4, including oil and pressure seals and a special thrust reverser seal. The fire resistant seal is reinforced with woven glass fibre and pressurized to form a seal around the fireproof bulkhead. Its design has been optimized to improve air flow and eliminate turbulence. The resulting improvement in fuel consumption is claimed to be in the order of 0.1% — enough to save several thousand pounds on the annual fuel bill of each engine in normal operating conditions.

Another aerodynamic seal of unusual design is the annulus filler seal. This is a multilayer composite of low temperature silicone rubber and polyester fabric. The seals are bonded to the base of the forged annulus filler component, totally embracing the new wide-chord fan blade, so as to control air flow, reduce turbulence and improve aerodynamic efficiency.

5.3 SCIENTIFIC

5.3.1 EXTRUDED POLYIMIDE IN THE SUPERCONDUCTING SUPER COLLIDER

The new Superconducting Super Collider to be built in Waxahatche, Texas, USA, is posing tough problems for the materials engineers. The Superconducting Super Collider is essentially an evacuated tube, 84 km in circumference, around which two opposing beams of protons travelling at almost the speed of light will be allowed to collide. The materials used for the tube must be able to withstand high radiation fluxes at cryogenic temperatures.

An important role will be played by a set of keys mounted every 45 cm along the tube, which have to maintain alignment of the trimming coils. Tests conducted at the Brookhaven National Laboratory have established that Rogers' 'Envex' extruded polyimide showed no signs of embrittlement, bloating or crazing after a radiation dose of 1000 Mrad. Under these conditions, most thermosets and thermoplastics are rendered extremely brittle or otherwise unserviceable.

5.3.2 RADOME TRANSPARENCY

Allied-Signal's extended chain polyethylene Spectron, is, the company claims, the only fibre available today having both a lower dielectric constant and a lower loss tangent (2.2 and 2×10^{-4} respectively) than the matrix resin. This property imparts a high transparency to most of the electromagnetic radiation spectrum, making it suitable for reinforcing radomes — protective domes for radar aerials.

Allied says that Spectra fibres are supplanting glass, aramid and quartz in this application, and has published a brochure, 'Spectra high performance fibres for radomes', which describes the material's characteristics and pinpoints some of its applications in the electronics field.

5.4 MEDICAL

5.4.1 POLYMERS FOR BIOENGINEERING

New polymers launched by BASF include Ultrapek (polyaryl etherketone), Ultrason E (polyethersulphone) and Ultrason S (polysulphone). All three polymers are stated to have high glass transition temperatures and excellent wear properties.

BASF is claiming applications for them in the medical sector. Ultrason S is said to be stable against hydrolysis and radiation, thus allowing effective sterilization. Ultrapek is semicrystalline and is said to have excellent chemical resistance, along with good biocompatibility.

Hip implants are being made from Ultrapek, with the shell consisting of the plain material and the supporting core made from Ultrapek reinforced with carbon fibre. It is claimed that this combination will give twice the life of conventional prostheses.

5.4.2 POLYMER FOR CARDIAC MONITOR

Monsanto's Triax has been chosen for the manufacture of a cardiac monitor by US manufacturer Omnica Corp. Weighing only 250g, this instrument is worn by a patient to record the electrical activity of the heart over a 24-hour period. It needs to be able to withstand both careless dropping and chemically aggressive cleaning, and is a difficult product to produce in quantity because some of its reference dimensions have very tight tolerances. Triax works well for the outer shell because it has low shrinkage and good mouldability, according to the manufacturers.

5.4.3 CUSTOM FORMULATIONS FOR ANATOMICAL MODELS

Educational and Scientific Products Ltd of Rustington, Sussex, UK, specializes in making anatomical models for educational purposes. These include life-size and miniature skeletons, teeth, various joints and bones, hearts and ears. More than 50% of the company's range is exported to the USA, Japan and elsewhere.

These models make extensive use of polyurethane resins supplied by Macpherson Polymers, including semi-rigid high impact materials for the skeletons and moulding grade elastomers for flesh. Polyurethanes are chosen because of their ease of use and their economics in short runs.

5.4.4 CUSHION BASE FOR WHEELCHAIRS

Supracor Systems Inc has won a contract to develop a new cushion base for wheelchairs which, it is claimed, will significantly reduce weight while retaining comfort and durability. The cushion will make use of Supracor flexible aerospace honeycomb based on advanced thermoplastic elastomers.

The high tear and tensile strength of Supracor give it resilience and resistance to compressive loading and good shock absorption. Other products based on Supracor include the Reebok range of running, tennis and basketball shoes.

5.4.5 THREE AWARDS FOR INSULIN PUMP

An injection moulded insulin delivery device for diabetics, NovoPen II, has won the BPF (British Plastics Federation) Horner Award for Design and gained second prize in the Forma Finlandia Second International Design Competition. It is also the winner of the BBC Product Design Award for 1990. The product was designed by a three-way venture team involving Advance Injection Moulders, Sams Design and Hypoguard Ltd.

NovoPen II makes life easier for diabetics who must regularly and unobtrusively inject themselves with insulin. The device comprises 26 individual parts made from a variety of materials including acrylonitrile-butadiene-styrene, nylon 6, polycarbonate, polyacetal and GF nylon 66. It is supplied in a carrying case which holds three spare needles and a refill cartridge of one of the new generation of longer lasting insulins.

NovoPen will perhaps face competition from an innovative design, from Montreal-based Techmire Ltd, which incorporates all components of the syringe barrel, including the needle and cap, into one single piece. The Techmire product resulted from an attempt to help Third World companies who do not have the resources to deal with the complex processes of conventional manufacture.

5.5 INDUSTRIAL

5.5.1 ALL-PLASTIC FLOWLINE COUPLING

A Lancashire firm has launched an all-plastic coupling with a quick-release locking mechanism, intended for air, gas or liquid flow lines. Designed for use with flexible tubes, the APC is made of Du Pont's Delrin acetal copolymer, with a Buna-N O-ring seal, and is suitable for chemical, medical, pharmaceutical and vending equipment where hygiene is important. Other applications include industrial cooling water and pneumatic lines, and laboratory instrumentation.

5.5.2 STEAM-WATER MIXER

Victrex polyetherether ketone has been specified by Caradon Mira for a vital component in a new steam and water mixer, the Rada 203. The material was chosen because of its combination of mechanical strength, high continuous working temperature and hydrolysis resistance.

Designed for cleaning industrial equipment, particularly food processing machinery, the mixer incorporates a simple safety device to prevent steam mixing with hot water in the event of component seizure — which can occasionally happen, particularly in hard water areas.

5.5.3 REINFORCED POLYPHENYLENE SULPHIDE BOLTS

Engineers based at a technical test station in Kanagawa Prefecture, Japan, have filed a patent for producing bolts made of polyphenylene sulphide reinforced with carbon fibre. The bolts are said to have excellent dimensional stability and sufficient strength for their purpose — the composite material itself has a tensile strength of 210 MPa — while weighing only one-fifth as much as the equivalent steel bolt.

5.5.4 POLYVINYL CHLORIDE FOAM KEEPS GAS RIGS BUOYANT

Plasticell, a rigid polyvinyl chloride foam, is being used to give extra buoyancy to a pair of offshore drilling and exploration platforms in the Morecambe Bay gas field, UK. The material is calculated to give an extra 300 tonnes of uplift to the platform when settled into the mud of the seabed.

5.5.5 GRINDING DISCS

Monsanto's Triax alloy has been used as a replacement for nylon in grinding discs. In this application, toughness is essential. During manufacture, abrasive powder is sprayed onto the disc at high temperature; in use, the disc undergoes repeated impact and flexure. It also may have to be stored for long periods in a dry atmosphere, under which conditions polyamide tends to become brittle. Triax 1120 was successfully used for this application, as it was for a new type of herbicide sprayer in which the material's ability to withstand prolonged contact with aggressive chemicals was of primary importance.

5.5.6 POLYAMIDE HEAT EXCHANGER

Du Pont Canada Inc has been granted Patent No. 4 859 265 for a method of manufacturing a polyamide heat exchanger.

The device consists of a labyrinth of fluid-flow passages sandwiched between two panels. The patented method involves moulding the panels, coating them with polyvinyl alcohol in a pattern corresponding to the network of fluid passages, and applying heat and pressure to bond them together. Other methods of manufacture are possible.

5.5.7 DISPENSER FOR REACTIVE RESINS

Chemical resistance is essential for reactive resin dispensers. The need to keep reactive resins separate, and yet convenient to use, has led to the design of a simple-to-use dispensing cartridge. The maker, Supermix Systems, of Corby, Northants, UK, is a leading supplier of two component reactive resin dispensing equipment for use in applications such as electronics encapsulation and mould making.

The cartridge is injection moulded by North West Plastics of Manchester in Hostaform acetal copolymer C27063, an easy flowing high impact grade, marketed by Hoechst's Polymers Division. The complex body moulding comprises two cylindrical cavities, one within the other, and a nozzle. The size of the inner cylinder is varied according to the dispensing ratio, which can range from 1:1 to 10:1. Apparently, Hostaform acetal copolymer was selected for the dispenser, rather than polyamide or polypropylene, because it was essential to use an easy flowing material which could nevertheless produce mouldings that were dimensionally, accurate, rigid, impervious to the resins being dispensed, and resistant to chemicals such as methyl methacrylate. The cartridge must not flex under the operating pressure, and must also pass the standard Air Transportation Drop Test of 1.5 m to a concrete floor.

The nozzle, made from polypropylene, has a static mixing element also moulded in Hostaform acetal copolymer. A simple on/off valve moulded in Hostaform closes the chambers and prevents leakage when the dispenser is not in use. The success of the dispenser is indicated by the fact that over 70% is sold for export.

This is one of a growing number of high performance applications for acetal copolymer, including gear trains, taps and couplings.

5.6 OPTICAL

5.6.1 GEL THAT SHRINKS WHEN EXPOSED TO LIGHT

Researchers at the Massachusetts Institute of Technology (MIT), USA, have developed a polymer gel that shrinks when exposed to visible light. Based on *n*-isopropylacrylamide, the material includes a light sensitive salt of copper chlorophyllin with a structure similar to that of chlorophyll in plants.

The MIT team claims to have found that a co-polymer of the gel and sodium acrylate shrinks to as little as one-thousandth of its original size when irradiated with laser light of 488 nm wavelength. Apparently the polymer changes from a one-phase to a two-phase material when irradiated, i.e. from a homogeneous co-polymer to a separated mixture of solvent and gel. The time taken to do this depends on the size of the sample, but is in the order of milliseconds for a specimen a few micrometres in diameter. The material's transparency and refractive index also change.

Possible applications for the gel include electro-optical switches and muscle prostheses.

5.6.2 PLASTIC LENS

Toray Industries Inc of Tokyo, Japan, has developed a polymer, described as "containing urethane bonds", for moulding spectacle lenses. The company claims the material is easy to dye and has a refractive index of 1.61.

The lens is coated with three layers: the first cuts out ultraviolet radiation, the second reduces reflection, and the third protects it against moisture.

5.6.3 LIGHT SENSITIVE POLYMER

Researchers at the University of Rochester, USA, have developed a thiophene polymer which not only has non-linear optical properties but has one of the fastest known response times to an electric field. It is far faster than polydiacetylene in this respect and is likely to find application in high speed communication devices.

5.7. ELECTRICAL AND ELECTRONIC

5.7.1 BENZOCYCLOBUTENES AND POLYCYANATES

A promising future in a number of critical applications is forecast for two new thermoset polymers — benzocyclobutenes and polycyanates — recently developed by Dow Chemical Co.

Benzocyclobutenes were discovered by Dow researchers in the 1970s, but it is only recently that they have come to be recognized as a useful alternative to polyimides for microelectronic applications. According to Dow, benzocyclobutenes have better electrical properties — in particular, a lower dielectric constant — than polyimides. This enables conductive components to be spaced closer together when benzocyclobutene is used as the insulating layer between them. Benzocyclobutenes are also more processible.

Polycyanates are already used as matrix resins for circuit boards, but Dow spokesmen predict an explosion of interest in these polymers for applications in aerospace. They reportedly have high temperature resistance, low moisture absorbency, good electrical properties and resistance to microcracking.

5.7.2 AN ULTRA-THIN BATTERY

Scientists at Northwestern University in Evanston, Illinois, USA, and at a number of other laboratories, have developed advanced polymer materials for use in a new generation of batteries. These materials can be used to create a thin, flexible battery — reportedly similar to a playing card — that could power an electric car safely and efficiently.

Incorporating conductive polymers, the polymer-charged batteries would replace the lead-based batteries currently used in small electric vehicles such as golf carts and milk floats. Lead-based batteries are environmentally detrimental, have limited recharging capability and are too heavy for use in more powerful electric vehicles such as cars.

Polymer electrolyte materials were discussed at a meeting of the Materials Research Society in Pittsburgh, Pennsylvania, at the end of November. A key paper was presented by Professor Mark Ratner of Northwestern University. Ratner believes that the rising cost of oil, coupled with more stringent anti-pollution requirements, will bring about the large scale production of electric cars in the near future. Already the California Air Resources Board has predicted that there will be 200 000 electric vehicles on the state's roads by the year 2000.

The Northwestern research team has created a laboratory model of polymer electrolyte behaviour which, Ratner says, can be adapted to develop efficient polymer-based batteries to power vehicles. Building on the work of a British scientist who, in 1975, discovered that an electrical charge carrier could be created by combining polymers and salt, the group has expanded on the discovery, experimenting with different host polymers and salt concentrations at varying temperatures. The result has been a whole new family of polymer electrolyte materials. These conduct electrical charges at a rate comparable to that of liquid conduction, and thereby permit development of a solid state, all-plastics battery.

The research group is currently trying to understand how and why the physical properties of the polymer change with temperature or salt concentration; how to blend polymers effectively; and how the electrolyte polymers will react with electrodes. "When we understand these things", says Ratner, "we will be able to design better, higher conductivity polymer electrolytes, and thus create even better materials."

5.7.3 TOWARDS THE PLASTIC MAGNET

Research at the University of Hokkaido in northern Japan has led to the development of a polymeric material that can be permanently magnetized at very low temperatures. This material is reported to have been made by dissolving a derivative of acetylene in ethanol and reacting it in the presence of a rhodium catalyst for 24 h. A yellow precipitate was produced which, when heated to 200°C, changed into a uniform plastic material that was non-magnetic at room temperature but showed magnetic properties at −180°C or lower.

Another approach to the plastic magnet has been made by researchers at Gunma University in collaboration with Osaka Gas Corp. In this case, magnetic material appears to have been mixed into a plastic tailored for this purpose. It is reported that an aromatic multicyclic substance based on a coal tar distillate was polymerized and crosslinked by the use of paraxylene glycol or vinyl acetate. Ultraviolet radiation treatment during polymerization resulted in dehydrogenation and gave a stable radical structure which facilitated the incorporation of magnetic material.

Meanwhile, *Soviet Weekly* has reported that researchers at the Moscow Institute of Chemical Physics have made an organic polymer with magnetic properties similar to those of iron. According to the Institute's director, Academician Alexander Ovchinnikov, it is theoretically possible to produce a polymer with magnetization 50% stronger than that of iron.

5.7.4 PACKAGING FOR SENSITIVE ELECTRONICS

Cabot Plastics announces that its conductive compound Cabelec 3464 is being used successfully to extrude high quality sheet suitable for making into packaging for devices sensitive to static electricity. The compound, based on a precise mixture of carbon black and a polypropylene copolymer, is being extruded into conductive twin-wall sheet by a number of specialist packaging manufacturers. It is used as a substitute for corrugated cardboard where this material is not suitable.

Twin-wall sheet made with Cabelec 3464 is claimed to be permanently free from both dust and static electricity, and to be highly resistant to oil, water and chemicals. It can be manufactured in thicknesses ranging from 2 to 5 mm and weights from 250 to 1500 g/m^2, after which it is cut, folded and stapled or welded to form boxes in various sizes, including compartmentalized boxes for printed circuit boards.

Corrugated Plastic Products (CPP) of Ringwood, UK, has launched an antistatic storage and handling system for electronics products based around a new lightweight extruded plastic material called Hydrostat. Applications include boxes to handle and protect printed circuit boards from electrostatic discharges during assembly, storage or transit.

Manufactured from carbon-loaded polypropylene, Hydrostat is permanently anti-static and features two outer walls separated by an internal rib structure which, it is claimed, combines strength and rigidity with low weight. Conductivity is specified at 1.1×10^8 ohm/square, and the material meets the USA Military Standard B-817 05B and the Federal Standard 101B, method 4046. It is available in a variety of grades from 250 to 1500 g/m^2, to match a range of applications.

CPP added the Hydrostat material and manufacturing service to its range after visiting electronics manufacturers in Japan. Handling systems based on extruded plastics materials are extensively used there, particularly in providing support for advanced manufacturing systems such as continuous flow and just-in-time, with dedicated boxes issued to suppliers so that components go straight to the point of use on delivery. Hydrostat is ideal for this application, claims CPP.

Hydrostat is easily formed, says CPP, using conventional box-making techniques such as die-cutting, ultrasonic welding, plastic riveting or taping. Tests on the material, it is claimed, show it to be extremely durable and resistant to impact damage, with negligible reaction to both water and many common industrial chemicals and excellent resistance to mould. Hydrostat is thus superior to cardboard packaging, and, being extruded, considerably cheaper than moulded plastics. Containers can be supplied pre-constructed, or in flat-pack form for assembly on-site when required.

The material is complemented by a non-conductive version called Hydroflute.

5.7.5 HIGH TEMPERATURE ENCAPSULATING RESINS

Bayer's 'Epic' range of resins for electrical and electronic components are claimed to have far better thermal properties than their 'conventional' counterparts. They have withstood laboratory testing at over 200°C and have a projected service life of over 20 000 h.

According to Bayer, some Epic resins have reached a glass transition temperature of more than 300°C and a heat deflection temperature of 250°C. They are intended for encapsulation of components subject to high service temperatures, and could therefore find applications under the car bonnet.

Three versions of Epic resins are available. A-stage consists of isocyanate with epoxy and is mainly recommended as an impregnating and dipping resin. The AB-stage is a resin of medium viscosity (2000 mPa.s) for many casting applications. The solid B-state resin (melting point 50–60°C) opens up further applications, as it is fusible or soluble in organic solvents such as acetone or methylethylketone, and is therefore suitable for transfer moulding or the lamination of circuit boards.

Reaction time can be varied between a few minutes and several days, depending on the choice of catalyst. A special reaction inhibitor allows mixtures to be kept for more than six months.

5.7.6 HEAT CONDUCTIVE STRIP FOR ELECTRONIC PACKAGES

Removal of excess heat from electronic component packages becomes increasingly difficult as these packages become smaller and higher powered. Enclosures are typically cooled using some combination of forced air fans, heat sinks and heat frame laminates. However, Boyd Corp of Pleasanton, California, USA, is marketing a heat transfer strip that is claimed to eliminate the need for these devices while meeting a variety of environmental conditions.

Each strip consists of a length of high density Poron cellular urethane made by Rogers Corp of East Woodstock, Connecticut, USA. The material, which has a UL (Underwriters' Laboratories) flammability rating of 94 HBF, is covered with a 25 µm thick layer of copper foil. The strips are manufactured in 305 mm lengths, and a layer of acrylic transfer adhesive is applied to the top of the strip.

When the chassis is assembled, it compresses the heat transfer strips between 10 and 60%, depending on the height of the components on the PC board. The Poron core is compressible, enabling the strips to conform to the heights of various components within the chassis. The copper foil covering the strips provides a direct conductive heat transfer path from the components to an outside cover plate.

Poron was selected for its resistance to compression set. It also provides vibration and shock protection for both the integrated circuits and the entire unit.

5.7.7 POLYSTYRENE FOR VIDEO CASSETTES

Dow Europe has introduced to the European market a polystyrene resin, Styron 8023VC, designed specifically for the manufacture of video cassettes.

Developed two years ago, Styron 8023VC commands a large share of the US video cassette resin market, and has become an industry standard. The overall balance of properties, such as impact strength, ductility and fast cycle time, has been chosen to meet the manufacturing needs for this particular application. Furthermore, says Dow, its inherent high gloss makes it appealing to the end consumers.

Styron 8023VC has a melt flow rate of 3.5 g/10 min, a Vicat softening point of 104°C, and Izod and Charpy impact values of 85 J/m and 6 kJ/mm respectively.

5.7.8 HIGH PURITY POLYCARBONATE FOR COMPACT DISCS

Dow Chemical Corp is preparing to market a high-purity polycarbonate for compact disc manufacture. Known as Calibre 1001, the material has a melt-flow rate of 73 g/10 min, a refractive index of 1.584, a light transmission of 91.9% at 3.2 mm thickness, and a glass transition temperature of 144°C.

5.7.9 UNDERSEA CONNECTORS

Undersea electrical connectors must function at pressures of up to about 10 MPa. Ocean Design Inc of Holly Hill, Florida, USA, now manufactures fluid-filled connectors which can operate over this pressure range.

A flexible chamber within each connector is filled with a dielectric silicone oil, the pressure of which varies to match that outside. A male probe with a conductor tip enters through a port in the chamber, is wiped clean of sea water and then plunged into the oil. The electrical connection is made inside the oil filled chamber — a process which can be done underwater without turning off the power.

For many years the design used a traditional epoxy resin which had good dielectric properties but was brittle and crack sensitive. New toughened epoxy resins from ICI Fiberite, designated TEM 9001 and TEM 9010, have now been introduced into several areas of the design, including the thin coating on the conductor tip. They will also be used for a cylindrical stopper — a piston-like part about 25 mm long and 2 mm in diameter — which is displaced as the male probe enters the fluid filled chamber, and for the grip sleeve which actuates the connector's latching mechanism; this latter part is some 9 cm long and 3 cm in diameter. The parts will be transfer moulded by Ocean Design.

The new resins have been designed to provide increased mechanical strength and toughness while still retaining the electrical properties and resistance to seawater of traditional epoxies.

5.8. SPORT

5.8.1 POLYETHYLENE IN TENNIS RACKET

Much of the weariness suffered by top tennis players in long matches can be traced to the shock forces transmitted to the player's hand and arm by the repeated impact of ball on racket. It follows that a high performance racket should punish the ball, not the player.

Ellipse Sport of Altamonte Springs, Florida, and of Redmond, Washington state, USA, has conducted dynamic testing of various rackets made with a range of modern, high tech reinforcing materials, including aramid, in the frame. The purpose was to develop a relatively shock-free racket — i.e., one with a high vibration damping coefficient — which would, nevertheless, stand up well to repeated impact and abrasion. According to Ellipse's vice president John Gugel, Allied-Signal's Spectra extended chain polyethylene fibre performed the best of all fibres tested, offering maximum strength at minimum weight. In addition to the frame, a non-woven fabric of Spectra fibre cushions the joints in the racket's throat, where the yoke meets the frame, creating what is described as a highly effective vibration trap.

5.8.2 PLASTICS HORSESHOE

A steel reinforced plastics horseshoe, designed by private inventor Steve King, has recently come on the market. Manufactured by Trisport Ltd with the support of the British Technology Group, a technology transfer organization, this horseshoe is claimed to offer horses the performance advantages that modern technology gives to athletes.

A number of other plastics horseshoes are available on the market, but they do not have the rigidity of traditional metal horseshoes. The Trisport product, it is claimed, offers a number of advantages over traditional metal and alternative plastics horseshoes:

- use of this horseshoe eliminates the need to use different types of shoe for training and racing;

- the nails cannot pull through the shoe. (Shoes can also be stuck on if necessary.);

- the shoes are laterally rigid, which means that the nail load is spread over the hoof;

- good grip is achieved on both metalled roads and soft ground;

- they are lightweight, which helps give a competitive edge in jumping, racing and eventing; and

- they are longer lasting than a conventional shoe, which means fewer nail holes in the hoof and less embrittlement of the hoof wall.

Trisport believes that the new horseshoe will be particularly welcomed by trainers. Samples of the Trisport horseshoe were distributed to farriers for testing throughout 1990.

5.8.3 RACING SKI GOGGLES FEATURE BUTYRATE–SILICA SANDWICH

A line of recreational and racing ski goggles from Uvex Winter Optical Inc features a layer of silica gel sandwiched between two impact resistant, ultraviolet protective lenses formed with Tenite butyrate resin from Eastman Chemical Products Inc. The butyrate lenses are coated with anti-fog material and fused to the middle silica layer.

The sandwich adds flexibility and thermal insulation to the goggles' optical properties. The goggles are claimed to be the first to incorporate a triple lens system.

5.9 BUILDING

5.9.1 PLASTIC HOUSE

GE Plastics' 'concept house' at their headquarters at Pittsfield, Massachusetts contains about 30% plastics, mainly in cladding, roofing, plumbing, electrics, windows and even the foundation. GE classifies these uses as '1', '2' or '3': 1 represents products that already exist; 2 signifies that the product has already been developed and awaits commercialization; and 3 means that the application is still innovative and needs further development before coming onto the market.

Naturally, it is the third group of products that are arousing most interest. These ideas include a new type of insulation foam, Gecet, made from a blend of polyphenylene oxide and polystyrene. This is used to fill a variety of components, including cladding panels, interior furniture, exterior lighting bollards and floor panels. Another innovation, in the second category, is the use of Azloy reinforced sheet to make fire resistant roof tiles; GE and its client manufacturers are planning to market these units aggressively in California, where the traditional roofing 'shingle' is likely to be banned as a fire risk.

Prefabricated wall panels are another area in which GE foresees a plastics alternative to traditional brick or concrete. US Gypsum Corp of Chicago is developing a movable, non-loadbearing interior wall structure of low density Noryl PPO/PS foam faced with Azmet sheet. A loadbearing structure is being attempted by Weyerhauser Co of Tacoma, Washington state, comprising Noryl foam faced with Azloy sheet and stiffened with corrugated wood fibreboard. These units can incorporate ducts, fasteners, and even vacuum inserts — increasing thermal insulation on the thermos flask principle.

The economics of these innovations must depend not only on the price of the materials but on the efficiency of the manufacturing processes. GE emphasizes in particular the various roles of blow moulding, reinforced sheet forming, foam extrusion and bead moulding in making the new developments a feasible proposition. The company has made a model to illustrate the concept of a factory producing building panels on a mass production basis.

5.10 PIPELINES

5.10.1 POLYETHYLENE APPLICATIONS

Using a proprietary Ziegler catalyst, Solvay of Brussels, Belgium, has developed a polyethylene — designated Eltex TUB 125 — which can be extruded to form a gas pipe with a pressure rating of 7 bar. In a trial run, British Gas is laying a 7.5 km length of this pipe, which has been manufactured by Stewarts & Lloyds Plastics of Huntingdon, UK.

In another application of this resin, Wavin Industrial Products of Durham, UK, recently supplied medium density polyethylene pipe to British Gas Northern for use at 7 bar pressure. A total of 1 km of 250 mm SupaGas 7 pipe, in 18 m lengths, was installed to improve the gas supply to the Sunderland region.

This constitutes another application for Solvay's Eltex polyethylene resin which was designed for 7 bar gas piping.

In a separate operation, Wavin supplied 66 free-standing coils, each comprising 100 m of 180 mm WavinGas medium density polyethylene pipe, for a medium pressure gas installation at a large air base in North Yorkshire. The choice of free-standing (drumless) coils — the largest single use to date in the region, it is claimed — was influenced by the difficult nature of the terrain through which the pipe had to be laid, and the considerable distance involved.

Wavin also makes blue medium density polyethylene pipe and fittings for water systems under the trade name WavinSure, and has recently supplied over £100 000 worth of material to the British Rail engineering works at Derby for a new fire main.

Wavin has also developed a new range of mechanical ChemiJoint fittings for use with the WavinSure range. The fittings are also compatible with Wavin SupaSure medium density polyethylene pipe for use with high pressure systems.

A fast, efficient technique for relining gas and water mains has won a national construction industry award for Subterra Ltd of Wimborne, UK. The company's rolldown pipe system won one of the 1989 Castrol Multiplant/Construction News R&D Awards.

The rolldown pipe system makes use of the fact that, when compressed and then released, polyethylene does not immediately return to its original dimensions. In this application, polyethylene pipes are fed between rollers to reduce their diameter, so that they can easily be slipped inside existing mains. When water pressure is applied, the polyethylene lining tries to return to its original diameter, thus fitting closely inside the main. Because of the low frictional properties of polyethylene, the lined main will actually have a greater capacity than the original.

Subterra developed the rolldown technique in conjunction with Stewart & Lloyds Plastics Ltd and BHRA, the Fluid Engineering Centre. So far, 186 km of relined pipe, up to 450 mm diameter, have been installed in the UK, and the process has been exported to Europe and North America through licensees. Now Subterra is developing the technique further to allow its use on larger diameter pipes, up to 600 mm, and to reduce the size and weight of the on-site machinery. It is also exploring new markets, especially the petrochemical industry.

5.10.2 VINYL ESTERS USED TO RENOVATE PIPE

Derakane vinyl ester resins, produced by Dow Europe, have been used to renovate some 2000 m of buried or inaccessible pipe using an innovative repair system known as the Insituform process. A flexible non-woven fibre tube (Insitutube), impregnated with a liquid resin, is inverted by water pressure through the damaged pipe, without any need for excavation, and is pressed firmly against its inner surface. The resin is then cured, either by hot water or high frequency light, creating a new and durable structural pipe within the old.

The first European application of this process in which Derakane was used was carried out in 1987 by the French licensee, Entrepose, which used 5.8 t of Derakane 411-45 vinyl ester resin to carry out repairs at a paper mill in Bordeaux. Since then the Dutch licensee Ricol Technieken has installed about 75 t of Derakane 470-36 vinyl ester resin at Dow's Terneuzen plant in the Netherlands. This particular resin is resistant to organic solvents and high temperatures and is therefore suited to the renovation of pipes and sewers carrying aggressive liquids.

The Insituform process was developed in the UK by the Insituform Group Ltd. Dow has worked closely with this group in developing special thixotropic grades of Derakane resins that have the same handling properties as the polyester resin systems currently used in the Insituform process.

The company is also looking closely at other possible applications where a demand for exceptional properties such as Class 1 fire resistance could be met by specifying a special grade of resin, such as Derakane 510 which is based on a brominated resin and usually contains a fire retardant filler. However, the biggest application areas are still in the corrosion resistance market — composite pipes and tanks for the chemical industry and other industrial processes where acids, alkalis and solvent storage and handling can cause big problems.

5.10.3 PROTECTING NORTH SEA PIPELINES

A polyurethane elastomer system has been developed by Macpherson Polymers to insulate and protect pipelines carrying hot crude oil from North Sea wells to central production platforms. The latest of these Hyperlast insulation/protection systems was used on two 7 km pipelines laid in 1989 for Marathon Oil, in the Central Brae field, in the UK Northern Sector, north-east of Aberdeen. More piplines protected in this manner are expected as the field is developed.

Undersea oil wells are often equipped with a sub-surface collection system which gathers oil from the well and delivers it by pipeline to the nearest platform with processing facilities. This puts a heavy demand on the pipeline, which must not only be corrosion and impact resistant but must maintain the temperature of the crude oil to 30°C or more. This is necessary to keep the oil's viscosity low enough to maintain the flow. Oil emerges from the wellhead at anything up to 110°C, but would soon cool down to little more than the ambient seawater temperature — about 4°C — if the pipe were not heavily lined with insulation.

The Marathon pipelines, laid in 1989, are protected and insulated by two layers of different Hyperlast polyurethane moulding compounds developed specifically for this application. The inner, 170 mm diameter pipe is made of stainless steel and is first treated with a Hyperlast primer. Next, a 12 mm layer of Hyperlast 407 build coat is moulded onto the pipe. The outer layer is 35 mm of Hyperlast 512, which combines a low thermal conductivity of 0.12 W/m.K with physical toughness. The pipes are prepared in 12 m lengths and jointed together on the laying barge.

5.11 MINERAL RECOVERY

5.11.1 HYDROGEL POLYMER FOR RARE EARTHS EXTRACTION

The demand for rare earths and other scarce metals is burgeoning by reason of their increasing use as catalysts. The process of extracting these metals often involves the recovery of metallic ions from dilute aqueous solutions by the use of ligands — molecules which form complexes with these ions and facilitate their extraction.

At the University of Strathclyde, UK, researchers have managed to fabricate a hydrogel polymer in the form of a porous network with an affinity for ligands. When used as part of the ore extraction process, this should enable rare earth ions to be extracted far more efficiently than by existing processes.

One basic problem which the research team had to solve concerned the nature of the polymer. A hydrophilic material would dissolve in the water carrying the metal ions, while a hydrophobic material would repel ligands rather than absorb them. The team solved this by forming a hydrogel from an emulsion of hydrophilic monomer and water, then polymerized the monomer. When the water was driven off, the resulting polymer was left in a highly porous state, forming an open structure on which ligands could readily be adsorbed.

5.11.2 POLYMER ABSORBENT TRAPS GOLD

The Moscow-based Institute of Geochemistry and Analytical Chemistry, USSR, has devised a simple electrochemical method for extracting nearly all the gold or platinum from ores or industrial wastes. The key element is a polymer-based organic absorbent placed between the electrodes. The absorbent can take up its own weight in metals, including palladium, indium, ruthenium and osmium, as well as platinum and gold.

The Soviet scientists claim that the use of only 1 mg of absorbent per litre of waste water can produce concentrates of 80–90% of noble metals. The process is already in use at some Soviet industrial enterprises.

5.12 POLLUTION CONTROL

5.12.1 FLOATING BARRIERS KEEP BEACHES CLEAN

Polychloroprene synthetic rubber and other materials have been used in floating anti-pollution barriers set up in 1989 along parts of the Adriatic coast of Italy. The purpose of the barrier was to exclude large blooms of algae from popular tourist beaches.

Sewage and industrial and agricultural pollution had combined with an unusually warm summer to cause a proliferation of algae which formed an unpleasant scum on the sea surface. Alarmed by this threat to the tourist industry, the local authorities governing a number of popular resorts — including Venice, Cattolica and Pesaro — hired some floating barriers of the type used to contain oil slicks.

One such barrier consists of 10 m inflatable segments made of high tensile rubberized fabric, protected on both sides by a layer of black polychloroprene rubber-based compound. The segments incorporate cast iron blocks to keep them upright, and are joined together by rubber coated stainless steel plates. The ends of the barrier, which may be 1 km or more in length, are anchored to the sea bed.

Another type of barrier consists of 25 m long segments of continuous polyvinyl chloride sheet, reinforced at the edges and supported in the water by polyvinyl chloride foam floats.

Given an adequate supply of barrier segments and the necessary expertise in installing them — and keeping one's fingers crossed to avoid the sudden occurrence of a violent storm — this may well provide an excellent short-term solution to the problem of algae blooms. The long-term solution is, of course, to reduce the quantity of pollutants entering the Adriatic — but, despite recent legislation, it will still be a few years before this happens.

5.12.2 ABSORBENT PLASTIC FOR OIL SPILLS

Nippon Shokubai Kagaku Kogyo of Japan has developed a plastic that can absorb up to ten times its weight in petroleum and 25 times its weight in some solvents. According to the company, the active ingredient is an acrylic monomer which absorbs the liquids into its molecular structure.

5.12.3 GLASS REINFORCED PLASTIC TANK KEEPS RIVER CLEANER

Klargester Environmental Engineering Ltd has installed a 45 000 litre glass reinforced polyester tank at a dairy farm in the west of England to collect the liquor draining off 4000 t of ensiled grass, stored as winter feed for 400 cattle. With the main aim of preventing river pollution, a tank of this type is claimed to be rot and corrosion proof, lighter and easier to install than the traditional underground concrete tank.

5.13 AGRICULTURAL

5.13.1 ARTIFICIAL TREES COULD ENCOURAGE REAL ONES

A report in *New Scientist* describes a plastics 'tree' invented by Antonio Ibañez Alba, an electronics engineer from Barcelona, Spain. Standing between 7 and 10 m tall, the tree is made from fire resistant polyurethane and phenolic foam. It imitates the stages of evaporation and condensation of a natural tree, but does not require artificial or natural irrigation.

The tree, like its natural counterpart, is divided into three parts: roots, trunk and leaves. The stiff tubular trunk is filled with a polyurethane material which is riddled with channels which absorb water by capillary action. The trunk contains several layers of polyurethane of different densities, designed to retain water and release it slowly during the course of the day.

After planting, liquid polyurethane is injected under pressure into the base of the tree and finds its way into natural channels in the soil, where, on hardening, it forms strong 'roots' which anchor the tree firmly to the ground.

The tree creates cooler air by absorbing the moisture that condenses onto its surface during the cold desert nights and retaining it in the body of the tree. During the heat of the day, the moisture is slowly released, cooling the air. The branches and leaves are made from phenolic foam and are moulded to look like the crown of a palm tree, which, according to the inventor, is the most effective shape to trap dew from condensation and encourage evaporation.

Ibañez Alba believes that a plantation of his trees could, in the course of a few years, trap so much moisture that the local weather pattern could change so as to allow real trees to grow in a region that was previously desert. The Libyan government is sufficiently impressed by his claims to be planning to plant between 30 000 and 40 000 of these artificial trees in its £600 million project for the creation of an artificial river between Tripoli and Sebha, in the south of the country. Other North African countries are also interested.

5.13.2 DEGRADABLE FILM BOOSTS MAIZE HARVEST

Several agricultural fields bear witness to a successful application of photodegradable plastics film. Aglex is the name given to a French-made, thin (12 µm) polyethylene film incorporating starch, which is laid between rows of a cereal crop — typically maize (sweetcorn) — and helps the plants to thrive by trapping heat and moisture. Sweetcorn grown by this method will mature earlier, contain a higher proportion of dry matter, and produce a larger eared crop, says UK-based Gromax-Plasticulture Ltd.

5.13.3 GIANT VALVE FOR IRRIGATION

Asahi Yukizai Kogyo Co Ltd from Japan has made what it claims to be the world's largest plastics valve. Installed at Sorachi, Hokkaido, the valve has a bore diameter of 1.5 m and is used for agricultural water distribution.

The body of the valve is made from glass reinforced plastic and the diaphragm from ethylene-propylene-diene monomer. The valve is claimed to have excellent chemical resistance, water tightness when closed, minimal flow resistance when open, easy maintenance and long life.

5.14 FLAME RETARDANTS

5.14.1 NON-HALOGENATED POLYPROPYLENE

Monmouth Plastics Inc of Asbury Park, New Jersey, USA, has introduced a non-halogenated, thermally stable, flame retardant polypropylene. Designated EMPEE PP 9000, the material can be thermoformed or extruded into sheet form.

Monmouth has also produced a flame retardant wire and cable jacketing material called EMPEE PE JR 165. It is available in a range of molecular weights with various additive packages for special needs.

5.14.2 POLYCARBONATE AND POLYCARBONATE/ACRYLONITRILE-BUTADIENE-STYRENE

GE Plastics of Pittsfield, Massachusetts, USA, has announced a range of polycarbonate and polycarbonate/acrylonitrile–butadiene–styrene alloys intended for computers and other business machines. Lexan ML-6000 has a UL rating of V-0 and a heat distortion temperature of 118°C.

In addition, three new polycarbonate/acrylonitrile-butadiene-styrene alloys in the Cycoloy range are now available. These are flame retardant grades, with a 5-V rating, and have heat deflection temperatures of 82–99°C.

5.14.3 NON-HALOGENATED FLAME RETARDANT FILM

Courtaulds Films & Packaging (Holdings) Ltd of Bridgwater, Somerset, UK, envisages an "almost infinite" variety of new commercial applications for its non-halogenated, flame retardant film, Technoflam, which it can now produce in rolls up to 1.4 m wide. The film is used as a decorative facing material on rigid composites, particularly phenolic, and other panels used in vehicles or wherever fire safety is paramount.

5.15 PACKAGING

5.15.1 ZEOLITE ADDITIVE KEEPS FOOD FRESHER

In recent months there have been several claims, mostly from Japan, that the addition of a zeolite to a plastic packaging material will keep the contents fresher by removing bacteria and other harmful constituents.

Zeolites are ceramics that have high molecular-level porosity. This property apparently enables them to entrap microorganisms and prevent them from spoiling food or other perishable material with which they are in contact. Gases such as ethylene are also absorbed.

Zeolite powder could well be the additive used in a low density polyethylene sheet developed by Japanese inventor Tadashi Ogawa. He claims that this sheet not only preserves food wrapped in it but actually enhances its flavour; vegetables stay greener, flowers last longer, bread remains fresh, the inventor states.

Ogawa's own explanation is that the ceramic additive absorbs infrared radiation and re-emits it at a specific frequency that is harmful to bacteria — but many scientists treat this theory with a degree of scepticism.

Nippon Unicar also markets a polyethylene packaging film incorporating an unidentified ceramic additive. In this case, the company claims that the additive absorbs ethylene gas emanating from the fruit or vegetables stored in the package, thus slowing the ripening process.

Another additive used by Unicar is a deodorant, reportedly extracted from green tea.

A further application of zeolite impregnated plastics — this time polypropylene — is for water storage tanks or pipes. Takuma Co Ltd of Osaka, working in collaboration with Shinanen New Ceramics Co Ltd and the brewing firm Suntory, says that in tests in which water stood for 24 h in a tank at 37°C, the plastic, 'Asepla', killed 99.999% of the bacterium *E. Coli* present. Asepla is made by mixing a powdered artificial zeolite called 'Zeomic' with the resin. Tanks made from Asepla can be maintained in a sanitary condition without the need for regular sterilization, the companies say.

5.15.2 EDIBLE POLYMER

Mitsubishi Rayon Co Ltd of Tokyo, Japan, has developed an edible polymer intended for use in the food and pharmaceutical industries. Said to be derived from natural sources such as pine trees and selected tropical seeds, the polymer can be made into film for food packaging, or into capsule coatings for slow-release drug delivery systems.

5.16 MISCELLANEOUS

5.16.1 POLYETHYLENE FISHING HAWSERS

Nylon, polyester, aramid and even steel ropes are beginning to be replaced by Allied-Signal's Spectra high performance polyethylene fibres in commercial fishing. With the introduction of larger factory trawlers and the accompanying increases in dimensions, drag and catching capacity of pelagic nets, it appears that conventional materials are no longer adequate for cost efficient, large scale fishing.

Spectra ropes are only one-seventh the weight of conventional wire ropes, but are claimed to be just as strong and more flexible. They do not corrode in the salty environment, and because of their comparatively light weight they are easier to handle. To control such nets now requires only one man when previously four were needed, it is stated.

Spectra lines, it is reported, decrease the number of backlashes on the winch and ensure a smoother and more balanced winding. Another major advantage is the safety offered by the lines which, should they swing loose, will not do as much damage.

Spectron 12, an ultra high strength marine rope made from Spectra, is being used experimentally for gilson winch lines by Nor'Eastern Trawl Systems. It was developed by Samson Ocean System which reports that sea trials have proved successful. This type of gilson winch line is now fully commercial and currently in use on four large commercial trawlers, including two factory boats which process and package the fish on board. It is also being recommended for other fishing applications such as topping liftlines, set nets, running and beach lines, high strength slings and chokers, as well as trawl ropes.

The Spectron 12 single braid is described as a 100% Spectra 12-strand Parallay construction with a Samthane polyurethane coating.

5.16.2 AWARD-WINNING PLASTICS SAFE

A British Design Award has been given by the Design Council for what is believed to be the world's first plastics safe. Resistant to fire and a wide range of physical attacks, it is made by Racal-Chubb of Wolverhampton, UK, and is based on a new material, Ellox, developed by Macpherson Polymers in cooperation with Chubb research specialists.

The Planet safe is said to represent a breakthrough in safe technology as it combines anti-burglary properties with fire resistance. This is made possible by the development of Ellox — an energy absorbing polymer composite tested to withstand a temperature of 1000°C.

The Planet safe underwent vigorous testing when it was placed in a furnace pre-heated to 1090°C for 30 min, then dropped 9 m to simulate a building burning and collapsing. It was reheated for a further 30 min and left to cool in the furnace. At this time the internal temperature was well below the stipulated level and the contents were in perfect condition.

The exact composition of Ellox, an elastomer, has yet to be revealed, but embedded in the polymer are nuggets of alumina that are almost as hard as diamonds and which, according to the company, withstand disc cutters, drills, percussion tools, oxy-acetylene and acid.

The safe is being manufactured at Chubb's plant in the Netherlands.

5.16.3 MANHOLES FROM HIGH DENSITY POLYETHYLENE

High density polyethylene manholes for storm and sanitary sewer systems have been introduced by Advanced Drainage Systems (ADS), based in Columbus, Ohio, USA. The product is made by rotational moulding of a high modulus, high strength grade of Quantum Chemical's Microthene high density polyethylene powder that reportedly provides good melt flow. The manhole is filled with concrete, weighs about 100 kg and is 50 mm thick. It is designed for use in non-traffic or light-traffic applications.

5.16.4 COMPOSITE HARP FRAME

A composite of glass and carbon fibres in an Araldite matrix provides the frame for a harp made by Stratline of Chartres, France. The harp follows a traditional design as regards size and string spacing, but is reportedly 30–40% lighter than its wooden counterpart, is unaffected by variations in humidity or temperature, and is claimed to offer outstanding sound quality and volume. Another innovation is that pedal movements are transmitted by aramid fibre cables.

5.16.5 ACETAL–NYLON INTERFACE CONTROLS GAS FLOW

The wear resistant, nearly friction-free performance of Du Pont's Delrin acetal and Zytel nylon sliding against each other is claimed to be a significant factor in the choice of these materials for a gas regulator. Made by Fisher Controls and designed to reduce the pressure of natural gas from 415 to 1.75 kPa for domestic use, the regulator is being installed at many homes in the USA.

6. PROCESSING

It is notable that a significant proportion of this short section is taken up by computer control and analysis.

6.1 FASTER PROCESSING

6.1.1 FASTER BLENDS FOR PHENOLICS

The ability to blend more plastics in less time is the claim made by Dynamic Air Ltd for its air blending system, as supplied to Perstorp Ferguson of Newton Aycliffe and to BP Chemicals, Barry, UK.

The system is based on blending heads which can provide a homogeneous mix of finished product at the rate of 4 tonnes every 15 minutes. Perstorp Ferguson, a company which blends plastics granules (including phenolic moulding compound) for use in electrical products, reports that the new blending heads have increased the production capacity. This is due, it is claimed, not only to the reduction in production time resulting from handling larger batches, but also because of greater consistency which requires fewer blends to be tested each day.

According to Dynamic Air, improved efficiency and consistency in quality are particularly important in application areas such as the automotive industry, where Cellobond phenolic polymers from BP Chemicals are used as binders in friction materials such as brakes and clutch facings. For this application it is necessary to blend fine phenolic powders with a variety of fillers and additives. Without a consistent and reliable air system, this can be a difficult operation because of the differences in bulk density between the various materials.

6.1.2 FASTER CURING FOR POLYURETHANE FOAM

Curing of polyurethane foam blocks can be cut from about 12 h to 40 min, according to Hyman plc of Poynton, UK. The firm has developed a rapid curing system called Hypercure, which purges foam blocks of noxious fumes such as methylene chloride or tolylene diisocyanate and recycles these chemicals through an activated carbon recovery system.

Use of the Hypercure not only saves on storage space but results in a less polluted and therefore safer working environment, claims Hyman. The firm has won an export order to the United States for its system.

6.2 BONDING METHODS

6.2.1 JOINING TECHNIQUE FROM JAPAN

The Japanese developers of a method for joining parts made from engineering plastics predict that its use will greatly expand the range of applications for these materials.

Osaka-based companies Tokushu Kogyo and Mitsui say their method consists of coating a nickel wire with resin, the wire then being wrapped around, or sandwiched between, the parts to be joined. A current passing through the wire then melts the resin, which flows around or between the parts without leaving a gap. The resin can be selected to give the best results depending on the materials to be joined.

As a direct result of the development of this method, the two companies are now selling a purpose built, computer controlled power supply. The computer can adjust the current and time the joining process, and its operation requires no specialized training. A large version is available for factory production lines and a portable one for use at work sites.

6.3 COMPUTERIZED CONTROL AND ANALYSIS

6.3.1 COMPUTERIZED QUALITY CONTROL FOR PULTRUSION INDUSTRY

A computerized quality control system for the pultrusion industry has been introduced by Pultrusion Specialties of Alum Bank, Pennsylvania, USA. The CAN thermal analyser is claimed to be light, compact and portable, and can be switched from analysis to control mode so that it can serve either as a laboratory instrument or a production control device. It is programmed to accept a full line of variables, allowing the user to fine-tune his testing procedures.

The analyser is said to be capable of verifying incoming resin standards, checking pultrusion formulation development, optimizing pultrusion process spread, checking resin batches for quality, documenting process quality control and delivering studies on process or product performance.

6.3.2 COOLING ANALYSIS SYSTEM FOR INJECTION MOULDING

Intergraph has announced a new plastics cooling analysis system for the plastics industry. I/Cool has been designed to help engineers in heat transfer analysis for cooling circuit layout in plastics injection moulds, and runs on Intergraph's RISC UNIX workstations.

According to Intergraph, I/Cool supplies a convenient interface to Moldflow's Moldtemp thermal transfer analysis program and allows users to quickly prepare models and evaluate results. The package is claimed to increase productivity through the direct operation of Moldtemp within the I/Cool environment.

I/Cool has been developed to meet various needs identified by Intergraph's Plastics Consortium, a group of Intergraph system users representing the plastics industry. It enhances Intergraph's engineering modelling system (I/EMS) by simulating and analysing the thermal properties of injection moulds during the moulding process. I/Cool, it is claimed, enables users to determine geometry patterns for uniform mould cooling, reduce warpage and initial stress problems inherent in the cooling process, improve the flow characteristics of the mould and minimize cost through lower tooling costs and reduced wastage.

I/Cool also complements Intergraph's injection flow analysis system (I/Flow), which predicts plastics flow for the mould injection process and automatically builds a specialized model for analysis of the cavity fill. I/Cool uses this model as a basis for creating and cooling circuits optimizing coolant circuitry for shorter cycle time and improved product quality.

7. EQUIPMENT

The trends towards ever more massive injection moulders, to produce bigger products, is well illustrated here. There is also an increasing sophistication in control, involving not only computers but also sensitive measuring devices to provide data.

7.1 INJECTION MOULDERS

7.1.1 INJECTION MOULDING RANGE IMPROVED

Sandretto recently launched a range of injection moulding machines. The 'Series Eight' range, says the company, has benefited from a total redesign. The machines have been given more robust clamp units, sturdier platens, longer opening strokes and thicker tie-bars to match their increase in performance. The machines have greater mould height capacity, and a longer stroke ejector system is now built into the construction of the moving platen.

A choice of three injection units is available for most machines; in addition, Sandretto offers a choice of up to three L/D ratios for each of the six different injection units. For any one clamp unit, therefore, seven alternative standard profile screws will be available; there are also special screws for processing rigid polyvinyl chloride, amorphous and crystalline polymers. Another feature is a choice between a standard and a larger platen. The 60 t machine, for example, can be supplied with platens taken from the 100 t machine.

Screw and barrel changing has been made easy with the use of fast-change couplings and connectors. Moulds can also be changed quickly, it is claimed, without the need for modifications to the machine, guards or control systems. Process control is through the Selec 'System 90' multiprocessor. The range includes machines with clamps of 60, 100, 150, 200, 270, and 360 t.

Meanwhile, at Farmington Hills in the USA, Sandretto has assembled an injection machine reported to be the largest of its type ever built. 25 m long by 17 m high, the 'Model 9000' features a vertical clamp with hydraulic control that rises between two injectors located on opposite sides. It is used to produce industrial polyethylene containers for use by the automotive industry.

The machine requires 1600 kW of power to operate, of which 500 kW is used to heat the plasticization barrels and the rest used to run the motors needed for the machine to operate. Each injection unit can operate independently, allowing coinjection of different plastic materials. Together they can inject 95 kg of material per cycle.

7.1.2 MODULAR 1200 TONNE MACHINE

The components of B.M. Biraghi's new 1200 t injection moulding machine are designed on a modular basis, allowing models to be custom built to suit the requirements of individual customers.

The machine is the largest in a series of eighteen presses ranging upward from about 40 t. It comprises four double toggle clamp units ranging from 5000 to 12 000 kN, and five injection assemblies of Euromap sizes 2900 to 9950. Features include: Sankyo self-lubricating bushing;

cowled hydraulic system; high speed barrel change; barrel delivery system with preheater; high speed clamping of moulds; automatic coupling of ejector; automatic correction of clamp force; and a hoist for lifting and replacing barrels and moulds.

Biraghi claims that the use of Sankyo bushings eliminates oil lubrication, reduces contamination, cuts noise emission and meets the production requirements for delicate items such as medical components. Barrel replacement is said to be foolproof, with automatic cut-off of fluids supply while the barrel is being hoisted.

7.1.3 TANDEM DESIGN INJECTION MOULDING MACHINES

There is a growing need for machines that are not only capable of moulding large items but can still be used economically to produce smaller objects. With some machines this can be done by using the two halves of the machine independently for smaller items, and both halves together, for the bigger products.

Two injection moulding machines recently supplied by the German firm Battenfeld use this technique and are of particular interest.

8000 tonne machine to make waste bins

What is claimed to be the largest automated injection moulding machine in the world, in terms of injection capacity, was recently supplied to Waste Management Inc of Oak Brook, Illinois, USA, by Battenfeld. Designated Unilog 9000, the press has a total clamp force of 81 800 kN and a shot weight of more than 90 kg. It can produce parts up to $1.85 \times 1.55 \times 1.25$ m in size, and will be used to make a variety of products, including waste bins with a capacity of 1.53 m^3.

Despite belonging to Waste Management, the press is operated by Cascade Engineering of Grand Rapids, Michigan. Cascade has the expertise necessary to operate the machine, and cooperated closely with Battenfeld in its development; during this process the projected capacity of the press was raised from 55 000 to 68 000 kN, again to 72 700 kN, and finally to 81 800 kN. The machine occupies some 30.5 m $\times$ 9.1 m in plan area, $\times$ 7.0 m in height.

One of the main design features of the Unilog 9000 is its capability to use the two mechanically separate clamping units jointly and simultaneously under comprehensive close-loop control to produce an ultra large part. For smaller parts, the robots can act independently. Integrated multi-axis robots are used to demould and transport the parts, adding additional functions such as deflashing, printing, hot stamping or palletizing.

Twin machine for storage tanks

The firm of Otto in Kreuztal, Germany, is one of the leading manufacturers of storage tanks and large industrial mouldings. This company needed an injection moulding machine with a clamping force of 60 000 kN to make plastics storage tanks with a capacity of between 240 and 360 litres. In addition, economic manufacture demanded that the machine could be left to run automatically, unattended, throughout the weekend.

Battenfeld's solution was to provide two integrated injection moulding machines of the type BA-Twin 30000/63000 Unilog 8000, with a screw diameter of 210 mm, a specific injection pressure of 2057 bar, a swept volume of 30 479 cm^3 and an aspect ratio of 25. The clamping units each have a force of 30 000 kN, the platens measure 2.05 m wide $\times$ 4.0 m high, and the distance between tie bars is 0.85×2.8 m. Mould height is adjustable from 1.45 to 2.4 m. The unusual shape of the platens facilitates mould changing, it is claimed.

Demoulding is accomplished by robots capable of handling up to 100 kg and the machines are designed to cope with many different demoulding situations. Available handling time and buffer capacity are doubled if the robots can demould alternately when the machines are used for twin operation.

7.1.4 INJECTION MOULDING WITH RETRACTABLE TIE BARS

Kuasy of Germany has developed a new range of large injection moulding machines claiming to offer significant advantages to users. These machines are now available in the UK through SSX Plastics Machinery Ltd of Sully, South Glamorgan.

A special feature of the design is that the machines incorporate fully retractable tie bars. These ensure side access for robotic or automated loading, apparently eliminating the need for overhead handling and allowing the moulded components to leave the machine at a more convenient height. Tool changes are simplified, and headroom requirements reduced, SSX Plastics Machinery states. Installation of three Kuasy 410/100 microprocessor controlled injection moulding machines at Gloster Plastics, Gloucester, UK, has increased its moulding capacity by about 60%, it is reported. In addition, set-up time has been reduced by an average of 20 min per batch, and improved process control has enabled the company to produce a better quality of moulding than was previously possible.

7.1.5. CONTROLLED MOULD ACTION

Battenfeld's new BK-T series of injection moulders is claimed to offer a firm but gentle mould closing and opening action, protecting the mould and machine and ensuring a better quality of moulding. This, says the company, is achieved by a double toggle system that ensures that the closing speed of the two halves of the mould is reduced to a low value as they come together.

Operational parameters are controlled by the Unilog 4000 system; these include injection speed, holding pressure and back pressure, screw speed, stroke length, injection temperature and many other variables. Optional extras include a melt cushion control, an automatic granule feed and a computer interface for setpoint storage and central production data acquisition.

The series comprises eight models, ranging from the BK-T 1000 (distance between tie bars = 410 × 410 mm) to the BK-T 4000 (735 × 735 mm).

7.2 EXTRUDERS

7.2.1 TABLE-TOP EXTRUDER

A new table-top extruder under the Yellow Jacket tradename features a double reduction, helical gear box to process engineering polymers that requires high screw torque. This type of gear box was previously exclusive to heavy duty extruders on the production floor.

Standard features include digital automatic temperature controls; ammeters and automatic air cooling for each zone; nitrided barrels; and safety rupture discs. Many types of dies and downstream equipment are available to form complete laboratory extrusion systems for tubing, blown film, sheet, rod, wire installation, and compounding applications, it is claimed.

7.2.2 TWIN SCREW COMPOUNDERS

A new range of twin screw compounders — the HP4000 series — has been launched by APV

Chemical Machinery. The machines are designed for processing advanced polymer blends to critical specifications.

The HP4000 series, it is claimed, features design and engineering developments in three critical areas — torque range, process optimization, and screw and barrel geometry.

APV has developed a twin motor drive with increased torque. A single gearbox drives the two screws, ensuring synchronous rotation to maintain the intermeshing action.

The machine can run at a higher level of fill than previous APV models. Quality control parameters at low temperature and minimum viscosity shift can be achieved at high output levels, it is claimed; it is also easier to control the temperature.

To achieve the rheological needs of a diversity of resins, often sensitive to heat and shear, APV is introducing refinements in screw and barrel geometries. The agitator and bore perimeters show opposite trends as the centre-line ratio is increased; at the intersection point the perimeters are equal. This is claimed to optimize viscous shear and achieve a high level of mixing.

7.2.3 ICI INSTALLS COMPOUNDING UNIT

ICI Advanced Materials has installed a Werner & Pfleider twin screw compounder at its 'Procom' plant at Wilton, Teesside, UK. It is used to produce mineral-filled and rubber-modified grades of polypropylene which are used in the automotive industry for components such as bumpers, fascias and interior trim panels and also in the domestic appliance, electrical and electronic industries.

The new unit incorporates automatic raw material selection from hoppers and IBCs; a control system; plant monitoring and operation; and full event logging, together with automatic packing and palletizing.

7.2.4 MOULDING MACHINE ENCASES OLD FOAM WITH NEW

With the ever-increasing emphasis on recycling and reuse of discarded materials, the Italian machinery firm Presma has introduced a device which allows virgin plastic foam to be coinjected with scrap foam. The resulting item has an inner core of old foam securely bonded to an outer shell of new foam.

The process is run on a Presma injection press with twin barrels — one for prime resin, the other for scrap. The key to the technique is a specially designed, bi-chambered nozzle, through which both melt streams pass. The process works by timed injection of prime and scrap resin through the nozzle ports, with the prime material going first. Each port opens long enough to inject the correct proportion of resin into the mould. Injection is controlled within precise limits by a microprocessor.

Presma says that scrap foam can comprise up to 70% of the part being moulded, although it recommends 50–60% as a working maximum. Economies result not only from the fact that scrap material is cheaper than prime, but that pigments and other additives are often only needed in the outer layer.

The process is adaptable to presses ranging from 300 to 4500 kN and with up to ten mould stations. The control system allows different parts to be moulded at each station. Presma offers a shuttle press with two injection units, four barrels, and the potential to use up to four different resins.

7.3 COATING TECHNOLOGY

7.3.1 RAPID CURING FOR COATED RESINS

Rapid-cure facility for coated resins and adhesives is now available from Rexham (UK) Ltd. The company has commissioned a 1.6 m wide coater/laminator machine incorporating two banks of 36 kW infrared lamps.

Located at both ends of the drying oven, these lamps complement the normal thermal drying capacity and generate high curing temperatures — particularly useful for crosslinked or water-based coatings.

Rexham has also installed a pilot ultraviolet curing line. This is intended for use with coatings containing a photoactive compound or photoinitiator. When ultraviolet light of a suitable wavelength and intensity impinges upon and penetrates the coating, the photoinitiator is broken down into free radicals, which are then able to initiate a rapid addition polymerization reaction.

This technology will allow Rexham to use environmentally friendly solvent-free coatings, and will also open up the possibility of using new resins and adhesives.

7.4 PROCESS CONTROL

7.4.1 CONTROL SYSTEM OF MODULAR DESIGN

What is described as a 'self-optimizing feature' could be a useful property of a new control system claimed to be particularly suited to plastics processing machinery. ABB Metrawatt's GTR9000 control system features a modular design which enables it to be configured for individual needs.

The GTR9000 is used in conjunction with Gercom systems equipped with 'intelligent' colour monitors. As many as six GTR9000 units with a total of 192 control loops can be connected to one Gercom /9000. Five different displays on the monitor provide information about the processes to be controlled. These include a protocol with texts; a file manager; two pages for free texts (e.g. for designation of control loops, marking of products, lots, identity numbers, and so on); and a bar chart. Flow chart displays are planned as an expansion of the system.

Hard copies of all important information can be obtained via a range of peripherals such as a protocol or other type of printer. The flow programmes for the GTR9000 system can be loaded from a disc drive via a standard RS232 interface.

7.4.2 PLASTICS AND RUBBER LINES CONTROLLED TO 0.5%

Automatic high precision control of plastics and rubber sheet and film production lines is claimed for a new version of the Statesman control cabinet, according to its manufacturer, Control Techniques Sussex Gauging. The Statesman II provides remote control of up to two extrusion or calendering lines with an accuracy of typically ±5 μm or 0.5%, maintaining stable extrusion by control of the line speed.

The basis of the Statesman II is an IBM computer with a 20 Mbyte hard disk, 35 cm high resolution colour monitor, membrane keyboard and custom designed software. The cabinet includes its own air chilling system and dehumidifier, allowing it to be used in hot or damp climates. An optional extra is an Epson type EX800 colour printer which can provide a permanent record. Software extras include records of reel length, weight and thickness for up to 30 000 reels, and batch production information records.

The system is claimed to be simple to operate and its use is said to lead to higher quality levels, improved quality control, reduced downtime and less scrap.

7.4.3 ON-LINE CONTROL FOR ICI AUSTRALIA

A Porpoise P5 on-line viscometer is being used as the prime feedback element in ICI's gas phase reaction for the polymerization of polypropylene. A direct sampling system takes a sample from the reactor and delivers it to the viscometer in a delay time of less than 10 s. The samples are called for on a just-in-time basis to coincide with the measurement cycle of the analyser.

The Porpoise rheometer is described as having a rapid melting and measurement cycle, a constant response time and a self-cleaning capability. A melt flow index (MFI) reading is produced every 7.5 min, with a total delay time of 10 min from extraction from the reactor, making the system ideal for statistical process control, it is claimed.

Both powder and granule Porpoise machines can handle products with MFI values in the range 0.05–600 at 190 and 250°C, covering all polyethylene and polypropylene products.

7.4.4 ABSORPTION GAUGES

A family of absorption gauges from Infrared Engineering is now available. Featuring microprocessor control, the gauges can undertake a wide range of measurements of, for example, moisture content, plastic film thickness or coating thickness.

Using five narrow pass band filters, measurements of selected infrared absorption bands can be made in transmission or, alternatively, in forward or back scattering geometries. The appropriate measuring head is connected to the MM55 system which has a record of gain, zero offset, alarm limits, measuring algorithms etc. for each head stored in memory. The system comes complete with RS232 serial interface to communicate with a printer, dumb terminal or control computer.

8. TESTING AND STANDARDS

Computer control and analysis continues to be introduced into routine and non-routine testing, as two of the items here illustrate. This can lead to more realistic simulation of service conditions in the test.

8.1 GENERAL

8.1.1 ASTM UPDATES STANDARDS ANNUAL

The American Society for Testing and Materials (ASTM) has published the 1990 edition of its annual in which standards relating to the same class of materials are grouped together.

Plastics materials and products are covered in Section 8, comprising four volumes: Volumes 08.01, 08.02 and 08.03 are devoted to plastics as such, while Volume 08.04 covers plastics pipe and building products. Section 9 (Volumes 09.01 and 09.02) covers rubber — materials, products and test methods.

8.1.2 COMPENDIUM OF GERMAN PIPELINE STANDARDS

No fewer than 88 German standards for pipes, pipe supports, connectors and linings, together with guidelines for pipelaying, test methods and chemical resistance requirements, are included in a handbook compiled by the German Standards Institution (DIN) and published by Beuth Verlag.

'Rohrleitungsteile aus thermoplastischen Kunststoffen' ('Thermoplastics pipeline components'), 4th edition, is priced at DM114.

8.1.3 PROCESSIBILITY ASSESSED BY RHEOMETRY

Increasing demand for quality control has set new targets for measuring and understanding the processibility of polymers in the context of efficient production schedules. This often has to be achieved without using trained personnel to evaluate test results and yet without introducing subjective judgements.

The Data-Processing Plasti-corder PL 2000 from Brabender claims to enable the user to simulate many production processes such as mixing, kneading, masticating, extruding or calendering on a laboratory scale by using different measuring heads. It generates data which can be used to assess quantitatively the processibility of thermoplastics, thermosets, rubbers and many other plastics to help processors and researchers.

An example of how this can be achieved is related to the plasticizing behaviour of polyvinyl chloride. The many different grades of polyvinyl chloride require several different types of processing machinery and processing conditions which have to be optimized in each particular application.

The Plasti-corder offers the possibility of achieving this using the measuring mixer head with roller blades attached. In this head, the material to be tested is exposed to a controlled temperature and a defined shear rate, simulating the processing procedure. The output, called a Plastogram, represents the plasticizing behaviour or fusion rate of the particular material. In the standard plasticizing test the measuring mixer is heated up to between 140 and 160°C, depending on the polyvinyl chloride mix.

A roller blade speed of 30 rev/min is used initially, but this can be changed after the first preliminary tests. A fixed mass (± 0.1 g) of the sample to be tested is fed into the mixer within a fixed time — say 20 ± 5 s — and the test is started. The Plastogram shows the increase of torque to a loading peak as the sample is loaded and the subsequent decrease as the sample is compressed and heated. As plasticization begins, the torque increases to a maximum when all the polyvinyl chloride particles are plasticized or fused.

The time from the loading peak to the plasticization peak is known as the plasticization or fusion time, beyond which further mixing and homogenization takes place in the mixer. Satisfactory processing of polyvinyl chloride requires an accurate knowledge of the plasticization time to calculate screw length at the given screw speed and barrel temperature. The peak torque is needed to deduce the power needed for the process.

By varying parameters such as temperature and shear rate over a wide range it is possible to create test conditions which simulate processing conditions in mixers, moulds, calenders, extruders etc. This makes it possible then to investigate the influence of stabilizers, plasticizers, lubricants and so on. Heat and shear stability of polymer compounds can also be investigated on a single machine, as can flow curing.

Special rheometric capillary die heads are available for measurement of direct viscosity on the processed polymer, both for research and production control processes.

The Plasti-corder comes with software which allows ready comparison of tests of different batches, or tests carried out under different conditions. Results can be compared with so-called Mastercurves which define upper and lower limits of acceptability. Mastercurves can be stored to build up a database with which to compare the properties of a given batch of material.

8.1.4 UPGRADED SOFTWARE FOR RHEOMETERS

Rheometrics, a manufacturer of rheological and on line process control systems, has upgraded the software package for two of its instruments. Two new test modes are now available on both the Rheometrics mechanical spectrometer 800 (RMS 800) and the Rheometrics dynamic spectrometer II (RMS II), and the programming has been simplified.

The multiple frequency (or multiwave) test runs up to seven different harmonic frequencies on a user-selected fundamental. Thus it is possible to run a frequency sweep in 2 min rather than in the previous 7–10 min.

The random waveform test allows the user to design his own waveform to deform the sample. According to the manufacturer, users are no longer limited to sine wave oscillation but can generate almost any waveform which is described by a mathematical equation.

Information display has been improved; data such as the maximum and minimum shear rates available for the particular specimen under test are now computed and displayed. The menu has been simplified to help the relatively inexperienced user.

8.1.5 NEW INSTRUMENTS

Single fibres can now be tested to greater sensitivity with the Rheometric solids analyser II (RSA II), the firm claims. This is an imposed oscillatory strain instrument capable of characterizing the viscoelastic properties of a range of materials, including polymeric solids, films, fibres, composites and elastomers. A newly designed fibre testing fixture enables fibres as thin as 5 μm to be characterized.

Another development by Rheometrics is an improved version of its dynamic analyser. Known as RDA II, this instrument characterizes the viscoelastic properties of polymeric materials in their fluid and solid states in dynamic shear using parallel plates, torsion, cone and plate geometries. A tension fixture allows dynamic tension/compression on film and fibres.

9. HEALTH, SAFETY AND THE ENVIRONMENT

Much of this section is concerned with environmental issues. The environment is a complex system not amenable to simplistic analyses; this is illustrated well by the debate on degradable polymers, which were originally supposed to be friendly to the environment and are now, in some cases, considered to be hostile to it. There are interesting development in polymers from vegetable and other non-petrochemical sources.

Agreement has been reached on chlorofluorocarbons, which are now known to be damaging to the ozone layer. Their use will continue to be reduced and, where possible, phased out. This is particularly relevant in the production of polyurethane foams.

9.1 DEGRADABLE POLYMERS

There has been considerable debate on the subject of degradable polymers. They were originally put forward as a solution to the mounting problem of waste plastic. There are several factors which make them seem now less attractive.

One of these is that they can be environmentally damaging, when looked at in the light of recent concerns with global warming; degrading polymers give off greenhouse gases. This objection is not valid for all degradable polymers, however. There are some, of which a number are mentioned in the following paragraphs, which originate from vegetable or other living matter, and their degradation is no more environmentally objectionable than that of a plant; the carbon released into the atmosphere during decay is that which the plant extracted from the atmosphere during its lifetime. But for polymers which originate from petrochemicals, this consideration can lead to the conclusion that degradables are more environmentally damaging than non-degradables; hence the recent news that environmentalists in the USA are calling for a boycott of the former.

A second objection is that they simply don't work. Recent studies, notably by W. Rathje of the University of Arizona, have shown that domestic refuse in landfill sites can take a very long time indeed to degrade. Digging to depths of up to 30 m, Rathje found that 20 year old newspapers were still readable and waste food still recognizable as such. There is little chance that any commercial plastics would degrade significantly in the anaerobic conditions usually found in such tips.

Thirdly, with the growing emphasis on recycling, with consumers increasingly being encouraged to hand in their waste plastics for this purpose, there is an increasing danger that batches of recyclable plastics will be ruined by the inadvertent inclusion of items containing a substance which promotes degradation.

Below we include summaries of two contributions on this subject in the scientific literature.

9.1.1 DEGRADABILITY AS A MISTAKEN GOAL

'Degradable plastics: a critical review' is the title of a recent article by Peter P. Klemchuk of the Additives Division, Ciba-Geigy Corp, Ardsley, New York, USA. Writing in *Polymer Degradation and Stability* (Vol. 27, pp. 183–203), the author reviews the applications of degradable plastics, describes the chemical mechanisms of degradation, discusses the commercial approaches, and concludes that degradable plastics are not a satisfactory solution to the problems of municipal solid waste.

Klemchuk begins by distinguishing between photodegradability and biodegradability — terms which, he says, are often used synonymously and incorrectly. He mentions the environmental pressures which are leading, in many cases, to the banning of non-degradable plastics for some applications; in particular, a United States federal law that requires that six-pack ring connectors for beverages should be photodegradable. The impetus behind such legislation is to reduce the problem of solid waste disposal, in respect of both the scarcity of landfill sites and — more visually — unsightly and sometimes dangerous litter.

Nearly all commercial plastics degrade, to some extent, on exposure to sunlight. Klemchuk details the photochemical reactions which lead to loss of integrity of the original polymer. He lists the patented technologies — the earliest dating from 1941 — by which the process of photo-degradability can be enhanced, many of them involving copolymers with either carbon monoxide or ketone. Biodegradability, on the other hand, mainly relies on the inclusion of starch as an additive to the polymer; the earliest patent for this technology dates back to 1977.

Klemchuk states categorically that none of the commercial packaging plastics are biodegradable. This is because their molecular weights are too high and their structures too rigid for assimilation by organisms; most of them also have substituents which prevent biodegradation by the enzymatic fatty acid oxidation mechanism. Linear polyethylene, he concludes, is the only commercial packaging plastic with potential for biodegradation — and then only when its molecular weight has been reduced drastically by photodegradation.

The author acknowledges that there are some plastics — notably caprolactone — which do biodegrade, given the right conditions, and that there are some worthwhile applications for these plastics. One of these is containers for tree seedlings; after burial in the soil, polycaprolactone planting containers were found to have undergone significant biodegradation, losing 48% of their weight in six months and 95% within a year. Photodegradability, too, has its uses, notably for agricultural mulch films that degrade into small particles after a few months. However, he warns that no hard data are available to support the manufacturers' claim that these particles eventually biodegrade into the soil, nor has the environmental safety of the degradation products, such as ketones, alcohols and acids, been established unequivocally.

Klemchuk claims that even a truly biodegradable packaging plastic would not help to solve municipal solid waste problems where the waste is being landfilled or incinerated. Recycling of municipal solid waste into a number of separate streams and composting would have to be implemented to dispose of biodegradable packaging plastics by biodegradation. However, such a biodegradable polymer is not yet available in the quantities necessary and at a favourable cost.

9.1.2 REASONS FOR, AND METHODS OF, DEGRADABILITY IN PLASTICS

An article by Gerarld Scott of Aston University, UK, (*Polymer Degradation and Stability*, Vol. 29, pp. 135–54) begins by highlighting the growing problem of plastics litter, particularly that discarded at sea and washed up on shore. Over a three-year period in the 1970s, he notes, a fivefold increase in plastic litter occurred in North-West Scotland in a remote area with almost no local population. It was concluded even at that time that the litter was sea-borne and wind-driven from the Atlantic shipping lanes. Despite an international convention, signed since then, to prohibit the dumping of plastics waste in the sea, there has been little evidence of improvement.

He argues that people will continue to dump litter, both at sea and on land, whether or not there is legislation to stop them. It is logical, therefore, to look for ways of pre-treating the material so that it will eventually decay to an innocuous form.

Scott lists the range of degradable plastics now commercially available, and discusses the chemistry underlying the processes of biodegradation, photodegradation and photobiodegradation. He pleads for a better understanding of the difference between these processes in order to facilitate the correct use of the alternative waste disposal strategies.

The packaging industry is often called upon to use degradable materials but has no economic reason to do so. Nevertheless, the industry's hand has been forced in many countries by the demands of 'green' consumers, or by environmental legislation which threatens to ban materials that do not show at least a token degree of degradability. The problem then is to produce materials which satisfy the legislation but are still useful for their initial purpose.

To make plastics inherently biodegradable is to make them less useful as packaging materials. An alternative stratagem, however, is to induce biodegradability by means of additives. Sometimes a combination of substances has the desired effect. A recent report indicates that the permeability of polyethylene filled with corn starch can be very much improved by the use of a polyethylene—acrylic acid copolymer as a plasticizer, making film products more accessible to microorganisms. Such water permeable materials are probably not very useful in packaging, but their use in agricultural mulch is being investigated.

Ideally, plastics should incorporate a 'trigger' mechanism in which the material performs just like normal plastics during manufacture and service but is triggered by some component of the natural environment to degrade rapidly after a controllable induction period. The ideal trigger has yet to be found, but already antioxidant-based photosensitizers are available which provide a predetermined lifetime for packaging which can be varied from a few weeks to several years.

Scott concedes that many useful products can be made from plastics waste consisting of one particular material without too much contamination. However, by far the major proportion of plastics waste is contaminated mixed collected waste. Household waste, for instance, is a mixture of all kinds of metallic, ceramic, biological and plastics waste and only the most intensive separation and cleansing procedures can produce mixed plastics of sufficient quality to recycle. Furthermore, even after these energy consuming procedures, the products that can be produced from the polymeric mixtures have properties greatly inferior to those from virgin polymer. Thus, for example, mixtures of the common packaging plastics are incompatible and give products with inferior mechanical performance.

It is unfortunate, he continues, that the major uses that have been proposed for recycled plastics are in outdoor applications (fence posts, park benches, boat docks, cladding etc.). To produce serviceable products from such waste will clearly require the use of substantial quantities of antioxidants and UV stabilizers. The auxiliary chemical manufacturers could find this profitable, but it seems unlikely that for most applications the value of the end application will match the cost of the additives involved.

It seems likely, Scott concludes, that for the foreseeable future most of the contaminated collected mixed plastics waste will continue to form inert landfill. Hopefully, the design of better incinerators will lead to increased energy recovery.

9.1.3 PLASTICS FROM VEGETABLE OILS

Having developed a starch-based material, Battelle in Frankfurt has recently succeeded in producing advanced plastic materials from vegetable oils; relevant patent applications have been filed.

According to Battelle, these materials have been made possible by a method of polymerizing the fatty acids of vegetable oils to form long molecular chains. A wide range of polymer properties is possible, depending on the chain length.

Raw materials that are particularly suitable for these plastics are vegetable oils in whose fatty acid composition one reactive acid is predominant. These are, for example, castor oil, the oil of a new (high oleic) sunflower variety and the oil from the seeds of *Euphorbia lathyris,* an oleaceous plant being grown experimentally. The two latter oils contain more than 80% oleic acid in their fatty acid fraction. They have been earmarked specifically as industrial raw materials.

The possibility of using these oils to produce plastics and thus substantially expand the classic area of oleochemistry (detergents, cosmetics, technical auxiliaries) may open up new, promising sales markets for agriculture and expand the raw materials basis for the European chemical industry.

According to Battelle, the new materials appear to be attractive from an environmental point of view, since they use renewable resources and are likely to be biodegradable. However, their success will depend critically on their price and performance. Battelle estimates that these plastics will cost about the same as comparable conventional plastics. As regards performance, laboratory experiments have yielded very promising results. Materials with a very wide range of properties have been produced, suitable for processing by established methods. Battelle says it has been possible, for example, to produce plastics of high ultimate strength and stretchability and materials with rubbery properties.

The developers of these materials, Drs Rainer Frische and Jürgen Volkheimer, believe that their plastics will be useful as packaging materials of excellent wet strength (films, bottles, etc.), as materials with special electronic properties, and as structural materials (cable-insulating material, injection moulding compound for equipment and casing production, fibres and tissues, etc.). Before the plastics can be used on a large scale, however, much more research and development will be necessary. Above all, market analyses will have to reveal the priorities for examining the various potential applications. Battelle is therefore looking for industrial partners to continue the work.

9.1.4 BIODEGRADABLE PLASTIC SHEET

A new type of completely biodegradable plastic sheet is being developed at the Japanese Government Industrial Research Institute, Shikoku. It contains a combination of polysaccharides, including cellulose and starch, together with chitinous substances such as those that occur in the exoskeleton of insects, in crustaceans and in the cell walls of some fungi.

The resultant plastic sheet is transparent and is said to be comparable in strength and flexibility to traditional plastics. The material becomes completely decomposed when buried in the ground due to the action of micro-organisms; for example, a plant pot will completely disappear after burial for 75 days. It will also slowly decompose on the sea bed, claims the institute.

9.1.5 A POLYMER FROM PURE STARCH

Warner-Lambert of New Jersey, USA, has patented a process for making a commercial polymer consisting entirely of starch and water. This has previously been difficult to achieve as the starch decomposes during processing; however, Warner-Lambert claims that this will not happen if the right proportion of water is added.

The company intends to market Novon, as the polymer is called, as a fully degradable plastic.

9.1.6 PLASTICS FROM POTATOES

Researchers at the Argonne National Laboratory in Illinois, USA, are developing a process for converting potato peelings and other food wastes into lactic acid, a raw material for biodegradable plastics.

9.1.7 BLENDING BIODEGRADABLES FOR STRENGTH

Lack of strength has often appeared to be a problem for polyhydroxybutyrate and polyhydroxyvalerate, which are bacterially synthesized biodegradable polyesters. This limitation has so far restricted their applications to supermarket bags or consumer-product bottles. Recent developments at the University of Lovell, USA, however, may significantly increase the scope of these materials.

A team led by Stephen McCarthy has established that blending polyhydroxybutyrate/polyhydroxyvalerate copolyesters with more conventional polymers such as polyvinyl chloride, styrene-acrylonitrile or polystyrene results in stiffer, stronger materials which still degrade as effectively as pure polyhydroxybutyrate. For example, 0.5 mm films made of 80% polyhydroxybutyrate and 20% polystyrene degrade in 120 days, as do similar thicknesses of polyhydroxybutyrate.

McCarthy, together with an associate, Richard Gross, head the Polymer Degradation Consortium at the University of Lowell, Massachusetts. This currently has six corporate members and reports considerable interest in five other organizations who have expressed interest in joining.

9.1.8 CHEAPER ROUTES TO POLYHYDROXYBUTYRATE

Conventional polyhydroxybutyrate has been slow to enter the market place, not only because of its poor strength (see item above) but also because of its cost. Typically, it is isolated from a bacterial culture grown on glucose — and glucose is not cheap, particularly in Europe.

Researchers at George Mason University, Virginia, USA, have isolated a species of bacterium which accumulates polyhydroxybutyrate when grown on plant cell wall polysaccharides such as cellulose, starch and xylan, as well as on glucose. Considerable work is still required to optimize this process before it becomes commercially significant.

Another route is being explored at the Biotechnology Research Institute in Canada, where they have discovered a strain of bacterium which produces polyhydroxybutyrate on a methanol substrate.

9.1.9 ICI SEEKS PARTNERSHIP TO DEVELOP BIODEGRADABLE PLASTIC

ICI is reported to be planning to enter a partnership for the further development of a new biodegradable plastic based on polyhydroxybutyrate. This material has already been produced in ICI's biological products division, and is now considered ripe for commercial exploitation. It is of a different type from the starch-based 'degradables' produced mainly by American companies.

The first area of application for the new plastic is likely to be packaging. Other applications may include disposable non-woven fabrics.

9.2 RECYCLING

9.2.1 REPORTS

Two reports have recently appeared on the outlook for the burgeoning recycling industry. Business Communications Co Inc (BCC) has published a study of the solid waste recycling business in the USA. Another study by Vladimir M. Wolpert restricts its subject matter to recycling of plastics, but within a wider geographical range.

BCC's report, GB-127 *New directions in the solid waste recycling business*, cites plastics as having the fastest growing recycling rate of all solid waste materials. By 1994, forecasts the report, plastics will account for 0.4 million tonnes out of a total of 35.2 million. The projected growth rate of 31% per year reflects the fact that "collection programmes are just beginning to include plastics, and . . . there is a great deal of legislative and public pressure on the plastics industry to recycle and find markets for the recycled materials", according to BCC analyst Dorothy Krell.

The report hints at a certain desperation among American plastics manufacturers. BCC anticipates some renewed interest in glass containers by end-users who previously replaced glass with plastic bottles that are now perceived as being less recyclable than glass.

A booming market is available for those able to develop the new types of equipment seen as necessary for bulk recycling to be economic. For example, equipment is needed that will automatically sort waste more accurately, remove contaminants and effectively remove labels from containers. Ways need to be found for recycling products or materials currently deemed 'non-recyclable', and consumers need to be supplied with containers for specific materials, together with appliances for cleaning, separating and crushing recyclables to make storage easier.

Copies of the 162 page report GB-127 'New directions in the solid waste recycling business' cost US$1950.

Following discussions in many countries with industrial firms, their clients, recyclers, and local authorities with responsibility for waste disposal, Vladimir M. Wolpert of Colemans Hatch, UK, has published a comprehensive report on recycling of plastics. One of a series of international techno-economic reports, it describes the latest developments and trends of the growing plastics recycling industry in Western Europe, North America and Japan.

Plastics account for 6–8% of total municipal solid waste. Most of this ends up as landfill — often a euphemism for dumping — but the cost of this option is escalating since existing sites are filling up and new ones are both further away from the source of the rubbish and likely to be opposed on environmental grounds. However, there are many ways in which plastics can be recycled, and the report shows possibilities both for firms already active in this field and for newcomers.

The report shows the growing market potential for manufacturers of plant and equipment such as balers, separators, grinders, granulators, shredders, washing plant and in-house recycling systems, describing these types of equipment and giving names and addresses of leading manufacturers. Special chapters deal with recycling of agricultural films, plastics bottles, mixed plastics waste and engineering thermoplastics, and also deal with reclamation of polymers from discarded lead batteries, cable and wire scrap, and discarded motor cars.

Another chapter describes the developments of incineration of municipal solid waste which contains plastics waste, with production of steam and/or energy. The latest advances in pyrolysis and high temperature gasification are described.

An appendix refers to the controversial issue of degradable plastics products.

The 192 page report 'Recycling of plastics' costs £380 or US$760.

9.2.2 CURRENT ACTIVITIES

A combination of consumer pressure, restrictive legislation and — in many cases — economic good sense means that the plastics industry is turning more towards recycling. The likelihood is growing that much of the material contained in any plastics article you pick up has already been used before.

As recycling activities increase, differences are becoming apparent between those materials and products that are easy to recycle and those that are marginal, difficult or impossible. Mixtures of materials are a problem, particularly if the materials have been combined as a composite or laminate in the original item (polyethylene terephthalate/high density polyethylene beverage bottles, for example, or fibre reinforced polycarbonate car bumpers). It has to be recognized that what may be the ideal technical answer to a problem may not be the best solution from the recycling standpoint.

It is a sign of the times that Germany's three main plastics companies, BASF, Bayer and Hoechst, have launched a joint organization to promote the recycling of plastics. Based in Wiesbaden, it will act as an advisory service for firms engaged commercially in collection, recycling, disposal or incineration of plastics wastes.

9.2.3 AUTOMOTIVE

At an exhibition, in Düsseldorf, Germany, a recycling programme for car bumpers was highlighted. The German companies Metallgesellschaft AG, Opel AG and Hoechst AG demonstrated how car bumpers, moulded in an impact modified grade of polypropylene, could be recycled.

The old bumpers are first shredded into coarse chips from which any foreign bodies are removed. After further grinding and analysis, additives such as stabilizers, release agents and pigments are added if necessary to ensure the quality of the material, which can then be used again for car components such as wheel arch linings and luggage compartment mouldings.

The Germans, along with the Danes, are under great pressure to reduce the amount of non-recyclable material ending up as waste. Hoechst and its partners hope that the scrap vehicle industry will cooperate in using this and other techniques to ensure that a junked car is regarded as a valuable materials resource rather than a wasteful eyesore.

The Dutch chemical group DSM is to recycle plastics car bumpers for the West German manufacturers Volkswagen AG. Volkswagen will use the recycled material for production of new bumpers for its cars. The bumpers will be collected in a car dismantling plant recently opened at Leer in Germany.

Meanwhile, Volkswagen's rival BMW is itself setting up a car dismantling plant at Wackersdorf, Germany. The aim is to separate all car components according to their construction material; a future possibility is to code-label all components to identify the type of material and thus to facilitate recycling.

Cookson Industrial Materials, located at Newcastle-upon-Tyne, UK, is collecting used car batteries and recycling the polypropylene casings which are then sold, in pellet form, to injection moulders. The product is made in eleven different grades at the rate of about 7000 t/yr, and, while meeting the BS 5750 quality assurance accreditation, is cheaper than virgin polypropylene.

9.2.4 DISTRICT SCHEMES

Recoup (RECycling Of Used Plastic containers) is a non-profit-making joint venture formed by about 30 British bottle makers to organize the collection and recycling of used plastics bottles from domestic waste. The group's first activity has been to set up a number of bottle banks in Leeds, Yorkshire, UK. The city was chosen because of its existing good record for recycling materials.

The bottles are being sorted, processed and recycled by Reprise Ltd, a joint venture of European Vinyls Corp and PVC Ltd. Reprise is collaborating with NRT Inc of Nashville, Tennessee, USA, to develop a fully integrated closed loop system for automatically separating and reprocessing mixed plastics containers to yield pure streams of recovered polymers for recycling. A pilot plant is being installed in the UK.

Recoup is planning to set up similar schemes to the Leeds venture in other British towns. The aim is to recycle 50% of used plastics bottles by 1995. However, a higher figure may have to be aimed for if the latest EC proposal comes into effect.

The Blue Box scheme is operated at Sheffield, UK. Householders have been enthusiastic in sorting out recyclable plastics containers and putting them into the blue boxes provided by the environmental pressure group Friends of the Earth. (The boxes themselves are donated by ICI and made of 'Propathene' polypropylene.) The scheme has been running since November 1989.

On the opening day of the PVC '90 conference held in April 1990 at Brighton, UK, J. Guignard of Atochem's Vinyl Products Division described the performance to date of Opération Pélican, the French equivalent to the UK's Blue Box scheme for collecting used plastics containers and other waste.

A typical example is Dunkirk. The municipality provides each household with a special blue bin to collect recyclable materials: paper, glass, metal cans and transparent bottles. Once a week these recyclable materials are collected alongside the more normal domestic waste. The materials are separated at a new sorting centre specially designed for the purpose, and are subsequently sold.

After only two months of running the scheme, the results were most encouraging with, for example, more than 80% of the available polyvinyl chloride bottles being returned.

The operation also has an educational function, with talks being given at schools. Other towns are following Dunkirk's lead and are being supported by GECOM (Groupe d'Étude pour le Conditionnement Moderne — Study Group for Modern Packaging), a group created by the French packaging industry in 1975.

9.2.5 PLASTICS PRODUCTS AND THE CONSUMER

Any visit to a supermarket will reveal the vast amount of plastics packaging used for domestic fluids such as fabric softeners, most of it in very substantial bottles which, sadly, are mostly used once only and then finish up as landfill. There have been a few recent innovations of refill packs for fabric softener which are not sufficiently rigid to serve as dispensers but are just strong enough to enable the bottle to be refilled. These refills present a very significant saving in materials and transport costs. They therefore appear to be a welcome trend in packaging.

Lever España SA has just launched its 'Mimosin' liquid softener in flexible 500 ml pouches. They are made using an ICI Melinex 813/high density polyethylene laminate. The combination provides enough strength and stiffness for the pouch to stand by itself, maintaining a good aroma barrier and being also readily printable. The material is supplied in reels to Lever, which produces and fills the pouch 'online'. The new pouches have already been adopted in France and several other European countries.

Supermarket giants Sainsbury's and Tesco are cooperating with European Vinyls Corp in setting up a collection scheme for used plastics containers. Customers will be encouraged to bring the containers back to the shop, whereupon European Vinyls Corp will collect them, sort them and, if possible, re-use them. The plastics involved include high density polyethylene, polypropylene, polystyrene, acrylonitrile-butadiene-styrene, polyvinyl chloride and polyethylene terephthalate, all of which are used for different types of food and non-food containers. Trials have already begun at selected stores in north-west England.

In the USA, the word 'Recyclable!' is to be emblazoned on bottles of Heinz tomato ketchup. The company is changing from a six layer polypropylene/ ethylene–vinyl alcohol copolymer (EVOH) sandwich to a five layer polyethylene terephthalate)/EVOH combination. The difference is that the layers do not have to be glued together and are therefore readily separable at the recycling plant — though not, the company hopes, while the bottle is in use. Polyethylene terephthalate constitutes 98.5% of the new bottle. The change reflects both the growing market for recycled polyethylene terephthalate and the premium placed on homogeneous materials rather than mixtures of plastics. Although the old bottle was, in theory, recyclable, there was virtually no market for it.

Two major firms, Dow and Huntsman, are cooperating with the US National Parks Service to convert useless plastics litter to useful items. They will place bins for waste plastics in three national parks and return the collected materials to the parks in the form of objects such as benches, picnic tables, road signs, car stops (bollards) and guard rails.

Other companies making bulky items of this type from mixed plastics waste include Cascades Inc of Drummondville, Quebec, Canada, Resource Plastics Corp of Brantford, Ontario, Canada, and International Plastics Corp of Lexington, Kentucky, USA.

9.2.6 POLYETHYLENE TEREPHTHALATE PROJECTS

John Brown Plastics Machinery Inc of Providence, Rhode Island, USA, has developed an automated system for dealing with polyethylene terephthalate beverage bottles with high density polyethylene base cups and aluminium caps. The key component is a hydrocyclone that rapidly spins a suspension of shredded bottles in water, separating out the components. The company says that the system will process 450–2300 kg of bottles per hour.

In contrast, Recycled Polymers of Madison Heights, Michigan, has managed to do the same job without the use of a cyclone, by using what it describes as standard bottling technology. Caps are automatically removed from the bottles, which are then filled with water on what could be described as a bottling machine run in reverse — the label and base cup are scraped off instead of being stuck on! The washed bottles are then granulated, dried and sent away for reprocessing. The company claims that the recycled polyethylene terephthalate is 99% pure.

This may still not be pure enough for re-use. Eastman Chemical Co has announced the success of a chemical method for purifying post-consumer polyethylene terephthalate bottle flake. The company's methanolysis process reverses the polymerization reaction and returns the polyethylene terephthalate to its raw materials, which can then be reacted to produce new polyethylene terephthalate for use in any number of applications, including new bottles.

To prove that recycled polyethylene terephthalate has uses other than bottle making, Eastman cites the example of Image Carpets Inc of Armuchee, Georgia, which is taking recycled polyethylene terephthalate bottle flake and extruding it into polyester fibre for carpets.

Day & Zimmerman of Philadelphia, Pennsylvania, is also in the business of recycling polyethylene terephthalate bottles. The firm's subsidiary Day Products has recently started to operate a plant for this purpose at Bridgeport, New Jersey, using materials separation technology developed by the Center for Plastics Recycling Research at Rutgers University, also in New Jersey. The plant incorporates a hydrocyclone that mixes the shredded bottles with water and spins them rapidly to separate the aluminium bottle caps and the high density polyethylene bases from the lighter polyethylene terephthalate bottle bodies.

Johnson Controls Inc of Milwaukee, Wisconsin, has begun to build a 6300 t/yr recycling facility for polyethylene terephthalate bottles. The plant will use technology developed by the DSM subsidiary Reko. Johnson is the USA's largest polyethylene terephthalate container manufacturer, and recently acquired the Belgian firm Polypack.

9.2.7 AUTOMATED RECYCLING IN THE USA

Automated waste recycling, in which materials are sorted by a variety of ingenious devices, is becoming big business in the United States. In what is believed to be the largest single purchase of plastic recycling equipment in the US to date, National Waste Technologies of New York City has bought 30 ET-1 plastics recyling machines for $10 million to turn commingled plastic waste into synthetic 'lumber' at a recyling plant in Islip, Suffolk Country, Long Island. The new material, Syntal, will be used for items such as park benches and picnic tables, and will reportedly contain plastics, paper, cardboard, glass, metals, tyres, and construction and demolition debris.

Union Carbide has announced plans to build and operate the first full scale multi-plastics recycling facility in the USA. With a capacity of nearly 20 000 t/yr, the unit will be located at the group's technical and manufacturing complex at Piscataway, New Jersey.

Herbold Granulators USA, of Seekonk, Massachusetts, has inaugurated a system comprising a shredder, material separator, wet granulator, de-watering screw, mechanical dryer, cyclone separator, hot air dryer and storage silo. Herbold's RM series has individual drive systems for each shredder shaft, which reportedly creates a greater shearing effect than in centrally driven systems.

Du Pont and Waste Management Inc are jointly building a plastics waste processing unit at Philadelphia, Pennsylvania, which will supply some 20 000 t/yr of recycled materials to the north-eastern United States. A similar scheme is planned for the mid-western states. Du Pont claims that the pelletized resins emerging from this plant will be indistinguishable from virgin materials and will cost about the same. Chemical additives will be used to upgrade the scrap into a higher quality material.

Waste Management is also establishing a materials recovery facility jointly with Eastman Chemical Co. Located at the Eastman works in Kingsport, Tennessee, the facility will have the capacity to recycle 24 000 t of plastics scrap per year.

9.2.8 RECYCLING PLANTS PLANNED

Cabot Plastics International, a subsidiary of US-based Cabot Corp, is planning to set up more than ten plastics recycling facilities over the next few years, as part of its Ecoplast scheme. Ecoplast currently operates a 6000 t/yr recycling facility at Herstal, Belgium, and is collaborating with the EC and other bodies on plastics recycling and separation technologies.

9.2.9 LAW AND POLICY ON PLASTICS RECYCLING

Reuse and refill of plastic bottles and containers is becoming a matter for legislation in many countries. Sweden is introducing a ban on non-refillable polyethylene terephthalate bottles in 1991 but in the meantime has a packaging tax which is lower for refillable containers.

Refillable bottles are also mandatory for beer and carbonated soft drinks in Denmark. In Germany there is a 19p (equivalent) mandatory deposit on all plastic beverage containers (except for milk) and there are targets for the use of refillable bottles, ranging from 90% for beer and bottled water down to 35% for still drinks and juices.

Similar incentives to refill or recycle have also been introduced in Italy, Austria, Switzerland and in several states in the USA and Canada.

In parallel with this, the current proposal for the amended European Commission directive 339/85 on one-way beverage packs requires the industry to aim at a recycling rate of at least 70% for the materials used in such packs.

In the UK, the situation seems to be improving — but not, perhaps, as fast as is happening elsewhere. The Government has set a target to recycle 50% of recyclable waste by 2000. In addition, it plans to introduce 'recycling credits' to encourage the diversion of waste materials into recycling.

9.3 THE OZONE LAYER

9.3.1 CHLOROFLUOROCARBONS TO BE PHASED OUT BY 2000

An amendment to the Montreal Protocol, calling for a complete ban on the manufacture of chlorofluorocarbons (CFCs) by the end of the century, was passed unanimously at a 93-nation conference held in June 1990 in London, UK.

This supersedes an earlier agreement which called for a reduction of only 50% in the quantity of chlorofluorocarbons being manufactured by 2000. Chlorofluorocarbons are suspected of damaging the stratospheric ozone layer, thereby allowing more harmful ultraviolet radiation from the sun to reach the earth's surface; there has recently been some evidence that this is happening.

Hydrochlorofluorocarbons are increasingly being used in industrial processes to replace chlorofluorocarbons. They are believed to cause much less damage to the ozone layer but are not entirely safe in this respect. Hydrochlorofluorocarbons, too, are due to be phased out under the protocol, though not until 2040.

In the plastics industry, chlorofluorocarbons have been mainly used to produce polyurethane and other foamed plastics. Most of the world's producers have now gone over to other gases, but it appears that the use of hydrochlorofluorocarbons for these applications, which many manufacturers favour, can only be a short-term measure and not the final solution.

9.3.2 ICI ORDERS PLANT FOR HYDROCHLOROFLUOROCARBONS

ICI has commissioned a pilot plant to produce hydrofluorocarbon 123, a replacment for the chlorofluorocarbons used in foam-blowing, solvent cleaning and air conditioning. The new plant, based on Merseyside, UK, will produce enough HCFC 123 to allow for its development as a blowing agent for rigid polyurethane foam and as a coolant for air conditioners.

9.3.3 OZONE-FRIENDLY FOAMS

Dow Europe SA of Terneuzen, The Netherlands, has now combined with Hyman plc of Poynton, UK, to develop an environmentally acceptable method of blowing plastic foam. Hyman is well known in the UK for upholstered furniture. This follows the banning of unmodified polyurethane foam by the British government — a rare case of the UK leading its European partners in the health and safety field, rather than lagging behind.

Dow and Hyman have developed a method of making furniture foam which eliminates the need for ozone-depleting chemicals. A special additive is added to water, producing a stable foam. Tests undertaken by Dow show that its use enables a significant range of foams to be produced without any major changes in foam quality and characteristics, as compared to those previously blown with the acid of chlorofluorocarbons.

Dow has also developed a blowing agent for solid foam sandwich panels. Lightness, strength and good thermal insulation are important but sometimes conflicting requirements for this application; Dow's polyurethane foam (Voracor CD 300) has a density of 40 kg/m^3, a compressive strength of 200 kPa and a thermal conductivity of 0.02 W/m K. The use of chlorofluorocarbons has been reduced by some 50% in this instance, so there is still some research to be done in this field.

However, Dow has also produced a foam, Voracor CL 100, which is completely chlorofluorocarbon-free. This is used for flexible-faced panels used in roof insulation. As with the Hyman product, foaming is achieved by the use of water containing a special (undisclosed) additive.

ICI, too, has found that it is not that easy to eliminate chlorofluorocarbons entirely from the production of rigid, as opposed to flexible, polyurethane foams. Research workers at ICI Polyurethanes found that it was possible to substitute about half the 'traditional' CFC-11 by carbon dioxide, produced by the reaction between water and isocyanate. A similar path is being followed by the Royal Dutch/Shell group. Current research at ICI is aiming to replace chlorofluorocarbons by hydrofluoroalkanes.

Additives are also used by Union Carbide to produce Geolite and Ultracel flexible foams for upholstered furniture.

An important breakthrough has been made by Plaschem, part of the BPB Building Products Group, based in Manchester, UK. Plaschem has replaced the chlorofluorocarbons in its 'Aerothane' foam by a proprietary mixture of gases that have no ozone-depleting effect and are not flammable. Despite the high insulation value of chlorofluorocarbons, Plaschem claims that the new foam has the same thermal conductivity as the old — 0.23 W/m K — and is just as strong. This is achieved, the company says, by very strict control over cell size and orientation. The technique is at present limited to the production of foam insulated board by continuous lamination, with a maximum thickness of 150 mm.

A new foamed polyethylene terephthalate called Petlite has been developed by Goodyear. It is aimed up-market from the fast-food chains and aims to out-perform conventional foamed polystyrene in many ways — including the increasingly important environmental factors.

Petlite is foamed with blowing agents which Goodyear describes as 'environmentally friendly', and the foam can readily be recycled using existing facilities, it is claimed. The foam sheet is generally extruded to a thickness of 0.5–0.7 mm with a 50% reduction in density, which according to Goodyear should lead to cost savings of 25% to the customer. Petlite can withstand modest oven temperatures (it has a melting point of 260°C), is claimed to be cut resistant and to be a good thermal insulator. Existing process equipment can be modified for about $50 000–$100 000 to include the foaming process, Goodyear states. Hoechst AG has introduced Polyclear polyethylene terephthalate foam into the USA, about one year since its initial launch into the German market. Polyclear is foamed with CO_2 gas and is considered suitable for use in food packaging. It is light in colour and is claimed to have good mechanical properties and chemical resistance. Polyclear, like Petlite, is said to be suitable for recycling.

Stephan Co of Northfield, Illinois, USA, claims that its new series of polyurethane formulations will reduce the amount of chlorofluorocarbon blowing agents used by 35%. The high density foam systems are claimed to have improved physical properties and are designed for high and low temperature applications. They are prepared using the company's aromatic polyester polyols, which are based on phthalic anhydride.

9.4 CONFERENCE REPORT — PVC '90

9.4.1 POLYVINYL CHLORIDE AND ITS ENVIRONMENTAL IMPACT

"Polyvinyl chloride has had an accident prone history", declared environmental consultant John Elkington at the PVC '90 conference held at the Brighton Metropole Hotel, UK. He listed some of the problems the manufacturers of polyvinyl chloride had had to contend with in recent years — the cancer risk of the vinyl chloride monomer, the toxicity (real or perceived) of some of the additives used, the toxicity and corrosivity of its combustion products, and the waste problem ...

Elkington, a director of Sustainability Ltd and co-author of *The green consumer guide* and *The green consumer's supermarket shopping guide*, was speaking on the opening day of the conference, a day dominated by topics with an environmental or public health dimension.

The conference lasted from Tuesday 24 to Thursday 26 April, the second and third days being taken up by the technicalities of PVC materials and manufacture. Extrusion, plastisols/calendering and alloys and novel polymers were discussed on the 25th, with additives occupying the final session on the 26th. Since these topics are of special interest to the PVC industry but not quite as relevant to high performance plastics in general, they will not be discussed here.

9.4.2 PART OF A GREENER WORLD?

Arnaud d'Aramon, head of Atochem's vinyl products division, asked whether polyvinyl chloride could be an integral part of the environmentally-conscious world. All materials in general use must not only serve the customer's needs but must respect the environment. He quoted, as examples, two applications of polyvinyl chloride for which there was a growing market: window frames and blister packs for pharmaceuticals. There was little or no opposition to window frames, but blister packs had attracted criticism because of their contribution to the waste stream and also because of the suspicion that plasticizers, or traces of vinyl chloride monomer, could contaminate the tablet or other product enclosed in the pack. However, said d'Aramon, there was no evidence of anything other than total safety for the user; in particular, the level of residual vinyl chloride was so low as to be undetectable by present methods. In the Netherlands the pharmaceutical industry had refused to accept a ban on polyvinyl chloride since no other material available at present could satisfy the combined needs of this type of packaging.

Many speakers made the point that polyvinyl chloride requires less energy to produce than many other materials, including polyolefins. A polyvinyl chloride drainage pipe, for example, consumes only one-seventh of the energy needed to produce the equivalent cast iron pipe — and it lasts longer. This is offset by the fact that polyvinyl chlorides's chlorine content gives rise to corrosive hydrogen chloride when the material is incinerated or pyrolysed. A speaker from the floor cited an independent environmental assessment which concluded that polyvinyl chloride was, overall, just as environmentally friendly as polyolefins. This had embarrassed the polyolefin producers, who read the report as indicating that their own materials were just as harmful as polyvinyl chloride.

9.4.3 CONSUMER RESISTANCE TO POLYVINYL CHLORIDE

John Elkington outlined the details of a general hostility towards polyvinyl chloride particularly in mainland Europe, culminating in a total ban on PVC packaging by the Dutch supermarket chain Albert Heijn. Suppliers had turned either to polyolefins or to other packaging materials such as paperboard or glass. However, thanks to a public relations campaign by the Packaging and Industrial Films Association (PIFA) and its European counterparts, Heijn and other retailers were now having second thoughts about polyvinyl chloride as it was becoming apparent that other materials had their environmental problems too. John E. Webb-Jenkins later described the progress to date, and the future prospects, of the PIFA campaign.

9.4.4 BALANCING RISKS AND BENEFITS OF POLYVINYL CHLORIDE

The fact that no material was entirely friendly to the environment was emphasized by W. Freiesleben, director of the European Council of Vinyl Manufacturers, who claimed that the use of polyvinyl chloride was an environmental gain versus possible non-plastic substitutes. He outlined what he saw as the priorities for action by the industry. Of top priority was the management of polyvinyl chloride waste, including recycling of post-consumer waste and minimization of chlorinated residues. Second priority was given to the elimination of cadmium compounds as stabilizers, and to the replacement of some plasticizers by others in certain polyvinyl chloride applications. Thirdly, a closer look should be taken at the consequences of accidental fires involving polyvinyl chloride, with risk/benefit analysis being applied where appropriate.

9.4.5 PLASTICIZERS — WHY THE WORRY?

D.M.P. Pugh, of BP Chemicals' Occupational Health and Environment branch, spoke about the plasticizers di-2-ethylhexyl or di-octyl phthalate and di-2-ethylhexyl or di-octyl adipate. These have been widely used in many applications, including food contact and medical tubing, but a scare had arisen following a report by the International Agency for Research into Cancer (IARC) which classified di-2-ethylhexyl as 'a probable human carcinogen' on the basis of animal studies carried out in 1980 by the (US) National Cancer Research Institute (NCRI).

While appreciating that any attempt to extrapolate test results from animals (rats and mice) to humans has to err on the side of caution, Pugh pointed out that these toxicity tests involved a dose equivalent, in human terms, to 500 g per day, whereas the Association of Plastics Manufacturers in Europe (APME) and the Conseil Européen des Fédérations de l'Industrie Chimique (CEFIC) had jointly reported, in 1987, that the intake of all plasticizers via food packaging was extremely unlikely to exceed about 2 g per annum. "Officially appointed expert groups", said Pugh, "have concluded, without exception, that these plasticizers do not represent a risk to man."

Nevertheless, regulations have been introduced in many countries, notably Italy and Switzerland, to limit the types or quantities of plasticizer used in some polyvinyl chloride products. In California, USA, the use of di-2-ethylhexyl has had to be notified to the consumer in a warning label attached to the product. In the EEC, the legal position awaits the publication of a Directive on additives. For di-2-thylhexyl, which is used in cling film, a tolerable daily intake has been proposed that is well above the maximum likely daily intake of this plasticizer via food packaging.

A biomedical aspect of the use of di-2-elhythexyl was covered by C.R. Blass of Hydro Polymers Ltd. In a joint paper with authors from the Bioengineering Unit, University of Strathclyde, and the Glasgow Royal Infirmary, Blass reported on tests to establish the blood compatibility of polyvinyl chloride tubing plasticized with di-2-ethylhexyl, trioctyl trimellitate and polymeric adipate. The investigation indicates that plasticizers other than di-2-ethylhexyl merit further consideration, but that the continued use of DEHP is acceptable.

9.4.6 FIRE-SAFE CABLING

Marcelo M. Hirschler gave a lively presentation of a paper produced jointly with two colleagues from the Geon Vinyl division of BF Goodrich. He described the performance in fire tests of a new series of wire and cable insulation and jacketing materials, based both on vinyl compounds and on new vinyl alloys. The results were compared with those of traditional vinyl materials.

The new vinyl compounds and vinyl alloys developed should be applicable for uses where much better fire performance is required. They show much lower flammability (in terms of heat release or flame spread) and much lower smoke emission than is normally found with flexible vinyl products. At the same time, he said, the physical and electrical properties of these products make them suitable for multiple uses.

Such developments are welcome in view of the harmful emissions from polyvinyl chloride cables overheating in an electrical fire. Discussion on this topic elicited the surprising information that copper cables can catalyse the formation of dioxins in such an event.

9.4.7 THE DIOXIN PROBLEM

Christoffer Rappe of the University of Umeå, Sweden, discussed the possible formation of dioxins and dibenzofurans during the incineration and pyrolysis of polyvinyl chloride. Present indications are that polyvinyl chloride is only one of many substances — including paper and wood — that could be producing these poisons on incineration, particularly in an inefficient incinerator using obsolescent technology. Assessments of any contribution made by polyvinyl chloride are contradictory and controversial. Research is continuing, hampered by the necessity of detecting dioxins and dibenzofurans in extremely small quantities, which makes cross-contamination of samples very difficult to avoid.

9.4.8 POLYVINYL CHLORIDE AND THE PROBLEM OF WASTE

It is perhaps inevitable that a material constituting about 30% of the total plastics used in the world should have produced some worrying problems for its producers. Many of them have been or are being solved. The carcinogeneity of vinyl chloride monomer, for example, is no longer much of a problem since it is now generally polymerized under enclosed conditions. Cadmium and lead pigments are now being phased out, to be replaced by harmless alternatives. Flame retardants of increasing efficacy are being used in cabling grades of polyvinyl chloride.

One big problem remains: the related questions of waste, incineration and recycling. In continental Europe, as in the USA, landfill sites are in short supply and a large proportion of domestic waste is incinerated. Polyvinyl chloride one of a number of materials producing corrosive gases when incinerated, a cause of local pollution and global acid rain. Several polyvinyl chloride producers state that their product constitutes only a small proportion of domestic waste, and that in a modern, well-run incinerator the flue gases can be scrubbed clean of any harmful emissions. However, many communities cannot afford to invest in modern incinerators, nor do they have the expertise to run them at maximum efficiency; and one can hardly blame the authorities if they decide to ban polyvinyl chloride and any other materials which, in their view, are contributing to the pollution.

The industry's answer must lie in an increased commitment to recycling. Polyvinyl chloride is technically one of the easiest plastics to recycle, yet — probably because of its cheapness — hardly any is being recycled at the present time. However, a positive feature has been the initiative shown by the APME and various national organizations to set up systems and encourage all users, especially the general public, to get into the recycling habit.

9.5 INDUSTRIAL SAFETY

9.5.1 PHILLIPS EXAMINES CONTRACT JOB SAFETY

Phillips Petroleum is sponsoring a study to help determine how the petrochemical industry's reliance on contractors and subcontractors affects job safety. Some 1500 companies will be asked for information on the extent that they use, and train, contract labour.

An explosion at Phillip's plant at Pasadena, Texas, USA, on 23 October 1989 killed 23 contract workers and injured many others.

9.5.2 SAFETY AWARDS

Both Petrokemya, a wholly-owned affiliate of Saudi Basic Industries Corp (SABIC) and the Saudi Petrochemical Co (SADAF), another of SABIC's affiliates, have recently received the US National Safety Council Award of Merit for achieving, respectively, three million and one million man-hours without any accident involving an injury requiring time off work. Petrokemya has been given Saudi Arabia's highest honour for industrial safety, the Shield Award.

9.5.3 FIVE KILLED IN GERMAN PLASTICS PLANT

Two explosions occurred on 9 February 1990 at the VEB Chemische Werke at Buna, Germany, killing five workers and seriously injuring at least 25 others. The plant makes polyvinyl chloride, synthetic rubber and a number of industrial chemicals.

Attention has focused on some ageing reactors in which calcium carbide is converted to acetylene as part of the synthetic rubber manufacturing process. These reactors were already under threat of closure because of pollution problems.

Production of synthetic rubber had also been halted in 1987 following an explosion at the plant.

9.5.4 ETHYLENE PLANT EXPLOSION KILLS 25

About 25 people were killed in an explosion at an Indian Petrochemicals' plant near Bombay. The disaster occured in an area receiving hydrocarbon gases for conversion to ethylene.

9.5.5 ULTRAVIOLET LIGHT METER

With the increasing use of ultraviolet curing systems, an important consideration is the maintenance of a safe working environment as some wavelengths of ultraviolet light can be harmful. UVA light has now made available to processors a compact, hand-held ultraviolet meter, powered by a 9 V battery, which measures the intensity of direct and reflected light from ultraviolet lamp installations.

The Dr Honle UV-C Meter incorporates a photocell filtered to produce maximum sensitivity in the wavelength region 200–280 nm. According to the manufacturer, the unit can be used to monitor the output of all types of ultraviolet lamps, or to calculate the transmission factors of glass, plastics and liquids.

10. RESEARCH

Here are included two topics of quite general relevance for the future of polymer research.

10.1 GENERAL

10.1.1 VAMAS ACTIVITIES ON POLYMER BLENDS

VAMAS (Versailles project on Advanced Materials and Standards) is an international scheme to stimulate the introduction of advanced materials into high technology products and engineering structures, with the overall aim of encouraging international trade in such materials. The organization seeks international agreement on codes of practice and performance standards. It encourages multilateral research aimed at furnishing the scientific and metrological base necessary to achieve agreement on standards.

Six years ago, VAMAS established a technical working party on advanced engineering blends with participation from associated members in each of the VAMAS countries: Canada, France, West Germany, Italy, Japan, UK, USA and the EEC.

Predicting material performance

The aim of the VAMAS technical working party on polymer blends is to establish processing, morphological and property relationships for representative types of immiscible polymer blends. The links between melt rheology, the resulting morphology, the thermomechanical characteristics and the mechanical properties are being studied in blends of polycarbonate/linear low density polyethylene and in high impact polystyrene.

The first step has been to identify the minimum set of test methods which can provide a meaningful characterization of a blend. The search is now on for a practical alternative to the 'box-dropping' test, using lessons learnt from fracture mechanics.

10.1.2 POLYMER RESEARCH IN JAPAN

Professors Ian Ward and Tony Johnson, Director and Associate Director of the Polymer IRC (Interdisciplinary Research Centre) based in the UK at Leeds, Bradford and Durham, have recently returned from a fact finding tour of Japanese polymer science groups. In a two week tour sponsored by the British Council they visited the Tokyo Institute of Technology, the Research Institute for Polymers and Textiles and the Universities of Kyoto, Osaka, Hokkaido, Nagoya, Kanazawa and Tokyo. They also visited Mitsubishi Corp, Asahi Chemicals, the National Chemical Laboratory and the ICI Japan Technical Centre.

The visits revealed a significant complementarity between Japanese and British polymer research. The Japanese universities are particularly strong in the areas of polymer synthesis and in polymer physical chemistry and chemical physics. Surprisingly, polymer physics and polymer engineering have a low profile in Japanese universities. This is perhaps a result of a difference in attitude as they see development as a task only for industry. By contrast, in the UK, the Polymer Engineering

Directorate (now the Polymer Engineering group of the SERC) was established specifically to foster more applied polymer research and development in universities and polytechnics — and this emphasis continues. Many UK university groups find it easier to obtain funding, from both SERC and industry, for applied science and technology rather than for fundamental polymer science.

11. INDUSTRY NEWS

Some of the activities reported here — expansions, takeovers, mergers and restructuring — are no doubt in preparation for the European single market, due to become a reality at the end of 1992. This influence is not restricted to companies in Europe, as the expected integration of the European industry will produce responses worldwide.

11.1 NEW COMPANIES

11.1.1 ENVIRONMENTAL SERVICES LAUNCHED IN UK

Faced by ever-increasing demands that their industrial processes should be environmentally acceptable, many manufacturers are turning to advisory agencies to help them. By coincidence, two distinct environmental services for the chemical industry were announced in the UK on 28 March 1990.

Lloyd's Register has launched Environmental Assurance. Lloyd's Register first asks its client company to supply information on, for example, its size, number of employees, product range and any safety or environmental policies already in operation. Lloyd's Register's team of chemical engineers then visit the site to conduct an itemized audit on each function within the plant. These audits include design and operation procedures, waste management, complaints handling, management communications and environmental management and involve Lloyd's Register talking to all relevant personnel.

After this detailed survey, a report is prepared, identifying organizational strengths and weaknesses and detailing recommendations for improving environmental management systems. This report is then discussed with the client in order to help the company develop or improve its environmental strategy.

Inspectorate Griffith's networked environmental testing service is aimed particularly at the plastics and rubber industry, where the health hazards posed by solvents, inks and airborne solids can often be serious. A key area of activity for the new service will be in aiding and advising companies on how to meet the new Control of Substances Hazardous to Health (COSHH) regulations, which became fully implemented at the beginning of this year.

The service usually involves an initial site visit by Inspectorate Griffith personnel to assess the scope and scale of the client's processes and to identify potential or actual problem areas in relation to the COSHH regulations. If appropriate, the company will then perform the on-site monitoring and laboratory tests on which remedial measures can be based to control or eliminate the occurrence of hazardous fumes and dusts. Periodic re-examination of the client's premises and maintenance of records (as required by the COSHH regulations) are also undertaken.

11.2 EXPANSIONS/INVESTMENTS

BP

11.2.1 INCREASES IN POLYBUTENES CAPACITY

BP Chemicals has invested around £15 million in modernizing and 'debottlenecking' its polybutenes plants in France and Scotland.

Both stages of the programme are now complete.

The programme equips both production sites with modern control technology and additional storage capacity, providing enhanced process control, product quality and production output. The investment includes provision for improved standards of technical safety and environmental protection.

BP's programme is being backed by the expansion of its ethylene crackers, particularly those at Grangemouth which will be completed in mid-1992. This will increase the supply of C_4 feedstock used for producing polybutenes.

11.2.2 NEW BARRIER RESINS PLANT

BP Chemicals is to step up its European marketing operations for Barex barrier resins. This follows the commissioning, in March 1990, of a US$30 million Barex plant in Lima, Ohio, USA, to meet increasing demand from the packaging industry. BP claims that Barex offers the highest barrier rating of any single plastics material.

11.2.3 EXPANSION OF ETHYLIDENE NORBORNENE PLANT

BP Chemicals is to increase the capacity of its ethylidene norbornene plant at Antwerp, Belgium, from 11 000 to 16 000 t/yr to meet a substantial increase in customer demand. Ethyliden norbornene is an important intermediate in the manufacture of ethylene-propylene diene monomer rubber.

11.2.4 POLYETHYLENE PLANT IN INDONESIA

BP has signed an agreement with the government of Indonesia to build the country's first polyethylene plant. Construction has already started on the 200 000 t/yr plant, which is located near Merak on the West Java coastline some 120 km from Jakarta. Using BP Chemicals' gas phase process, the plant will be able to produce either high density or linear low density polyethylene. Ethylene feedstock will be imported and the plant's output sold in the domestic Indonesian market.

ICI

11.2.5 OPENING OF POLYURETHANE RESEARCH CENTRE

A £4.5 million investment in ICI Polyurethanes' world R&D centre at Everberg, near Brussels in Belgium, provides almost 4000 m^2 of additional space — including laboratories, an expanded machine hall, offices and storage. A purpose-built, 3500 m^2 three storey block now houses the fundamental research group and provides extra facilities for chemical analysis and physical testing. Newly acquired analytical instrumentation includes advanced systems for nuclear magnetic resonance and gas chromatography/mass spectrometry.

The opening of the Everberg centre underlines ICI's belief that polyurethanes are among the most versatile and fastest growing plastics. The company has been conducting R&D at Everberg since 1983, but the volume of the research effort has more than quadrupled since then.

ICI has increased the scope of its European test capability by installing a 700 tonne polyurea press in the machine hall. This complements a similar press at Sterling Heights (Detroit), the main centre for automotive work in the USA.

Speaking at the official opening of ICI Polyurethanes' R&D centre at Everberg, principal executive officer Alan Pedder called for the wider adoption of optimum insulation standards in all buildings. "Energy conservation is an area in which polyurethanes have a key role to play", he said. "Rigid polyurethane foam provides a highly effective insulant to cut fossil fuel consumption — a major contributor to the greenhouse effect."

11.2.6 EXPANSION IN ACRYLICS

ICI's Chemicals and Polymers Group is planning to spend about £450 million, worldwide, on an expansion of its acrylics business. Key sites will be located in the UK (Teesside), the Far East and North America.

The use of acrylics is currently increasing at 5–10% per annum, and ICI's expansion programme represents a major vote of confidence in this family of materials.

Another division of ICI, Advanced Materials, has embarked on a joint programme with the Southwest Research Institute (SwRI) of San Antonio, Texas, USA, to evaluate the fire, smoke toxicity and corrosivity performance of polymeric composite materials and systems used in aircraft interiors, naval vessels, mass transit vehicles and building products.

11.2.7 OPENING OF EUROPEAN CENTRE

ICI Advanced Materials has opened a European Application Centre close to the ICI Fiberite plant at Oestringen in West Germany. Consisting of a computer aided design office, a laboratory and a machine hall, the new facility will allow ICI to work together with targeted customers on innovations and to focus on the business's development in the automotive, aerospace, defence, electrical and teletronics markets.

11.2.8 POLYESTER PLANT IN JAPAN

ICI Films, part of the ICI group, has opened a plant at Tamatsukuri, near Tokyo, to produce a variety of 'Melinex' polyester films for use in the photographic, reprographic, printing and heavy electrical goods industries. The plant will have an annual capacity of 7500 t and will enjoy close links with ICI's Technical Centre at Tsukuba.

DSM

11.2.9 EXPANSION IN ANNIVERSARY YEAR

1990 marked the fiftieth anniversary of the foundation of the DSM research laboratories at Heerlen in the Netherlands, although the firm itself is nearly forty years older. The success of the research effort is apparent in the current expansion of DSM's activities both in Europe and in North America.

DSM Kunstharze GmbH, a subsidiary of DSM Resins based in Meppen, Germany, has expanded its resin production to 12 000 t/yr with the installation of a computer-controlled batch production line for advanced materials such as polyurethane and saturated polyester resins. Representing an investment of DM25 million, the new reactors will each be able to produce a batch of 5–30 t of resin, depending on type.

Back in The Netherlands, DSM has inaugurated new plants at Geleen for producing Stanyl (polyamide 4.6) and its principal feedstock diaminobutane. To cope with its growing diversity, DSM is forming a new division, with headquarters in Brussels, Belgium, to be named DSM Engineering.

11.2.10 US MARKETING SUBSIDIARY

In the USA, a new marketing subsidiary — DSM Engineering Plastics North America Inc — has been launched. Products offered to date include Stanyl, Stapron S (a rubber-modified styrene maleic anhydride copolymer) and Faradex, a range of conductive thermoplastic compounds for EMI/RFI shielding and electrostatic discharge protection.

DU PONT

11.2.11 NYLON PLANTS IN ASIA

The Du Pont Co has announced plans to spend $1 billion over the next decade to build plants for nylon (polyamide) intermediates and downstream manufacturing in the Asia Pacific region. The object of this investment strategy is to position Du Pont as a major nylon 6.6 supplier in the area in the 1990s and to strengthen its position as a leading worldwide supplier into the next century.

The first plant to be constructed will be a $200 million facility in Singapore for 'Adi-Pure' high purity adipic acid. Scheduled to go on stream in 1993 with an initial annual capacity of more than 100 000 t, it will increase Du Pont's worldwide adipic acid capacity to more than 700 000 t/yr.

The Singapore plant will be the company's first adipic acid manufacturing site outside North America — Du Pont currently operates two adipic acid plants in the USA and one in Canada. Adi-Pure is a key ingredient in the production of nylon 6.6 polymers.

Also scheduled to begin operation in 1993 is additional polyamide polymerization capacity to support compounding plants currently being built or expanded in Korea, Japan and Singapore. These plants will serve the region's automotive, electronics, appliance, electrical and industrial markets.

Longer range plans are to construct a world-scale plant in the late 1990s for the production of the nylon intermediates adiponitrile and hexamethylenediamine.

11.2.12 TECHNICAL CENTRE OPENS IN EUROPE

Du Pont International has opened a £36 million European Technical Centre at Geneva, Switzerland. The company's aim is to strengthen its technical presence in Europe by consolidating its research and technical support work at the new centre.

Computer aided design will feature prominently in the R&D work carried out at the centre. Du Pont has developed a solid modelling system, called Somos, which electronically converts CAD input into sequential cross-sectional formats. These are imaged, layer by layer, through controlled irradiation by ultraviolet laser of a liquid photopolymer, building up a precise three-dimensional model of solid or hollow form.

The company also intends to build a plant for 'Nomex' meta-aramid fibre in Asturias, Spain, as part of a worldwide capacity increase of 50%. The $200 million project is due for start-up at the end of 1992.

11.2.13 US OPERATIONS STREAMLINED

Meanwhile, back at the Du Pont headquarters at Wilmington, Delaware, USA, the company has been forced to 'streamline' its polymer products businesses. This will involve dropping some new ventures, selling or leasing small product lines, and eliminating 900 jobs worldwide. Du Pont hopes to save US$100 million in 1990 by this action.

Against this background came the announcement that Du Pont will invest more than US$1 million to build an ethylene propylene diene monomer pilot plant in Beaumont, Texas, USA. The plant will focus on improving the quality and processing technology for 'Nordel' hydrocarbon rubber.

EXXON

11.2.14 BOOST IN BENZENE AND XYLENE CAPACITY

Exxon Chemical has announced plans to increase capacity of both benzene and paraxylene at its aromatics complex in Baytown, Texas, USA. The additions will increase the company's capacity for both chemicals by about 70 000 t/yr, with an eventual capacity of over 480 000 t/yr for benzene and 450 000 t/yr for paraxylene. Construction has already begun on both projects.

11.2.15 DELAYED EXPANSION IN FIFE

The company decided to delay further work to expand its ethylene plant at Fife in Scotland, UK. Costing more than $400 million, the expansion will enable the plant to run on propane and butane from the North Sea oilfield as well as ethane. Production of ethylene will increase from 650 000 to 900 000 t/yr, with an extra capacity for 200 000 t/yr of polymer-grade propylene. The project, which was due to start up in the first half of 1993, is now expected to be delayed by about two years.

The delay was blamed on the current unsettled world economic outlook and the expected slower growth in chemicals.

11.2.16 PLANNED REACTOR AT ANTWERP

Exxon Chemical is investing in a new autoclave reactor at its Antwerp, Belgium, polymers plant. The reactor, with an initial capacity of 85 000 t/yr, will be installed in parallel with an existing tubular reactor and will not immediately increase the total capacity of the plant. Its main goal will be to support and further expand Exxon Chemical's strong position in coating polymers for the packaging and photographic paper industries.

The planned investment amounts to 400 million Belgian francs (US$11 million). Detailed engineering has already begun, and Simon Carves (UK) has been chosen as the contractor for the project. The start-up date for the project is scheduled for the second half of 1991.

11.2.17 CRACKER EXPANSION IN FRANCE

Exxon Chemical has announced that it will expand its steam cracker capacity at Notre-Dame-de-Gravenchon, France. This is a necessary move to cope with the demands of its projected linear low density polyethylene plant (a joint venture with Shell — see section on Joint Ventures) and a newly announced polypropylene plant.

Ethylene capacity will be increased from 315 000 to 400 000 t/yr and propylene capacity will be increased from 220 000 to 270 000 t/yr. The expansion, scheduled for late 1991, will mainly consist of modifications to existing equipment and the replacement of some compressors.

The polypropylene project builds on a successful business in the United States. In Baytown, Texas, Exxon Chemical operates the world's largest polypropylene plant with a capacity of 470 000 t/yr. Three of the four lines in this plant are based upon technology originally licensed from Sumitomo Chemical of Japan but extensively improved upon by Exxon since startup in 1983. The French plant will duplicate the Exxon–Sumitomo process.

Exxon Chemical already operates two steam crackers in the UK, and has recently announced a project to recover 50 000 t/yr of chemical-grade propylene at the Esso refinery at Ingolstadt, Germany.

With the completion of these projects and the current expansion under study at its plant in Fife, Scotland, Exxon Chemical would have access in Europe to a total ethylene capacity of 950 000 t/yr and more than 500 000 t/yr of propylene.

SHELL

11.2.18 LOW DENSITY POLYETHYLENE CAPACITY UPGRADED IN UK

Shell Chemicals UK Ltd is to upgrade its low density polyethylene capacity at the Carrington complex near Manchester. The new plant will replace two small lines and raise total production from 70 000 to 170 000 t/yr.

11.2.19 SHELL SUBSIDIARY IN MALAYSIA

A new Shell subsidiary, Shell Polystyrene (M) Sdn Bhd, has been granted a licence to manufacture expandable polystyrene in Malaysia. With an initial capacity of 30 000 t/yr, the plant will cater for the needs of the Malaysian and other markets in the Asia–Pacific region. It will be located at Pasir Gudang, near Johore Baru.

BASF

11.2.20 EXPANSION IN BRAZIL

German chemicals group BASF is to invest $20 million in expanding capacity at two plants located near São Paulo, Brazil. One plant produces plastic dispersions, the other Styrofoam.

11.2.21 ACRYLIC COPOLYMER DISPERSION PLANT IN USA

BASF Corp of Parsippany, New Jersey, USA, is to build a $29 million plant for acrylic copolymer dispersion at its existing complex at Monaca, Pennsylvania. Completion is scheduled for mid-1992.

BAYER

11.2.22 POLYCARBONATE AND POLYPHENYLENE SULPHIDE AT ANTWERP

Bayer's Belgian subsidiary Bayer Antwerpen AG has commissioned a second plant for the production of Makrolon polycarbonate resin. Capacity for polycarbonate at Antwerp has now reached 50 000 t/yr, and Bayer plans to increase this figure to 70 000.

Also at Antwerp, Bayer is now producing 4000 t/yr of Tedur poylphenylene sulphide, using its own polycondensation process.

11.2.23 EXPANSION IN KOREA

Bayer has set up a subsidiary — Bayer Korea Ltd — to market polymers, organics, industrial chemicals and agricultural chemicals in South Korea.

11.2.24 HOECHST

Building work has started at the Walton Manor site in Milton Keynes, UK,for a new Customer Technical Centre for the Polymer Division of Hoechst UK Ltd. The new building complex will provide space in 1991 for the existing activities of the Engineering Polymers and Standard Polymers groups, and there will also be space for further expansion.

11.2.25 SABIC

Saudi Basic Industries Corp (SABIC) will expand ethylene production at its wholly owned affiliate Arabian Petrochemical Co (Petrokemya) in Al-Jubail industrial city. The new plant will produce 500 000 t/yr of ethylene, with additional production of propylene, butadiene and benzene. Manufacture of ethylene glycol and polyethylene is envisaged. The plant is expected to go on-stream in 1993.

11.2.26 RHONE-POULENC

The engineering plastics department of the French firm Rhone-Poulenc has invested £5 million in the construction of a new production unit of nylon 6.6 (Technyl) in Taiwan. The unit will start operating in mid-1991. This will serve a Taiwanese market for engineering plastics that is growing at 20% per annum.

11.2.27 AKZO

Akzo Chemie is to build a rubber chemicals plant at LeMoyne, Alabama, concentrating on the production of mercaptobenzo-thiazole, a vulcanization accelerator. Akzo already produces mercaplotensothiazole at Hillhouse, UK, and Termoli, Italy, and the new plant will bring its capacity up to about 20 000 t/annum.

11.2.28 IDEMITSU PETROCHEMICAL

Growing demands for polypropylene film and fibres have prompted Idemitsu Petrochemical Co to set up a new plant in Chiba Prefecture which is expected to produce 50 000 t of polypropylene per year. Sumitomo Chemical Co, Tosoh Corp and Ube Industries Ltd also plan significant expansions to their polypropylene production facilities.

11.2.29 NESTE OY

ABB Lummus Crest BV of Voorburg, The Netherlands, has been awarded the contract to build a polyalphaolefin plant for Neste Oy of Finland. The plant, said to be the first of its kind, will have a capacity of 20 000 t/yr and will be located at Antwerp, Belgium.

11.2.30 SOLVAY

The Belgian company Solvay has decided to build a high density polyethylene industrial plant with a capacity of 100 000 t/yr at Antwerp. Start-up is expected for the end of 1991.

11.2.31 PHILLIPS PETROLEUM

Phillips Petroleum Co has approved construction of a new plant which will enable the company to restore its annual US polyethylene production capacity to 820 000 t. This was Phillips's production capacity before the accident in October 1989 which killed 23 workers and destroyed the polyethylene plant at Pasadena, near Houston, Texas, USA. Start-up of the new facility is scheduled for mid-1992.

11.2.32 MACPHERSON POLYMERS

Macpherson Polymers has announced a £1 million investment plan for its production and administration site at Birch Vale near Stockport, UK. The money will be spent mainly to provide new production equipment, improve handling and storage facilities for raw materials and finished products, and install advanced process control and monitoring systems throughout the plant.

Macpherson Polymers, UK, has also opened a technical and commercial centre at Dordrecht in the Netherlands. The centre will enable Macpherson to provide a high level of technical service and commercial support for its customers across Europe, and will serve as a demonstration facility.

11.2.33 EASTMAN CHEMICAL

Eastman Chemical Co has announced plans to expand production facilities for Kodapak polyethylene terephthalate at its Ectona Fibers Ltd plant in Workington, UK. The expansion project, expected to be completed in 1992, will increase Eastman's 500 000 t/yr capacity for polyethylene terephthalate by an additional 50 000 t/yr.

11.2.34 PLASTIMAT

Battenfeld and other German companies have erected new buildings and supplied equipment worth over DM40 million to create what is claimed to be Europe's most modern plastics processing plant. Located at Liberec, Czechoslovakia, the Plastimat enterprise supplies bumpers, dashboards and other components for the Skoda car plant, which is expanding its production capacity and hopes to increase its exports to Western Europe.

11.2.35 ATOCHEM

Atochem has built a plant located at Lavéra, in the south of France, for the production of hydroxylated polybutadiene. Costing FF200 million and having a capacity of 4000 t/yr, the plant will complement Atochem's existing facilities in the USA and Japan.

The new plant will specialize in producing Poly BD R45HT, which is mainly used for the production of solid propellants for the aerospace industry, including the Ariane V rocket programme. Outside the aerospace industry, hydroxylated polybutadiene is used in formulations for various uses including mastics, impregnation and encapsulation lacquers, telephone cable jointing compounds and waterproof membranes.

11.3 JOINT VENTURES/AGREEMENTS

11.3.1 EXXON CO-OPERATES ON POLYETHYLENE PLANT WITH SHELL

Exxon Chemical and Shell Nederland Chemie are proceeding with a project to construct, in France, a joint linear low density polyethylene plant with a capacity of 220 000 t/year. Based on Unipol technology, the plant will also be able to produce high density polyethylene.

Costing about $170 million, the plant will be located at Notre-Dame-de- Gravenchon and is scheduled for start-up in 1991. It will be owned by a 50:50 Groupement d'Intérêt Economique (GIE) and operated by an Exxon Chemical subsidiary. Marketing and product application will be conducted separately by each of the parties.

Exxon Chemical will supply ethylene to the plant from its adjacent steam cracker, and in return Shell Nederland Chemie will supply ethylene from Moerdijk to Exxon Chemical's plastics plants in Belgium.

In related news, Allied-Signal Inc and Exxon Chemical Co have become partners in the operation of a high density polyethylene plant at Baton Rouge, Louisiana, USA. The plant will continue to operate with the current Allied-Signal management and staff, but ethylene will be supplied by Exxon from its various refineries.

The facility at Baton Rouge employs about 450 people, has a capacity of over 540 000 t/yr and boasts an annual sales figure of US$450 million. The new business is to be called Paxon Polymer Co Ltd.

11.3.2 BP IN INDONESIAN POLYETHYLENE VENTURE

A joint venture company, PT Petrokimia Nusantara Interindo (PT Peni), has been formed involving BP Chemicals, PT Arseto Petrokimia, Mitsui & Co and Sumitomo Corp. It plans to set up a new plant by 1992 for the production of up to 200 000 t/yr of linear low density and high density polyethylene, and high density polyethylene using BP's gas phase process. The output is destined for the domestic Indonesian market in order to reduce the substantial drain on foreign currency.

BP Chemicals has also signed a licence agreement with Samsung General Chemicals Company Ltd for the use of the BP fluid bed polyethylene process in the Republic of Korea. The plant is to be constructed as part of Samsung's new petrochemical complex at Daesan. The design capacity will be 80 000 t/yr of linear low density polyethylene.

11.3.3 BASF–MITSUBISHI LINK ON ENGINEERING PLASTICS

BASF Japan and Mitsubishi Petrochemical Co Ltd, Tokyo, have established a new basis for marketing engineering plastics made by the former company in Japan. They have created a joint venture, BASF Engineering Plastics Co Ltd, with BASF Japan holding an 80% share in the venture, and Mitsubishi Petrochemical the remaining 20%. The new company will take over the marketing of BASF's polyacetyl and polyamide grades, currently the responsibility of the Mitsubishi Yuka Badische Co, owned equally by the parents.

Plans include the construction of a plastics applications centre at BASF's Yokkaichi site. A compounding plant will be added later.

11.3.4 EASTMAN AND TORAY IN INJECTION MOULDED PLASTICS AGREEMENT

Eastman Performance Plastics, USA, and Toray Industries Inc, Japan, have announced an agreement on injection moulded engineering plastics. Toray will distribute Eastman's products in Japan and Asia, and the two companies will work together to develop new products.

Possible further cooperation will include joint compounding of engineering plastics and manufacture of polyesters and copolyesters.

11.3.5 HOECHST AND DAICEL COLLABORATE ON ACRYLONITRILE-BUTADIENE-STYRENE COMPOUNDS

Hoechst AG of Germany and Daicel Chemical Industries of Japan have set up a joint venture company to compound and market flame retardant acrylonitrile–butadiene–styrene copolymer compounds and alloys in Europe, Africa and the Middle East.

The new company, Hoechst Daicel Polymers Ltd, is based in the UK with its headquarters initially at Watford. It is responsible for the compounding and marketing of Cevian flame retardant acrylonitrile-butadiene-styrene compounds and Novalloy acrylonitrile-butadiene-styrene alloys. Similar materials are currently being produced by Hoechst UK's subsidiary Whitefield & Son (Plastic Compounders) Ltd, located in Telford.

A new compounding facility, costing £3 million and with an estimated output of 4000 t/yr, is due for completion in 1991.

11.3.6 GOODRICH AND ORKEM LINK ON HIGH PERFORMANCE RESINS

The French group Orkem has concluded a 50/50 joint venture agreement with BF Goodrich of Cleveland, Ohio. The new firm, Télénor, will be based at Paris and is expected to develop, make and sell high performance liquid resins for structural components used in the agricultural, electrical, heavy vehicle and factory equipment industries.

The agreement provides for the creation of a technical assistance centre at Drocourt in the Pas-de-Calais, which could also be the location of a second production unit.

11.3.7 GOODRICH AND HYDRO COMBINE ON VINYL COMPOUNDS

The US and Scandinavian giants BF Goodrich Co and Norsk Hydro AS have agreed to negotiate the formation of a joint venture to provide specialized compounds for the European market.

The new company, probably to be called Hydro Goodrich, will market vinyl-based injection moulding compounds for use in markets serviced by flame retarded engineering thermoplastics, such as business equipment components, appliance parts and telecommunications equipment. The products to be marketed will include Goodrich's 'Geon' compression injection moulding vinyl compounds and Hydro's 'Vinakon' range.

Hydro Polymers Ltd, a subsidiary of Norsk Hydro located in north-east England, will provide the joint venture with manufacturing facilities and services, together with a technical centre for formulation and applications development.

11.3.8 DU PONT STARTS UP TURKISH NYLON PLANT

Du Pont has announced the start-up of what it calls its first European high tenacity nylon plant, at Izmit, Turkey. The plant, a 50/50 joint venture with Turkey's Sabanci group, will have a capacity of

22 000 t/yr and employ approximately 330 people. The high tenacity nylon 6.6 is aimed at the tyre and other rubber reinforcement markets in Europe and the Middle East.

The Izmit plant is the latest in a major European expansion programme for engineering fibre systems, launched by Du Pont in 1985. Production facilities for Kevlar para-aramid fibre in Maydown, Northern Ireland, and for Tyvek spunbonded olefin in Luxembourg, were brought on-stream in 1988. A plant for Nomex meta-aramid fibre is now under construction in Asturias, Spain, and is scheduled for start-up in late 1992.

Du Pont hopes to double its European business by the mid-1990s.

11.3.9 GE AND HUNTSMAN TO MAKE POLYSTYRENE

GE Plastics Ltd and Huntsman Chemical Co Ltd will form a 50/50 joint venture to add a 45 000 t/yr impact polystyrene production line at Huntsman's plant at Carrington, UK. The venture will supply most of the polystyrene to enable GE to produce Noryl modified polyphenylene oxide in Europe.

11.3.10 NESTE OY AND COMBUSTION ENGINEERING IN SIBERIA

Neste Oy and Combustion Engineering Inc have signed a joint venture agreement in Moscow. The partners will build a $2 billion complex in Siberia to make styrene—butadiene—styrene thermoplastic elastomers and polypropylene for the medical, automotive and consumer goods markets.

Neste and Combustion Engineering will each own 15% of the joint venture, but the rest will belong to Tobolsk Petrochemical Co, an agency of the Soviet Ministry of Chemical and Oil Refining Industries. Combustion Engineering will be responsible for overall project management, including design, procurement, supply of process control instrumentation and automation systems, as well as foreign currency financing. Neste will oversee worldwide export marketing.

Neste, of Espoo, Finland, has a 40 year history of profitable dealings with the Soviet Union. US-based Combustion Engineering supplies products, systems and services to the energy industry worldwide.

The three-plant complex at Tobolsk is expected to be completed in 1993 and will have a production capacity of about 100 000 t/yr. A further expansion is planned for the future.

11.3.11 MONSANTO–EXXON TIE-UP ON THERMOPLASTIC ELASTOMERS

Monsanto Chemical Co and Exxon Chemical Co have agreed in principle to merge their thermoplastic elastomers businesses, creating an 'independent' company to develop, manufacture and market these materials.

11.3.12 ICI–UBE COLLABORATE ON POLYIMIDE FILM

ICI Films and Ube Industries of Japan have signed a technical collaboration agreement under which they will jointly develop Upilex polyimide films.

The two companies will also carry out a feasibility study for a manufacturing plant for Upilex R or Upilex S in the USA. One location under consideration is ICI's site at Hopewell, Virginia.

11.3.13 DU PONT–TEIJIN LINK ON POLYESTER FILM

The US company Du Pont has concluded a joint venture agreement with the Japanese firm Teijin Ltd to manufacture and sell polyester film for audio and video applications in Europe, the USA and other regions, excluding Japan. The film will be made in Luxembourg, the USA and Japan and marketed under the trade names 'Mylar', 'Cronar' and 'Tetoron'.

11.3.14 EASTMAN NEGOTIATES EUROPE–INDONESIA TIE-UP ON POLYESTER FIBRE

Eastman Chemical Co, acting through Eastman International Technology BV of the Netherlands, has reached an alliance agreement with P.T. Eastindo Polymertama, and Pan European Fibres Ltd, companies that are part of the Texmaco Group. Under the agreement, Eastman can market up to 180 000 t/yr of additional polyester staple fibre from new polyester staple plants to be located in Europe and Indonesia.

The European plant will be one of the largest single-site polyester production facilities in Europe and should come on stream in mid-1992. The projected start-up of the Indonesia plant is also in mid-1992.

In addition to the marketing alliance, Eastman has a minority interest in these two manufacturing companies and will supply synthetic fibre technology.

Primary markets for the fibre will be Europe, Africa and the Far East.

11.3.15 SHELL ENTERS EUROPEAN POLYETHYLENE TEREPHTHALATE MARKET

Shell has concluded a joint venture agreement with the Italian plastics manufacturer Mossi & Ghisolfi, with the aim of setting up a £50 million factory for polyethylene terephthalate at Frosinone, near Rome. Start-up of the plant, with a capacity of 60 000 t/yr, is scheduled for late 1991. This will be Shell's first move into the booming polyethylene terephthalate market.

11.3.16 HIMONT AND HOECHST CO-OPERATE ON POLYPROPYLENE TECHNOLOGY

Himont Inc of Wilmington, Delaware, USA, has licensed its Adipol process technology to Hoechst AG of Frankfurt, Germany. This will make Hoechst the world's second largest supplier of stabilized, non-pelletized polypropylene.

11.3.17 MELTON MEDES IN HUNGARIAN VENTURE

Melton Medes, an industrial group based at Nottingham, UK, has announced details of a joint venture agreement to establish a plastics extrusion business, Profilplast MKFT, in Hungary. Under the terms of the deal, Melton Medes controls 58% of the voting capital of the new activity which will produce profiles, in polyvinyl chloride and other materials, for the domestic housing market.

The group's German subsidiary, Bolta Industrie und Bauprofile GmbH, is providing know-how, plant and equipment. In its first year the new business is projected to have sales of £1 million; thereafter, turnover will accelerate, with some of the production being sold to adjacent West European markets.

11.3.18 DSM TO HELP CANADIANS MAKE UREA

Stamicarbon bv, the licensing subsidiary of the chemical group DSM, is to provide Saferco Products Inc (Canada) with the technological know-how for a urea plant to be located at Belle Plain, Saskatchewan, Canada. The unit, with a capacity of 2000 t/day, is expected to come on stream early in 1992.

CHAPTER 4

CONCLUSIONS

4. CONCLUSIONS

4.1 CERAMICS

The positive tone of market reports published during the last year point to a healthy future for the advanced ceramics industry.

Those reading through the ceramics chapter will gain clues as to which sectors of the industrial will be healthiest. The consensus of the analysts is that electronic applications are, and will continue to be, the largest market for ceramics over the coming decade. However, market growth will be faster in biotechnology, in coatings and in cutting tools; the relevant order of importance of these latter three applications being dependent on the particular report to which you give most credence.

The shifting fortunes of these various markets over the next decade will depend of course on the technological advances which are made. The materials section reflects the continuing commitment to develop all forms of advanced ceramic. This Source Book covers only developments over the past year but most material types are represented. Nevertheless, a cursory glance is enough to realize that development of high temperature superconductors remains high on the industry's agenda.

High temperature superconductors featured prominently in the previous Source Book, published in 1989. However, at that time, the emphasis was on making new materials with ever higher transition temperatures. Now the focus is on refining the best of the materials (particularly the yttrium-based, thallium-based and strontium-based oxides) and in producing them in practical forms. Researchers, most notably those in Japan, have made progress towards usable thin films, wires and cables. There are even a few claims for practical applications.

More generally, advanced ceramics found a large number of novel applications during the year, as the relevant section in this Source Book clearly shows. But again it is worth noting that the majority of reports concern Japanese organizations. The role of Japan in the development of the ceramic engine, an older but still high profile goal of the industry, is also obvious in that section. It may be some time before the target is finally reached, but the coming decade will see increasing numbers of ceramic components in engines.

Whether or not these applications develop into significant markets will depend to a large extent on the reliability of the final ceramic components. Two factors related to this continue to be addressed: the search for new and improved processing routes, and the development of accepted standards. Among the new processes reported here, it is worth mentioning the use of microwaves. During the last year, two groups have formed with the intention of investigating microwave processing of ceramics; one in the UK and one in Australia.

The formation of research groups has also been part of the strategy to develop standards and testing methods for advanced ceramics. Several national programmes have begun and it is to be hoped that these will ultimately lead to international agreements. The formation of a single European market by the member states of the European Community at the end of 1992 will increase the need for such standards if advanced ceramics are to find wider applications.

The creation of the single European market has clearly been a high priority for the Japanese. Much of the industry news section is concerned with Japanese companies moving into Europe in anticipation. US organizations have responded as well and commercial activity is high, particularly with regard to expansions and investments. Worldwide the picture is similar, even in a time of deepening recession for the world's economies, reinforcing the optimism of the market reports.

4.2 COMPOSITES

Market reports on the advanced composites industry published during the year also make optimistic predictions for the decade ahead, despite the worldwide recession and the additional threat of reduced sales to the military as a result of improving East-West political relations.

Military aerospace markets have traditionally supported many developments in advanced composites. However, while in the short term it may be necessary for the industry to consolidate and prepare for better times ahead, in the long term a decreased reliance on this sector could prove advantageous as advanced composites manufacturers are forced to seek more diverse markets.

Civil aircraft could prove to be one market for the future, with several reports predicting large sales over the coming years. But perhaps more significantly, in-roads are being made into other transport industries including rail, marine, and most importantly, automotive.

At present, success in the automotive industry is largely through the increasing use of sheet moulding compound (SMC) in body panels. The desire for further progress is stimulating the search for high volume production techniques, such as preform technology, and the search for new material forms which will reduce manufacturing costs.

However, the move towards the high volume production of composites will force the industry to address environmental questions. Lightweight composite automobiles can make a positive contribution in terms of reducing the amount of fuel which must be used to power them, but how will the materials be recycled, can the necessary recycling technology be made cost-effective, and even so will a composites recycling industry emerge to take over from its scrap metal counterpart? Already there is competition from aluminium which is replacing plastics in some automobile applications because it is simpler to recycle.

Automotive applications, in this case for the engine, are also one of the reasons for developing ceramic and metal matrix composites. However, generally metal matrix composites are struggling and there is a need for improved processing techniques, such as liquid metal infiltration. Before ceramic matrices realize their full potential the industry must find higher temperature fibres.

It is encouraging to note, therefore, that research into fibres is active. The focus is on modifying the microstructure and on investigating different shapes in order to gain a better balance between strength and stiffness. Another promising feature for the industry is that carbon fibre sales have stood up well and increased production of carbon fibre continues.

If the optimism expressed above is to be translated into a thriving industry the question of testing and standards will have to be resolved; as is the case with advanced ceramics. Increasingly the responsibility for developing accepted standards is being taken by the larger organizations, such as the Suppliers of Advanced Composite Materials Association (SACMA) and the American Society for Testing and Materials (ASTM). Their work could be crucial.

4.3 PLASTICS

Environmental pressures are exerting a significant influence on the plastics industry. For instance, the future for polyurethane foams is unclear now that the major industrial nations have agreed to extend the Montreal protocol with the ultimate aim of eliminating the production of chlorofluorocarbons.

An earlier response of the industry to environmental concern was to develop degradable polymers, but this is now being questioned as the degradation processes usually involve the release of greenhouse gases. The alternative is to recycle. However, while this Source Book clearly shows that such activity is high, there is a threat to plastics in the automotive industry, still the most important end-user, from aluminium. As was noted above, aluminium is finding its way back into automobiles, at the expense of plastics, because it is considered to be easier to recycle.

Nevertheless, market reports still predict that the automobile industry will continue to be the largest end-user of high performance plastics. Several reports say that within this market electronic applications will show the fastest growth, and this reflects the growing importance of the use of plastics in the electronics industry as a whole.

The growth of the electronics sector is also illustrated by the development of many new conductive polymers. Other new materials notable in this book are liquid crystal polymers, polymer blends and polyarylether ketone (PAEK). PAEK has a higher melting point than polyetherether ketone (PEEK) and it will be interesting to read the next version of this Source Book to discover what progress it makes.

Development is not restricted to new materials, there are also reports of new grades of older polymers and even existing grades are finding new applications as a result of novel processing techniques. Labour saving is a major motivation in developing processing methods and many examples quoted here involve moulding large parts in one step. In turn, this has provided the stimulus to produce large injection moulding machines and develop automated control for them.

While the industry has been dealing with the problems described above it has also been preparing for the single European market. Much of the activity reported here is related to this and it should be noted that this is not restricted to just European companies. Others, notably from Japan, have been active. However, Europe is not seen as the only opportunity and several of the reports concern the launch of environmental services, showing this issue is as an opportunity as much as a threat.

CHAPTER 5

INDEX

The index has been divided into three parts corresponding to each of the major chapters of the Source Book; ceramics, composites and plastics. Each part is further divided into four sections; materials and properties, applications, processing and equipment, and organizations. This approach has been taken to allow the reader to refer easily between the Source Book and the Directory.

Each entry in the index consists of a page number followed in bracketts by the reference number(s) corresponding to the items on that page where the indexed phrase can be found. Source Book references to organizations are also included in their Directory listings.

5.1 CERAMICS

5.1.1 MATERIALS AND PROPERTIES

ZrC . . . . (see zirconium carbide)

ZrO_2 . . . . (see zirconium oxide)

$ZrSiO_4$ . . . .(see zircon)

5.1.2 APPLICATIONS

abrasive . . . . 19(3.2.6),109(11.2.24)

aerospace . . . . 33(4.5.9),51(5.1),77(6.3.3),78(6.4.1),
111(11.2.30,11.2.32)114(11.2.44)

agricultural . . . . 110(11.2.28)

analytical . . . . 21(3.2.10)

antenna . . . . 60(3.2.10)

armour . . . . 22(3.3.1),46(4.13)

automotive . . . . 23(3.3.2),52(5.2),53(5.2.1),54(5.2.4),
77(6.3.3),78(6.4.1),104(11.1.3),108(11.2.19),
114(11.2.44),117(11.3.10)

batteries . . . . 19(3.2.6)

bearing . . . 18(3.2.2),55(5.2.9),81(6.5.5),104(11.2.1)

bioceramic . . . . 16(3.1.1),29(4.3),55(5.3),104(11.2.2)

biological . . . . 29(4.3)

biomedical . . . . 55(5.3),56(5.3.1),111(11.2.30)

biotechnology . . . . 13(2)

bone . . . . . . 57(5.3.6,5.3.7),74(6.1.6),104(11.2.2)

cable . . . . .14(2),68(5.10.3),100(10.1.4)

capacitor . . . 16(3.1.1),19(3.2.5),59(5.5.4),100(10.1.4)

carbon fibre . . . . (see fibre)

carbon film . . . . (see film)

carbon matrix composite . . . . .105(11.2.4)

carbosilicon fibre . . . . (see fibre)

catalyst . . . . . .19(3.2.6),34(4.7.1),47(4.14),73(6.1.5)

catalyst support . . . . .15(3.1.1),19(3.2.6)

catalytic convertor . . . . 108(11.2.20)

ceramic fibre . . . . (see fibre)

ceramic matrix composite 18(3.2.2),19(3.2.4),120(11.4.5)

chemical . . . . . .19(3.2.6),47(4.14,4.14.1)77(6.3.3)

circuit . . . . .21(3.2.10),46(4.12.30)

clearance control . . . . 24(3.4.4)

CMC . . . . (see ceramic matrix composite)

coating . . . . .13(2),14(2),19(3.2.6),21(3.2.10),
24(3.4.3,3.4.4), 30(4.4,4.4.2),32(4.5.7),53(5.2.1),
55(5.2.10), 56(5.3.2,5.3.4), 57(5.4,5.4.1),
58(5.4.2,5.4.3,5.4.4), 70(5.12.2),79(6.5),
80(6.5.2,6.5.3,6.5.4), 81(6.5.5,6.5.6),
87(7.4),103(11.1.1), 114(11.2.43),118(11.3.15)

thermal barrier . . . . .18(3.2.2),24(3.4.4),53(5.2.1)

communications . . . . 21(3.2.10),60(5.5.5)

component . . . .58(5.5),67(5.10),72(6.1.2),74(6.2.1),
75(6.2.4),101(10.1.7),107(11.2.12),108(11.2.19),
116(11.3.7)

composite . . . . 19(3.2.6)

computer . . . . 65(5.7.3)

consumer . . . . 21(3.2.10)

cutting tool . . . . .13(2),16(3.1.1),18(3.2.2),21(3.2.10),
30(4.4.1,4.5), 57(5.4,5.4.1), 58(5.4.3),87(7.3.5),
104(11.2.1), 109(11.2.24)

cylinder component . . . . 53(5.2.1)

detector . . . . .59(5.5.2),66(5.9.3)

diamond coating . . . . 32(4.5.7)

diamond film . . . . (see film)

diamond-like coating . . . . 56(5.3.2)

diaphragm . . . . 33(4.5.9)

die . . . . 69(5.11.2),110(11.2.25)

diesel engine . . . . .53(5.2.1),54(5.2.5),109(11.2.21)

diode . . . . 60(5.5.8)

DLC . . . .(see diamond-like coating)

drills . . . . 58(5.4.3)

dust collection . . . . 35(4.9.1)

electrical . . . . 19(3.2.6),22(3.2.12),68(5.10.3)

electric motor . . . . (see motor)

electrode . . . . .46(4.13),47(4.13.1)

electronic . . . . . . 15(3.1.1),18(3.2.4),19(3.2.5,3.2.6),
21(3.2.10),22(3.2.12),27(4.1.1), 30(4.5),51(5.1),
58(5.5),61(5.5.10,5.5.13),74(6.1.6,6.2.1), 75(6.2.4),
77(6.3.3),78(6.4.1),100(10.1.4),117(11.3.13)

electronic material . . . . 16(3.1.3)

electronic package . . . . (see package)

electro-optic . . . . .46(4.12.29)

5.1.3 PROCESSING AND EQUIPMENT

5.1.4 ORGANIZATIONS

5.2 COMPOSITES

5.2.1 MATERIALS AND PROPERTIES

5.2.2 APPLICATIONS

5.2.3 PROCESSING AND EQUIPMENT

5.2.4 ORGANIZATIONS

5.3 PLASTICS

5.3.1 MATERIALS AND PROPERTIES

5.3.2 APPLICATIONS

5.3.3 PROCESSING AND EQUIPMENT

5.3.4 ORGANIZATIONS